MONTGOMERY COLLEGE LIBRARY
GERMANTOWN CAMPUS

# Steam Power and
British Industrialization to 1860

# Steam Power and British Industrialization to 1860

G. N. von TUNZELMANN

1978

CLARENDON PRESS · OXFORD

*Oxford University Press, Walton Street, Oxford* OX2 6DP

OXFORD LONDON GLASGOW NEW YORK
TORONTO MELBOURNE WELLINGTON CAPE TOWN
IBADAN NAIROBI DAR ES SALAAM LUSAKA
KUALA LUMPUR SINGAPORE JAKARTA HONG KONG TOKYO
DELHI BOMBAY CALCUTTA MADRAS KARACHI

© *Oxford University Press* 1978

*All rights reserved. No part of this publication may be reproduced, stored in a retrieval system, or transmitted, in any form or by any means, electronic, mechanical, photocopying, recording, or otherwise, without the prior permission of Oxford University Press*

**British Library Cataloguing in Publication Data**
Von Tunzelmann, G N
 Steam power and British industrialization to 1860.
 1. Steam-engines—History. 2. Steam-engines—Economic aspects
 I. Title
 338.4′7′621160941    HD9705.G72    77–30200

ISBN 0–19–828273–7

*Text set in 11/12 pt Monotype Baskerville, printed by letterpress, and bound in Great Britain at The Pitman Press, Bath*

To Carol

*Colleague and Companion*

So many hundred Hands in this Mill; so many hundred horse Steam Power. It is known, to the force of a single pound weight, what the engine will do; but, not all the calculators of the National Debt can tell me the capacity for good or evil, for love or hatred, for patriotism or discontent, for the decomposition of virtue into vice, or the reverse, at any single moment in the soul of one of these its quiet servants, with the composed faces and the regulated actions. There is no mystery in it; there is an unfathomable mystery in the meanest of them, for ever.—Supposing we were to reserve our arithmetic for material objects, and to govern these awful unknown quantities by other means!

CHARLES DICKENS, *Hard Times*

# Preface

THIS BOOK is an adaptation of my thesis 'Some Economic Aspects of the Diffusion of Steam Power in the British Isles to 1856, with special reference to the Textile Industries', submitted for the Degree of Doctor of Philosophy at Oxford University in November 1974—hereafter simply referred to as *Thesis*. I am indebted to many for help with its preparation. Professor Gary Hawke and Carol Dyhouse, for instance, read early drafts of Chapters I and II of *Thesis*, corresponding very roughly to Chapters 2, 3, 5, and 6 of this book. My understanding of the technology of cotton-spinning relevant to Chapter 7 was greatly assisted by Dr. H. J. Catling of the Shirley Institute, Manchester, though I should not wish to hold him responsible for the conclusions I come to. An earlier, more extended, version of Chapter 9 was published as 'Technological Diffusion during the Industrial Revolution: The Case of the Cornish Pumping Engine', in R. M. Hartwell (ed.), *The Industrial Revolution* (Nuffield College Studies in Economic History, no. I; Basil Blackwell, Oxford, 1970). I am grateful to Basil Blackwell (Publisher) for permission to re-use the diagrams which appear as Figures 9.1 and 9.2 of this book. Different sections of Chapter 7 first emerged as papers read to the Second Nuffield Conference on Quantitative Economic History in May 1971 and to the Second Anglo-American Conference on the New Economic History of Britain in September 1972. My paper 'Steam Power and Textiles in Britain to 1856', presented to the panel on New Methods of Research in Economic History at the Sixth International Economic History Conference in Copenhagen in August 1974, allowed a brief interim summary of some of the more novel points. Seminars at Oxford University, Cambridge University, and the University of Canterbury, New Zealand, have all acted as instructive sounding-boards for one or other portion of the work. I am glad to be able to take this opportunity of thanking all those who contributed in this way to the end result. Most of all I owe my thanks to the several people who supervised the thesis at some stage, Dr. R. M. Hartwell, Professor R. C. O. Matthews, Professor P. A. David, Mr. H. J. (Sir John) Habakkuk, Professor P. Mathias, Professor J. A. Mirrlees, and Professor S. L. Engerman. Stanley Engerman in addition read the completed thesis and made extensive comments, including suggestions for the design of the present work. Of the many other individuals who have assisted me, I should like to pay special tributes to my parents, who accommodated me whilst writing up the final drafts of *Thesis*, and to Dr. G. E. J. Llewellyn and his family. Finally, may I express my gratitude to the Warden and Fellows of Nuffield College, Oxford, and the Master and Fellows of St. John's College, Cambridge, for providing the conditions and creature comforts which made the completion of the study more enjoyable. Final preparation of the typescript for publication was assisted by the Nuffield Foundation's Small Grants Scheme.

As well as commenting on drafts of the *Thesis*, my wife more than anybody saw it through to this stage, and to her the book is dedicated.

G. N. VON TUNZELMANN

*Cambridge,*
*January 1976*

# Contents

LIST OF FIGURES — xii

## PART I. THE CONTEXT

1. Introduction — 1
2. The Engine: Technology and Scale — 15
3. The Economic Issues — 38

## PART II. THE EVALUATION

4. The Diffusion of Best-Practice Techniques — 47
5. The Backward Linkages — 98
6. The Social Savings — 116

## PART III. STUDIES IN THE APPLICATION OF STEAM-POWER

7. Steam-Power and the Cotton Industry — 175
8. Textiles, Steam, and Speed — 241
9. Cornish Mining and the Cornish Engine — 252

## PART IV. PERSPECTIVES

10. The International Context in the Early Nineteenth Century — 265
11. Conclusions — 283

BIBLIOGRAPHY — 299

INDEX — 321

# List of Figures

3.1. Schematic cost curves for transportation with and without railways — 40
4.1. Distribution of coal prices in England and Wales 1842/3 — 65
6.1. Price relatives: oats and coal, 1771–1856 — 121
7.1. Employment on each grade of cotton yarn, based on 178 firms in the SELNEC region, 1833 — 183
7.2. Cost of yarn per lb spun according to spindle speeds — 208
7.3. Isoproductivity schedules and a possible expansion path for U.K. cotton-spinning, 1835–1856 — 212
7.4. Average value-added in cotton yarns, 1833–1856 — 213
9.1. Percentage of all reported engines performing at duties over 35 millions, 1816–1838 — 259
9.2. Monthly duties of three major engines, August 1827 to August 1829 — 260

# PART I. THE CONTEXT

## 1. Introduction

AN EIGHTEENTH-CENTURY 'ENERGY CRISIS'?

FOSTERED BY schoolday familiarity, the association between the adoption of steam-power and the rise of modern industry has enjoyed a full life. Propositions concerning industrialization and economic growth have been erected on it—and not just for Britain. As a measure of how the tale has thrived, few historians have demurred from spotlighting the role of the steam-engine in 'the first industrial revolution'. Frequently, their views have brought wider concurrence of opinion. To quote the Nobel Prize winner, Simon Kuznets: '. . . one could argue that of the three major technological inventions associated with the Industrial Revolution—in the fields of cotton textiles, iron, and the steam engine—the last was by far the most important and fundamental to subsequent economic growth. . .'[1] Often, however, factors other than innovation, or the steam-engine in particular, have been seen as holding the centre of the stage, with the advent of steam-power in at most a shadowy supporting role. In part it could be argued that this indecision stems from the lack of any extended treatment of the evolution of steam-power within the context of its links with industrialization; if so, that ought to be reason enough for this study.

Each generation of historians has seen fit to underline the importance of the steam-engine. Contemporaries applauded the inventions of James Watt as the greatest the (Western) world

---

[1] S. Kuznets, *Modern Economic Growth: Ra.e, Structure, and Spread*, New Haven and London, 1966.

had seen—excepting possibly the ship. This was the belief of the engineer John Farey,[2] amongst many. In no sense can he be thought of as one wild-eyed at, but basically ignorant of, technical accomplishment in the early nineteenth century. There is a strong, perhaps overwhelming, case for seeing Farey's volumes as the finest monographs on technology produced during the Industrial Revolution. Without the preposterous gusto that vitiated the wilder extravagances of many of his Smilesian-minded contemporaries, he carefully arrived at the conclusion that the Watt engine represented the greatest step ever made by a single inventor 'in the history of the arts'.[3]

The rise of economic history as a discipline later in the nineteenth century saw the key position of the steam-engine reaffirmed, though by men less awestruck at its mechanical wonder. Arnold Toynbee, in popularizing the very notion of an 'Industrial Revolution', saw the engine as the lynchpin of invention: mechanization in far-ranging areas could have been retarded had the prime mover itself not been mechanized. Of four great inventions that he saw as altering the character of the cotton yarn manufacture in the course of Britain's Industrial Revolution (the jenny, the water-frame, the mule, and the self-acting mule) 'None of these by themselves would have revolutionized the industry',[4] unless Watt's patents had been applied to the cotton industry. This interpretation, of a kind of descending hierarchy of innovations—with the energy source at the top and perhaps minor alterations of design of the cloth (or something like that) at the bottom—was implicit in some of the panegyrics indulged in by Toynbee's predecessors, and has come down to us in several forms today. In his textbook *Britain Yesterday and Today*, for instance, W. M. Stern writes, 'The Industrial Revolution of the eighteenth century derived eminently and decisively from steam. Not until James Watt had perfected a steam engine to drive productive machinery, could that revolution transform cotton manufacture.'[5]

On the other hand, these authors—or others—often qualified their statements by remarking on an obvious chronological

---

[2] J. Farey, jr., *Treatise on the Steam Engine*, vol. i, London, 1827.
[3] Ibid., p. 473.
[4] A. Toynbee, *Lectures on the Industrial Revolution in England*, London, 1884, p. 90.
[5] W. M. Stern, *Britain Yesterday and Today*, London, 1962, p. 54.

difficulty. If the Industrial Revolution was to be dated from around 1760, as Toynbee believed, then the Watt engine can hardly have triggered off industrialization, since it was not being marketed commercially until the mid-1770s. Without altogether abandoning the notion of the engine being central to other innovation, there were at least two historical possibilities. One was to attribute a greater importance to the atmospheric engine. But Watt's 'decisive' breakthrough is usually supposed to have been the extension of rotative motion in early factories, and the atmospheric engine was ill-adapted to provide that kind of motion. As T. S. Ashton wrote in his famous essay, *The Industrial Revolution 1760–1830*:

Before their patents expired in 1800, Boulton and Watt had built and put into operation about 500 engines, of both types, at home and (in a few instances) abroad. The new forms of power, and, no less, the new transmitting mechanisms by which this was made to do work previously done by hand and muscle, were the pivot on which industry swung into the modern age.[6]

A second possibility was to date the Industrial Revolution from the 1780s rather than the 1760s. In the first edition of his celebrated book *The Stages of Economic Growth* W. W. Rostow dated the crucial take-off stage as having taken place between 1783 and 1802 in Great Britain. His hypothesis has since been subjected to some highly critical scrutiny, although the criticism itself has not escaped unscathed.[7] Rostow has reformulated his own argument in the second edition of his book, published in 1971. In an extensive Appendix, entitled 'The Critics and the Evidence', Rostow reiterates the choice of the years 1783–1802 as designating the take-off in Britain, but recasts his reasons.

. . . why was the surge in production at the end of the century so critical, not only for Britain but for the world? Why were these years (accepting the importance of their antecedents, from at least 1688) the beginnings of modern economic history in an important sense? The answer is that:

[6] T. S. Ashton, *The Industrial Revolution 1760–1830* (1948), O.P.U.S. edn., 1968, p. 58.
[7] The most extensive critique of the Rostovian take-off in Britain is by Phyllis Deane and H. J. Habakkuk, 'The Take-off in Britain', in W. W. Rostow (ed.), *The Economics of Take-off into Sustained Growth*, London, 1963. For a criticism of part of Deane and Habakkuk's rebuttal of Rostow, see S. D. Chapman, *The Cotton Industry in the Industrial Revolution*, London, 1972, Ch. 6.

the convergence of developments in spinning machinery, steam engines, and the efficient cleaning of cotton produced a radical decline in the cost of manufacturing articles with a high price and income elasticity of demand, and, therefore, yielded extraordinary rates of increase in output;

the modern system of factory production emerged in England on a relatively massive scale due to its leadership in the process and command over foreign trade;

the improvements in steam-power and machinery manufacture generated laid the basis for the technology that was to lead on to the railway . . .[8]

Steam-power is mentioned specifically in the first and third of these 'answers', and hinted at in the second, the rise of factory organization. The matter re-emerges in his most recent work, *How it all Began*, when he states that by 1800: '. . . modern industry had acquired its most essential long-term foundation. The short-run effects of this radical reduction in the cost of power and its almost complete locational mobility had revolutionary consequences over a wide range of industrial processes.'[9]

In contrast to this stress on the early dominance of steam-power and the radical cost reductions that resulted, a succession of authors have pointed to the persistence, and perhaps even extension, of the use of water-power. This reservation can be traced back at least to Cunningham, who said of the steam-engine that 'Its full effect was only gradually felt, and water continued to be economically the better agent during the first quarter of the nineteenth century; but eventually as a consequence of Watt's invention, water-falls became of less value . . .'[10] Not all those who would subscribe to the main argument would, I imagine, be prepared to accept the direct contradiction of Rostow's 'radical reduction in the cost of power' implied by Cunningham's early views. Nevertheless the view that in general the use of steam-power became massive in the economy only during the second quarter of the nineteenth

---

[8] W. W. Rostow, *The Stages of Economic Growth*, 2nd edn., Cambridge, 1971, p. 204.
[9] W. W. Rostow, *How it all Began: Origins of the Modern Economy*, London, 1975, pp. 166–7.
[10] W. Cunningham, *The Growth of English Industry and Commerce in Modern Times* (Part II: *Laissez Faire*), Cambridge, 1913, pp. 626–7.

Introduction 5

century has been common ground for Clapham, the Hammonds, Redford, and a host of later writers;[11] the bulk of present-day opinion would probably agree. In the last decade or so some attention has also been directed at the continuing use of horses in early mills and collieries. Little has been done in the twentieth century, however, to evaluate *how* massive the build-up of steam-power was between 1825 and 1850.

Some authors have managed to maintain both the innovational-hierarchy view and the delayed-diffusion view, e.g. Stern in the text quoted.[12] A cautious summary that seems to me to come closest to whatever consensus now exists is that given by Eric Roll (as long ago as 1930!):

The question as to whether steam power alone was responsible for the industrial revolution was debated even before the series of profound economic changes of the last half of the 18th Century began to be know by that term. It is now generally recognized that the factory system, together with all the forces that moulded the modern industrial age, originated before and apart from the application of steam power to industry. At the same time it is admitted that without the existence of some such agent neither the rapidity nor the completeness of the development of the modern economic system can be explained. The evolution of the steam engine occupies, therefore, an important place both technically and economically in all discussions of the history of the industrial revolution.[13]

With a vastly wider perspective, David S. Landes in *The Unbound Prometheus* comes to conclusions that are by and large similar to those expressed in the last sentence quoted from Roll's book. Landes regards the steam-engine as one—but only one—of three interlinked innovational trends:

. . . the substitution of machines—rapid, regular, precise, tireless—for human skill and effort; the substitution of inanimate for animate sources of power, in particular, the introduction of engines for converting heat into work, thereby opening to man a new and

---

[11] J. H. Clapham, *An Economic History of Modern Britain*, vol. i, Cambridge, 1930, pp. 441–5; J. L. and B. Hammond, *The Rise of Modern Industry*, 9th edn., ed. R. M. Hartwell, London, 1966, p. 130; A. Redford, *The Economic History of England 1760–1860*, 2nd edn., London, 1960, p. 123.
[12] Op. cit. cf. p. 61.
[13] E. Roll, *An Early Experiment in Industrial Organisation: being a History of the Firm of Boulton & Watt, 1775–1805*, London, 1930, p. 1.

almost unlimited supply of energy; the use of new and far more abundant raw materials, in particular, the substitution of mineral for vegetable or animal substances. These improvements constitute the Industrial Revolution.[14]

After examining the second of these, steam-power, in further detail he concludes:

> The development of mechanized industry concentrated in large units of production would have been impossible without a source of power greater than what human and animal strength could provide and independent of the vagaries of nature.... Coal and steam, therefore, did not make the Industrial Revolution; but they permitted its extraordinary development and diffusion. Their use, as against that of substitutable power sources, was a consideration of costs and convenience... the steam-engine could be relied on in all seasons; but the initial outlay was higher and it was costly to operate.[15]

To a degree, therefore, the current orthodoxy depicts a kind of eighteenth-century 'energy crisis'; not necessarily one that actually eventuated, but at the least one that lay in store for the economy at no very great time into the future. It does not seem possible to explain the amount of attention accorded to the steam-engine by economic historians in any very different terms. To be more precise, the argument might go as follows: one of the potentially most constrictive bottlenecks to industrialization during the Industrial Revolution period was a shortage of convenient sources of power. The limit to the use of water-power, animals, or other traditional prime movers may or may not have been reached by, say, 1782. But even if it was not then actually attained, it would seem likely that the pace of economic expansion would have slowed considerably for want of relatively cheap and accessible sources of power, had it not been for the steam-engine, and Watt's major improvements in particular.

This argument appears to unify the views contrasted above. We find it in Rostow, the archproponent of the idea that

---

[14] D. S. Landes, *The Unbound Prometheus*, Cambridge, 1969, p. 41.
[15] Ibid., pp. 95, 99. Similar views are expressed by P. Mantoux (*The Industrial Revolution in the Eighteenth Century*, London, 1928; rev. edn., ed. T. S. Ashton, 1966, pp. 311, 337–8).

steam-power opened up vastly expanded production possibilities:

Arkwright's and Crompton's inventions lent themselves to steam power; and, as Edward Baines observed, progress in cotton 'would soon have found a check upon its further extension, if a power more efficient than water had not been discovered to move the machinery. The building of mills in Lancashire must have ceased, when all the available fall of the streams had been appropriated.'[16]

But we also find it, for example, in Cunningham who, it will be remembered, appeared to be implicitly contradicting Rostow by contending that 'water continued to be economically the better agent during the first quarter of the nineteenth century':

The only check to the indefinite expansion of the [cotton] trade lay in the limited supply of water-power available; that cause for apprehension was removed, however, by the invention of Boulton and Watt.[17]

Possibly the most explicit formulation is provided by Phyllis Deane in her textbook *The First Industrial Revolution*:

The most crucial and general of the bottlenecks limiting the expansion of the British economy on the eve of the British industrial revolution (that is in the middle of the 18th Century) were two: they were the shortage of wood and the shortage of power. These were closely related problems . . . If we were to try to single out the crucial inventions which made the industrial revolution possible and ensured a continuous process of industrialization and technical change, and hence sustained economic growth, it seems that the choice would fall on the steam-engine on the one hand, and on the other Cort's puddling process which made a cheap and acceptable British malleable iron.[18]

---

[16] Rostow, (1975), p. 164. The quotation is from E. Baines, jr., *History of the Cotton Manufacture in Great Britain*, London, 1835; 2nd edn., ed. W. H. Chaloner, London, 1966, p. 220. Certainly Baines, along with Farey, can be adjudged one of the most reliable and level-headed contemporary commentators on industrial and technological history. Note also my findings on this issue, in Ch. 6, below.
[17] Cunningham, op. cit., p. 626.
[18] P. Deane, *The First Industrial Revolution*, Cambridge, 1965, pp. 129–30. Wood shortage will not be investigated further in this book, but compare the results of C. K. Hyde ('Technological Change in the British Wrought Iron Industry 1750–1815', *Econ. Hist. Rev.* 2nd Ser. xxvii, May 1974) for Cort with mine for Watt.

## OBJECTIVES OF THIS BOOK

The primary aim of the present study is to present the first detailed critique of the 'energy crisis' interpretation that has been shown to be implicit or explicit in many of the best-known economic histories of the Industrial Revolution. Needless to add, I am in the fortunate position of being able to build on the endeavours, in monograph or article, of many predecessors, even though in many cases their main objectives were rather different from mine (Lord on the steam-engine and capital formation, Roll on the steam-engine and industrial organization, Musson and Robinson on the steam-engine and the role of scientific discovery, and so forth).[19] Most of these studies, including all of those just named, concentrate on the last third of the eighteenth century. By contrast, economic historians have hesitated to probe into the evolution of the engine in the first half of the nineteenth century. Apart from the odd more or less uncritical count of horsepower and engines, there is one notable exception—Karl Marx.[20] His analysis is spare and succinct, encapsulating what emerge in my study as the truly significant links between steam-power and cotton. Had his work acted as a springboard for serious research in economic history, not to speak of countless other disciplines, rather than for polemic and counter-polemic, there is little doubt in my mind that the years 1800 to 1860 would not have remained the dark ages of the stationary steam-engine. As it is, his points have been forgotten, although they were central to his exegesis of industrial capitalism. Since his work in this area has thus never entered the mainstream of economic history, I feel free to put forward my present findings (particularly in Part III) as essentially a new contribution.

A glance at Parts II and III will reveal that, naturally, this work is quantitative in emphasis, in that the steps of each argument require one to impose some numerical values (not

---

[19] J. Lord, *Capital and Steam Power, 1750–1800*, London, 1923 (2nd edn., ed. W. H. Chaloner, 1966); Roll, op. cit.; A. E. Musson and E. Robinson, *Science and Technology in the Industrial Revolution*, Manchester, 1969.

[20] K. Marx, *Capital*, vol. i, Moscow edn., 1965, Ch. XV. After my work on this had been completed, I encountered a similar statement by Nathan Rosenberg ('Karl Marx on the Economic Role of Science', *J. Pol. Econ.* 82, no. 4, July–Aug. 1974).

necessarily exact ones). Nevertheless, the statistical procedures employed have been kept as straightforward as seemed possible; almost nothing in this book ventures beyond the territory surveyed by Roderick Floud in his *Introduction to Quantitative Methods for Historians*,[21] and most of it is straight arithmetic.

Bearing that standard in mind, I have set out to confront the figures with the interrelationships that I particularly desire to bring out between history, economics and technology. The interaction of the latter two deserves fuller consideration at this stage. The economics of technological change is one of the most under-developed branches of economic theory. Through limited understanding of the technical advances involved there is confusion over which set of assumptions ought to stand. To give an example that arises in the course of this book: a large body of recent work in economic theory has made clear the significance of the problem that in the real world it is inappropriate to think in terms of a single malleable capital good, and, a little less obviously, to apply the standard theoretical results obtained with such an assumption to the practical situation. Perplexing and sometimes perverse conclusions might be drawn in a theoretical world in which capital goods were heterogeneous.[22] However, attempts to gauge the importance of these possibilities in practice have been few and far between, to say the least. Economic historians have long been familiar with one persistent type of heterogeneity, between fixed and working capital; indeed, the process of capital accumulation in British industrialization could hardly be understood without bearing this distinction in mind.[23] The present work emphasizes another dichotomy, hitherto rarely emphasized, between two types of fixed capital: namely the engines that drove the machinery on the one hand, and the machines being driven on the other.

A more fundamental criticism is that economic theory in isolation may be too abstract a way of interpreting historical problems of technological change. To relegate all productivity

[21] Roderick Floud, *An Introduction to Quantitative Methods for Historians*, London, 1973.
[22] See G. C. Harcourt, *Some Cambridge Controversies in the Theory of Capital*, Cambridge, 1972, and refs. therein.
[23] As brought out, e.g., by S. Pollard, 'Fixed Capital in the Industrial Revolution in Britain', *J. Econ. Hist.* xxiv, 1964.

growth to the category of, say, 'learning by doing' may be obscuring as much as one ought to be clarifying. It would seem especially inappropriate—*ex ante*, anyway—to apply this approach to the steam-engine. To give a different illustration relating back to the question of heterogeneity of capital just alluded to: before beginning some of the present work I had imagined that one way to gauge technical advance in steam-power was by calculating the number of machines that could be driven by a unit of power.[24] But after doing this for spindles in the cotton industry I was surprised to find that the number of spindles per horsepower seemed to have fallen drastically over the observation period. The neo-classical economist might retort that this would result if the price of power was falling relative to the price of machinery. Indeed, this appears to be the case. However, the question is, what meaning can be given even so to the decline in the number of machines driven by a unit of power? Is this in historical fact some form of regress?

My intention, therefore, is to help illuminate aspects of economic theories of technical change; though it will be obvious from the Conclusion that my achievements in this regard are modest. Looking at it from the other side, one cannot fail to notice the extent to which discussion of the technology under consideration here has hitherto been couched in purely engineering or scientific terms. This has severe limitations for understanding the historical pattern of development, for several reasons. The 'Two Cultures' separation of the scientific and the literary consideration of the steam-engine has meant not just a meagre cross-flow of information, but more importantly an over-simplified and sometimes misleading interpretation of industrialization by each school of thought. Even when appraised in its own terms the engineering literature looks to suffer from an implicit belief that by and large the chronology observed was inevitable. Despite extremely careful studies of the influence of each theorist on his successors, and of the influence of both on technology and vice versa one is given the impression that history could have evolved only in this particular way. By contrast, the economist tends to think automati-

---

[24] This view is implicit in the works of several authorities on nineteenth-century industrial technology.

cally of a range of choices open to the inventor at any stage. Without necessarily adopting the 'innovation possibility frontier'[25] (whether implicitly or explicitly), the economic historian who ventures into tracing patterns of causality must, in explaining why a particular step was taken, be saying certain things about why some alternative, counterfactual step was not taken. For all its shortcomings, this approach is likely to lead to far greater insights than a study which presupposes that what actually happened resulted mainly from fortuitous events and, in due course, would inevitably have occurred.

It is obvious why an economic historian should be reluctant to envisage economic determinants as fortuitous. One may also doubt whether even scientific discoveries should be so regarded. James Watt's invention of the separate condenser has looked less like a classic instance of a quite exogenous scientific application since the careful and comprehensive research of D. S. L. Cardwell has established that the theory that Watt's breakthrough arose out of hearing from Joseph Black of latent heat is substantially erroneous, and that Watt of all inventors in this field was continuously guided by cost considerations, i.e. by economic factors.[26] In any case, whether or not the separate condenser represents manna from a Glaswegian heaven, the Cornish improvements to be examined in Chapter 9 clearly owed nothing to *a priori* scientific reasoning.

An economic historian can correct the bias of historians of technology in another way, and that is through emphasizing the contribution to economic growth (and welfare, where applicable), rather than the contribution to mechanical knowledge. It is conceivable that quite inconspicuous technical changes could have disproportionately large effects on the pace of economic expansion. Modern studies have shown that this

---

[25] P. A. Samuelson, 'A Theory of Induced Innovation along Kennedy-Weizsacker Lines', *Rev. Econ. Stats.* xlvii, Nov. 1965; W. D. Nordhaus, 'Some Skeptical Thoughts on the Theory of Induced Innovation', *Quart. J. Econ.* lxxxvii, May 1973.
[26] D. S. L. Cardwell, *From Watt to Clausius*, London, 1971. Watt himself repeatedly denied the legend that he had learnt of latent heat in this connection from Black before developing his engine. More convincingly still, Cardwell shows that proper deduction from the theory of latent heat would have led Watt to perfect the Newcomen engine rather than invent his own. The statement that Watt's investigations were always prompted by cost considerations appears on p. 50.

is more than mere conjecture.[27] If it can happen in the mid-twentieth century, with our currently huge outlays on Research and Development (both in the public and private sectors), then it could also have happened in the world of a century ago when the influence of organized industrial research was so much smaller.

PLAN OF THE BOOK

In principle the ground that might be covered in a study such as this is immense. I have limited it in two ways. First, I have restricted the analysis almost exclusively to the stationary steam-engine. There is no denying that forward linkages to the locomotive and marine engines represent some of the most substantial avenues along which steam-power could be visualized as becoming 'massive' in the second quarter of the nineteenth century, and, to the extent that they have been ignored here, the story is incomplete. However, not only would it be a gigantic undertaking to follow them up, but both fields have been previously scrutinized by quantitative economic historians. Professor G. R. Hawke has assessed the economic contribution of the railways in England and Wales up to 1870, while Professors Hughes and Reiter have looked at the early British steamships (to the mid-1850s).[28] This is not to say that I can fully accept the over-all judgements in either case. (I shall be interpreting Hawke's procedures briefly in Chapter 3, along lines suggested by some reviewers of his book.)

Second, I have concentrated on the interconnections between steam-engines and the cotton industry, though other textiles and mining will be given some attention. Related work has been done for some of the other industries that adopted steam-power at an early stage, e.g. by C. K. Hyde on iron. The emphasis on cotton here seems in keeping with the general evaluations of the role of steam-power by Kuznets, Toynbee,

[27] E.g. S. Hollander, *The Sources of Increased Efficiency*, Boston, 1965, and the whole field of 'learning by doing' pioneered by K. J. Arrow.
[28] G. R. Hawke, *Railways and Economic Growth in England and Wales, 1840–1870*, Oxford, 1970; J. R. T. Hughes and S. Reiter, 'The First 1945 British Steamships', *J. Amer. Stat. Assoc.* 53, 1958. At the time of writing, Dr. R. Craig and Professor R. C. Floud are engaged in re-examining the work of Hughes and Reiter.

Rostow and others, summarized on pp. 1–7 above. But it should be stressed that the economic impact of the engine computed in Part II is for *all* stationary steam-engines, regardless of industry.

The geographical coverage attempted is for the whole of the British Isles. For one reason or another the focus in some sections is narrowed to the most important subregion, e.g. the Lancashire cotton industry, though the final assessment is intended for the British Isles as a whole. Comparisons between different regions are made in Chapter 8 for the woollen and linen industries. The time scale is explained in Chapter 3, in discussing the nature of linkages.

Chapter 2 sets out comparatively familiar material on the evolution of engine technology and on the total horsepower installed, but gives both a rather unconventional interpretation in keeping with the findings unveiled later. Perhaps most important is an upward revision in the total capacity installed as Watt engines when their patent rights terminated in 1800.

Chapter 3 sets forth the basic framework within which the later empirical results will be expressed. Although I have borrowed the concept of 'social savings' from equivalent work on English and American railways, there are significant differences in the way I apply it, that I justify in this chapter. Wider consideration of backward and forward linkages leads into a brief preview of the principles of speed-up of machinery which constitutes the major forward linkage to textiles examined in Part III.

The bulk of the new empirical work is presented in Parts II and III. In Chapter 4 costs of building, purchasing, and erecting steam-engines of the various kinds encountered are brought together and made consistent. A simplified threshold model of relative profitability of each type (Newcomen, Watt, Cornish, etc.) assists in a tentative first step towards explaining diffusion patterns. Related information, on weights rather than costs of the metals used in constructing engines, is used in Chapter 5 to calculate backward linkages arising out of the steam-engine. It is shown that there was a much larger proportionate effect on coal than iron.

The cost data, when adapted to deal more realistically with the major mechanical limitations of earlier forms of the engine than Watt's, can be used to compute the social savings accruing to Watt's inventions, as in Chapter 6. Possibly of wider interest, though inevitably more speculative, are the social savings on all steam-engines, also given in Chapter 6. This computation requires cost information on alternative forms of technology (water-wheels, horses, etc.) which is given in Chapter 6 and its appendices.

Part III studies the forward linkages in a more detailed investigation of the adoption of steam-power in selected industries. An examination of the economics of mechanization at the level of both the individual firm and the sector as a whole in the case of cotton (in Chapter 7) enables the role of cheaper sources of power to be evaluated. In the spirit of Marx's work to which I have already called attention the rise of the high-pressure engine is shown to be associated with the first major reduction in power costs during the Industrial Revolution, and consequently with the further spread of mechanization. A similar model is employed to interpret the experiences of other textiles in Chapter 8, though an important modification becomes necessary. In Chapter 9 the original diffusion of the high-pressure engine within Cornish mining is studied in detail.

The conclusions follow in Part IV. Chapter 10 illustrates that the points raised in Part III have significance in explaining the diffusion of textile machinery not only within the British Isles, but also across countries and continents. Chapter 11 derives some lessons for history, for technology, and for economics.

## 2. The Engine: Technology and Scale

THIS CHAPTER presents the technical and quantitative information about the stationary steam-engine necessary to keep this book as self-contained as possible. Little of the content claims to be original. The first section, on the development of the technology, could readily be omitted by those conversant with that subject. Compared to the usual histories of technology, however, it does devote greater attention to what may at first sight seem rather minor points of interest, because of the clash between economic importance and mechanical importance noted in Chapter 1. It would have to be complemented by a thorough study from the scientific side[1] to give a complete picture of the technology. Lack of space precludes that here.

### TECHNOLOGICAL CHANGE AND THE STEAM-ENGINE

Historians of technology have sometimes dismissed the Savery engine of 1695 as not a true steam-engine, but rather a steam-pump. This is a harsh verdict, because in Savery's own design the engine worked as both pump and engine. Water could be drawn up some distance—theoretically 32 feet, but in practice little more than 20 feet—through suction created by condensing steam in a receiver by pouring cold water over the outside of it. However, with proper manual operation the water could be forced higher by the use of steam of greater than atmospheric pressure. According to Desaguliers,[2] Savery contemplated pressures of as much as 8 to 10 atmospheres in the engine he proposed for the York Buildings Waterworks. As this is much higher than in most of the high-pressure engines of the nineteenth century to be described below and in later chapters, the Savery engine ought to be regarded as a legitimate power source, at least in conception. Had it been successful, the York Buildings engine should have been capable of lifting

---

[1] e.g. H. W. Dickinson, *A Short History of the Steam Engine*; Cardwell, *From Watt to Clausius*.
[2] J. T. Desaguliers, *A Course of Experimental Philosophy*, London, 1744, vol. ii, pp. 466, 467, 472.

water some 300 feet. The engineering niceties are less important here in any case: simply by acting as a pump the Savery engine could supply motive force.

Savery-type engines were still being erected late in the eighteenth century, when they were generally used to pump water to the top of a water-wheel, for the latter to provide the rotary drive for a factory. In this function Savery engines were normally used at low pressures working purely by suction. High-pressure propulsion had placed explosive strains on the rather ill-constructed boilers then in existence and anyway had proved uneconomic in coal consumption when used in this way.[3] In the early 1770s Joshua Wrigley was able to reduce the function of skilled labour by getting the water-wheel to open and shut the regulator and injection cock at the right intervals.[4] But even this could not keep the Savery engine competitive indefinitely.

From the erection of the first of its type in 1712 the Newcomen engine predominated. High-pressure steam, with all its uneconomic (and explosive) consequences, was eschewed. In place of the receiver of Savery stood a cylinder, with piston inside. The top of the cylinder was left open, and the piston descended under the pressure of the atmosphere when the steam below the piston was condensed (unlike the original Savery engine but like Wrigley's, water for condensation was injected inside the cylinder through a cock). This descent was irregular, partly because of having to overcome inertia at the start of the stroke and partly because of the diminishing resistance offered to atmospheric pressure as it proceeded. The return of the piston was accomplished once condensation was completed by the weight of the large wooden beam through which the power of the engine was transmitted, as steam was being admitted into the cylinder. Most of the work was thus

---

[3] Farey notes that the steam was usually kept at about 217 °F. in the boiler for such engines in the late eighteenth century (*Treatise* i, 119–122; also J. Bourne, *A Treatise on the Steam Engine*, London 1846). The technology of steam boilers and its advance is set out more fully in *Thesis*, pp. 24–7.
[4] Farey, i, 131–3. J. Aikin (*Description of the Country from Thirty to Forty Miles Round Manchester*, London 1795, pp. 174–5) considered Wrigley a public benefactor "who never applied for a patent, but imparted freely what he invented to those who thought proper to employ him": see R. L. Hills, *Power in the Industrial Revolution*, Manchester, 1970, p. 140.

exerted in the down stroke only, and this added to the irregularity of motion of the engine. Since the pressure of the steam could not substantially affect the power produced by this engine, it was usually set for economy at about half an atmosphere.[5]

Though in Newcomen's hands the steam-engine emerged as recognizably the same machine that it was when the Industrial Revolution drew to a close about 150 years later, his design did not involve any dramatically new component. 'When we look into the matter closely, the extraordinary fact emerges that the new engine was little more than a combination of known parts . . .'[6]—though it is by no means certain that Newcomen would have been aware of all of his predecessors' work. The brilliance that he displayed was in putting these components together (especially the injection of water into the cylinder, and the valve combinations) plus his success in correcting the many defects encountered in the early years of operation, to a degree which makes even James Watt look something of a sluggard. For example, Newcomen compensated for the inaccurate boring of the cylinders of his engines by playing water over a leather seal fitted around the piston.

After Newcomen and his immediate associates there was little that could be described as a breakthrough on the technical front, but costs were steadily reduced by improved methods of manufacturing the components of atmospheric engines. The best known of these is the work done at Coalbrookdale in producing inexpensive cast iron cylinders to replace the costly brass cylinders of Newcomen's own time, although the use of iron meant a slight initial loss of efficiency.[7]

There was, however, investigation of a less dramatic kind which nevertheless had substantial effects on the economy of the steam-engine. Much of this can be attributed to John Smeaton of Austhorpe, who used a 1-horsepower model steam-engine (as he used model windmills and water-mills) for

[5] Late in the eighteenth century the steam pressure was increased, to allow a puff of steam to enter the cylinder at the beginning of the cycle, in order to displace the remaining air and water more effectively; see Farey, i, 422.
[6] Dickinson, op. cit., p. 29.
[7] Desaguliers therefore counselled against the use of iron and in favour of brass (e.g. op. cit., p. 468). See also A. Raistrick, *Dynasty of Iron Founders: The Darbys and Coalbrookdale*, 2nd edn., Newton Abbot, 1970, pp. 128–30.

experimentation.[8] He thought that power was being lost on existing colliery engines by having the beam reciprocate too rapidly, since its momentum was swamped each time it changed direction. The optimal load obtained from his model was 7·8 lb per sq. inch on the piston. To help produce a slower motion he used a longer stroke, an off-centre pivot on the working beam, and larger pumps. He was among the first to 'compound' the beam, i.e. construct it of layers of timber and sometimes iron.[9] Other aspects of good economy he stressed (whether or not he was the first responsible for them) were clothing the cylinder and piston to minimize heat loss, enlarging the steam passages, improving the boiler and air cock, and using the cataract (brought from Cornwall) to govern speeds. 'The improvements introduced by Smeaton chiefly resolve themselves into greater care in the construction of the engines, and a better proportion and arrangement of the boiler, and involve neither the application of any new principle nor any great expenditure of energy.'[10] The latter comment seems unduly tart. Through these technically minor developments Smeaton raised the duty achieved on his improved engines (i.e. lb of water raised 1 ft high by the consumption of 84 lb of coals) to 9·45 millions, e.g. in his large engines at Long Benton and Chacewater mines.

This duty of $9\frac{1}{2}$ millions was a useful advance from the average observed on fifteen engines surveyed on the Newcastle coalfield in 1769 of 5·59 millions (best engine, 7·44 millions).[11] Even so, it was hardly perfection as far as either customers or the engine capabilities were concerned. Because the steam was condensed in the cylinder, the temperature in the hot well (into which the eduction pipe drained) could not fall below 152 °F. if the engine was working properly; so that the uncondensed steam remaining in the cylinder still had an elasticity of nearly 4 lb p.s.i., i.e. about half its initial value.[12] This

[8] Farey, i., 166–72.
[9] *The Engineer and Machinist's Assistant*, Glasgow, 1850, p. 14; Dickinson, p. 87. Cf. the Cornish engine in Ch. 4.
[10] Bourne, op. cit., p. 12. Farey's verdict seems fairer. "Mr. Smeaton's improvements on Newcomen's engine consisted only in proportioning the parts, but without altering any thing in its principle; it was still Newcomen's, though perfected" (i. 307).
[11] Farey, i. 234.
[12] Ibid. i. 331; H. Reid, *The Steam Engine*, Edinburgh, 1838, p. 109.

## The Engine: Technology and Scale

in turn limited the load that could be placed on the piston at any economic level of operation.

The air entering the cylinder through atmospheric pressure when the cylinder was depressed, and the action of the water to seal the piston, both acted to cool the cylinder excessively. Above all, the alternate heating and cooling of the cylinder during the cycle resulted in the loss of a great deal of steam; about three-eighths of the fuel was wasted in this way, according to Farey.[13]

It remained for Watt to remedy the situation by inventing the separate condenser (patented in 1769), as is very well known. It would have been possible to do no more than fit the separate condenser to an atmospheric engine; this was sometimes done by those pirating Watt's invention, and occasionally even on Watt's own suggestion, but in this case the engine remained essentially atmospheric.[14] Watt achieved further gains in economy by enclosing the cylinder with a lid, and by fitting the whole inside an outer steam case so as to maintain the temperature of the cylinder and maximize the gains from his invention. The piston entered the steam case through a stuffing box, packed round with leather. However, the Newcomen method of sealing the piston inside the cylinder by a film of water across the top of the piston could not then be used. Watt instead experimented with a variety of oils, animal fats, and greases in conjunction with hemp or other packing to perfect the seal.[15] Even with these, though, the seal was probably not so effective as in the Newcomen engine. Moreover (and partly as a result), the cylinder itself had to be bored much more truly than for the 'fire engine'.[16] Even the method Smeaton had devised at the Carron ironworks for the improved atmospheric engine was inadequate, but the difficulty was eventually

---

[13] Farey, i. 307–8.
[14] Watt at least once recommended a customer to convert his atmospheric engine, because the capital cost of the Boulton and Watt engine would be too high (Roll, *An Early Experiment*, p. 36). In general, however, Watt set himself very much against this course of action because he believed it would damage the reputation of his company; see Farey (quoting letter from Watt to Smeaton), i. 329–30.
[15] Farey, i. 326–7. See also H. W. Dickinson and R. Jenkins, *James Watt and the Steam Engine*, Oxford, 1927, pp. 223–4, and Apps. II–III.
[16] Farey, i. 317–18. It is not always pointed out in historical studies of the steam-engine that the enclosure of the cylinder *compelled* Watt to find a superior precision in boring his cylinders if he were to retain a high level of efficiency.

overcome by John Wilkinson's new kind of boring mill, patented in 1774.

The separate condenser and Smeaton's air-pump were both encased in a hot well, with the temperature kept at about 102°F. This allowed the steam pressure to be lowered to as little as about 1 lb p.s.i., permitting a greater load on the piston, etc.

The enclosure of the cylinder enabled the piston on its return stroke to provide further power. The idea of double action had been mooted before Watt, e.g. by Dr. Falck in 1779 using two cylinders in opposition, and Papin had envisaged the use of two cylinders in 1690.[17] By suitable design and working of valves and pipes, Watt was able to obtain the same result in a single cylinder, by admitting steam on alternate sides of the piston (patented 1782). This involved some loss of fuel economy, since the power obtained in the single-acting engine from the return of the beam's own weight was sacrificed, but under appropriate cost conditions this could be offset by a decrease in capital costs, in doubling the power obtained from an engine with the same dimensions.

At his partner's (Boulton's) insistence, Watt developed means of rotative propulsion. The story of how Watt was pre-empted in patenting the crank and flywheel for direct drive by Wasborough and Pickard is familiar, if often misrepresented.[18] To avoid contesting their patent (in view of his own very vulnerable position on the score of patent rights), Watt developed the epicyclic gear he called the 'sun-and-planet motion' to achieve the same ends. It gave some advantage in driving the shafts twice as quickly as a crank without intermediate gearing, and this was some help in early textile-mills.[19]

[17] Hills, op. cit., pp. 145–8; see Farey, i. 658 ff., and C. Matschoss, *Die Entwicklung der Dampfmaschine*, Berlin, 1908, vol. i, pp. 330–1, for later developments of these ideas.
[18] Far and away the clearest account of the situation is provided by Dickinson and Jenkins, op. cit., pp. 148–58. They argue that the idea of the crank was an old one, and Watt probably rightly thought it would be unpatentable. What Wasborough actually might have obtained from Watt's workman by deceit was Watt's idea of a revolving weight rather than the crank itself. Later Watt himself was to give credit to Wasborough for first realizing the *importance* of having a flywheel; Watt had of course considered the problem without appreciating at this stage its crucial significance.
[19] J. Kennedy, 'Observations on the Rise and Progress of the Cotton Trade in Great Britain', *Mem. Lit. and Phil. Soc. of Manchester*, 2nd Ser. iii, 1819. After 1800 Boulton and Watt discarded the sun-and-planet mechanism because of its expense and mechanical shortcomings.

## The Engine: Technology and Scale

The double-acting engine required a rigid connection between the piston rod and the indoors end of the driving beam, in place of Newcomen's equalizing chain. For this purpose Watt in 1784 patented his parallel motion, of which he later declared 'I am more proud of the parallel motion than of any other mechanical invention I have ever made.' This statement may be as valid in economic as in engineering terms, as will be hinted at in subsequent chapters. Parallel motion paved the way for the cast-iron beam (*c.* 1797). To regulate the drive still further, Watt borrowed the centrifugal governor from wind-powered flour-mills, replacing the Cornish cataract which Smeaton and he had earlier promoted (1787).

Finally, Watt included in his patent of 1769 the idea of expanding steam before condensation, although like the governor it was probably not a completely new idea, and others were perhaps developing it independently at much the same time.[20] Because of the low steam pressures that Watt held to adamantly, comparatively little gain in efficiency was possible at this stage. Moreover, the amount of power delivered declined as expansion progressed; thus to avoid uneven motion Watt did not employ expansion at all in his rotative engines, i.e. those intended for the mills of the early Industrial Revolution. Indeed he actively discouraged the use of higher pressures. Like his scepticism about steamboats (and his earlier disbelief in the viability of rotative motion) this was to prove one of his major misjudgements: it was through the use of higher pressures that substantial gains in fuel economy were to be made in the early nineteenth century.

The Briton most closely associated with such a move towards higher pressures was Richard Trevithick, first with a high-pressure non-condensing engine familiarly referred to as the 'puffer'. Up to 1810 Trevithick built quite a number of these,

---

[20] "It is not peculiar to Mr. Watt's engine, that the steam passage is shut before the piston arrives at the conclusion of its stroke, and that the parts are brought to rest by expanding the steam. The atmospheric engine is regulated in that manner (see pp. 148, 179, 181 and 210). In fact it must be so, or else the catch pins would strike the springs every time." Farey, i. 363 n. Cardwell believes that Hornblower consciously hit on the expansive principle independently of Watt, though later (op. cit., p. 83). In any case, as he also shows, the water-pressure engine, to be described hereafter in App. 6.2, utilized a kind of expansive action before even Watt had patented his rather vague notions.

all apparently double-acting engines employing Leupold's four-way cock.[21] The steam entered the cylinder at 25 lb p.s.i., and was cut off after impelling the piston through part of its stroke, the rest of the stroke being completed through the expansion of the steam already in the cylinder until its pressure fell to about 4 lb p.s.i. above the atmosphere.

In 1824 they were stated to be only about four-fifths as efficient in fuel consumption as condensing engines.[22] Their advantage lay chiefly in their compactness. The elimination of all the apparatus for condensing, the air-pump (which created extra friction), the beam, and supporting rods and framework meant that they were (a) portable (which lowered transmission charges and allowed some decentralizing of power) and (b) considerably cheaper in first cost than Boulton & Watt engines. The cost per horsepower of small Watt engines was comparatively high,[23] so that even if Trevithick engines consumed 25 per cent more coal they could work out cheaper *in toto* where appropriate factor-price conditions prevailed.

Development along each of these paths continued through the first sixty years of the nineteenth century. Even before Trevithick, Smeaton had recognized the advantages of portability; other eighteenth century engineers such as Edward Bull followed.[24] Matthew Murray developed a portable engine in 1805 which, roughly speaking, was a beam engine turned upside down. This engine became known as the 'side-lever', and was built in large quantities at Soho in the early nineteenth century, mainly for boats or sugar extraction. One of the most suitable of the portable engines for many purposes was Maudslay's table engine, patented in 1807.[25] In 1856 Morshead[26]

---

[21] A. Titley, 'Trevithick and Rastrick and the Single-Acting Expansive Engine', *Trans. Newcomen Soc.* vii, 1926–7.

[22] R. Stuart (Meikleham), *A Descriptive History of the Steam Engine*, 2nd edn., London, 1824, p. 167. Matschoss, op. cit., p. 422. Simon Goodrich was cautious about advocating the Trevithick engine on grounds of fuel economy after inspecting one in Tottenham Court Road in 1807; see E. A. Forward, 'Simon Goodrich and his Work as an Engineer", Part II. *Trans. Newcomen Soc.* xviii, 1937–8, p. 8.

[23] See Table 4.11.

[24] J. Smeaton, *Reports of the late John Smeaton*, i, London, 1812, pp. 223 ff. (describing a portable fire-engine, 1765).

[25] Elijah Galloway, *History and Progress of the Steam Engine*, London, 1830, pp. 194–6.

[26] W. A. Morshead, jr., 'On the Relative Advantages of Steam, Water and Animal Power', *J. Bath and West of England Soc.* iv, 1856.

considered that the disadvantages of moveable as opposed to fixed engines in agricultural uses were: (a) consuming more fuel in proportion to the work done, and less able to run on inferior-quality fuels, (b) suffering a high rate of wear and tear (he thought a good fixed engine would outlast two or even three portables), (c) not being able to make use of the waste steam, e.g. for heating. Against that, they involved a smaller capital outlay, not because the engines themselves were cheaper[27] but because they saved the expenses of engine- and boiler-houses and stacks, which together nearly doubled the cost of 4-horsepower engines.

Dispensing with the beam and substituting direct action also went to reduce capital costs. The direct-acting engine had been used on his boat, the *Charlotte Dundas*, by Symington in 1802, by laying the cylinder on its side. Locomotive engines were, of course, direct-acting, so that progress went ahead on this simplified layout once the railways and steamboats attained greater prominence. Morshead further noted that '. . . as the beam-engine is heavier and more expensive than direct-acting ones, it is seldom employed for engines under 10 horse-power, unless condensing, in the case of which it affords facility for working the air-pump'. The oscillating engine, in which the cylinder is mounted on trunnions and rocks to and fro, was originally designed by William Murdock, but his employers Boulton & Watt did not encourage him to take it further. However, the company was producing later versions, for use on land as well as in steamboats, by the 1840s, and they too were usually less expensive than beam engines.[28]

As I have hinted above, economy of space and fixed cost did not suit all customers' requirements, certainly in so far as it

[27] Morshead's figures showed that, compared to fixed condensing engines, portables were £40 more expensive for a 4-horsepower engine and £25 more for one of 8 horsepower.
[28] The cost situation can be verified from the Boulton & Watt accounts. In 1842 they quoted J. Barrow £16,300 for 2 185-HP beam engines, compared with £15,350 for direct-acting engines of the same capacity. Later in that year they quoted C. Buschek of London £9,700 for a pair of 100-HP beam engines, compared to £8,200 for equivalent direct-acting engines. Similarly, the Herne Bay Co. was quoted £5,300 for 2 60-HP oscillating engines (including boilers), i.e. £44 per horsepower, and £5,950 for 2 65-HP beam engines (£45 per horsepower). All these from the 'List of Estimates rendered by the Boulton & Watt Coy., between the 30th Sept. 1840 and 21st Jany. 1841 [*sic*]'. Boulton & Watt MSS.

meant a sacrifice in economy of fuel. The high-pressure engine could, however, be made thermodynamically efficient in a number of ways, of which the two most important were: (i) still higher pressures—Trevithick used up to 120 lb p.s.i. in his 'plunger-pole' engines,[29] whilst Jacob Perkins attempted as much as 1500 lb p.s.i. in foreshadowing the Uniflow engine—but these standards were generally unacceptable in the days before steel boilers; (ii) condensing the steam after it had fulfilled its work (through expansion) at high pressures, and thus mating the Trevithick-type high-pressure engine with the Watt-type low-pressure one. This second option embraced the most important advances that I shall be concerned with in Part III of this book.

The condensation could be conducted either in the same cylinder or in a separate one. The latter principle, compounding, had a longer history: Jonathan Hornblower attempted to patent a compound engine in 1781, but the patent was ultimately revoked because Watt detected a disguised form of separate condenser in the apparatus, and that infringed his own patents of 1769–75. In fact there was not a great deal of advantage in Hornblower's practice, because, like Watt, he continued to use steam at low pressures, so that the gain from expansion was comparatively slight. His Cornish compatriot, Arthur Woolf, began building similar engines early in the nineteenth century, but it was not until he began erecting engines of this type in Cornwall after 1814, and imitated Trevithick by raising steam pressures to 3 or 4 atmospheres, that he achieved full success. Subsequent developments in the Cornish engine are dealt with at length in Chapter 9. More rough-and-ready solutions to compounding were adopted in manufacturing districts such as Lancashire and Glasgow, and some of these will be described in Chapter 4.

It is therefore misleading to see the pattern of progress as linear and inevitable: in explaining the direction and chronology of 'technical progress' in the economist's sense, it is vital to keep this diversity in mind.

[29] Titley, op. cit.; J. Gaudry, *Traité élémentaire et pratique . . . des machines à vapeur* Paris, 1856.

## The Engine: Technology and Scale

### ESTIMATES OF TOTAL HORSEPOWER

One dimension in which these technical improvements were translated into higher performances was in increasing the horsepower of the most powerful engines. (We shall find in Chapter 4 that, as suggested in the previous paragraph, this was by no means the sole objective, and often had to be sacrificed to others.) Savery engines were generally very small, and proved inefficient when sizes much above 4 to 6 horsepower were tried (see Chapter 4). The size of the largest Newcomen engines at each date has been estimated by S. B. Hamilton (see Table 2.1).[30] It is not clear that his list was intended to be complete,

TABLE 2.1

*Cylinder diameters and power of the largest extant Newcomen engines*

| Locality | County | Date | Cylinder diameter (in) | Estimated horsepower |
|---|---|---|---|---|
| Dudley | War. | 1712 | 21 | 5 |
| — | Various | 1720s | 15–24 | — |
| Heaton | Northumb. | 1733 | 33 | 15 |
| Heaton | Northumb. | 1734 | 42 | — |
| Horsehay | Staffs. | 1755 | 48 | 30 |
| Horsehay | Staffs. | 1757 | 60 | 45 |
| Walker Colliery | Northumb. | 1763 | 74 | 52 |
| Whitehaven | Cumb. | 1810 | 82 | 100+ |

and practically certain that it is not,[31] but it helps to give a general impression of the growing capabilities of the largest engines. Boulton & Watt soon had engines of greater capacity: the Hawkesbury Colliery engine with a 58-inch cylinder in 1776 and then the Wheal Maid engine of 63-inch in 1787 were

---

[30] S. B. Hamilton, 'Papers on Newcomen Engines in Previous Volumes of "Transactions"', *Trans. Newcomen Soc.* xxxviii, 1963–5. The 1810 Whitehaven engine employed a separate condenser.
[31] Note also the early 47-inch engine at Wheal Fortune tin mine (1720—Lord, *Capital and Steam Power*, p. 38). There was a 76-inch engine noted at Wednesbury in 1775 by La Houlière (see J. Chevalier, 'La Mission de Gabriel Jars dans les mines et les usines britanniques en 1764', *Trans. Newcomen Soc.* xxvi, 1947–9).

successively rated as the most powerful in the world.[32] By the second decade of the nineteenth century Watt engines in Cornwall could do even better: Stoddart's engine at United Mines were said to be capable of 250 horsepower, though for reasons that will emerge in the next chapter it performed on average at less than half this.[33]

One difficulty with rating the size of engines is that the unit of 'horsepower' was devised by Watt to describe his later rotative engines. His own reciprocating engines, as well as those of the atmospheric or Cornish type, were rarely designated in these units. Generally one has to approximate the horsepower (i.e. lb lifted 1 ft high per minute, or work done in unit time) by the size of the engine's cylinder, as is partly done in Table 2.1. In smaller sizes there is little difficulty in doing this, because Watt's measure is of 'nominal horsepower' (NHP) and takes no account of the (in)efficiency of the particular engine in practice. All engines with the same cylinder diameter and length from Soho are of the same NHP.

A greater difficulty is that Watt's NHP, which had always been a somewhat moderate estimate of the real work done by his best engines,[34] came less and less to reflect 'indicated horsepower' (IHP) in the nineteenth century. By the 1850s NHP is still useful for, say, estimating the backward linkages from steam-engines, since it remains an estimate of the physical size of engines, but it is less adequate for gauging the total real power supplied, as would, for example, be necessary if the social savings were to be computed at that date. By the 1860s Fairbairn noted that two engines of 100 NHP at Saltaire were working at 228 and 235 IHP.[35] Rankine estimated IHP as

[32] W. H. B. Court, *Scarcity and Choice in History*, London, 1970, p. 228; Farey, i. 339, 428–9.
[33] *The Cyclopaedia* (ed. A. Rees), London, 1819, xxxviii, puts Stoddart's engine at 250 HP. But cf. Farey, vol. ii, pp. 100–1, for actual performance.
[34] Farey, i. 576; ". . . all good engines, with a suitable allowance of fuel, are capable of exerting half as much more power, as that at which they are rated; for instance, a 20 horse engine can exert 30 horse power; a 40 horse engine 60 horse power, and so on; see p. 486. The nominal horse power by which each engine is rated in the table, is that exertion which it is competent to overcome in the most advantageous manner for a continuance, considering all the attendant circumstances, such as expense of fuel, wear of the machinery, and consequent stoppage and expense for repairs, and the first cost of the engine." Cf. the model explaining machine speeds in Ch. 7 below.
[35] W. Fairbairn, *Treatise on Mills and Millwork*, vol. i, London, 1861, pp. 233–5.

## The Engine: Technology and Scale

being anywhere from one and a half to five times the NHP in the 1850s;[36] the former ratio is probably applicable to Watt-type engines, and the latter to the best high-pressure compounding engines.

In his book John Lord was able to list 321 Watt engines erected by 1800. Dickinson showed that Lord's figure was far short of the total, and substituted one of 496 engines. However, this errs slightly the other way, through double-counting in cases of re-erection. If known cases of re-erection are deleted the number of Watt engines built (or converted from Newcomens) is brought down to 490 or possibly less (12 of them exported).[37]

Much the most serious attempt so far to collect and reconcile the many and diverse accounts of atmospheric engines in the eighteenth century is that of Professor J. R. Harris.[38] Harris splits the eighteenth century into three subperiods. In the first of these, the years of the Savery–Newcomen patent to 1733, he finds at least 52 and perhaps as many as 60 engines installed. Thereafter reliability slumps. The minimum count for 1733–81 is 223 engines, though this is clearly incomplete, and 300 or more may be more realistic. After 1781 there are the Watt engines to contend with, plus those Savery and Newcomen engines which continued to be built. On the basis of regional extrapolations from counties at present surveyed, Harris puts the minimum for the latter at about 220, and suggests that more than double this (470) may be appropriate, i.e. virtually as many non-Watt engines as Watts. From these results the minimum number of engines that made their appearance at some stage of the eighteenth century in Britain was 985; Harris thinks his expanded total of 1330 more reasonable.

However, Harris's figures were intended for rather different purposes than mine. He notes that his figures cannot exclude double-counting, especially of engines that were originally erected in one county (or even on one site), then moved to

---

[36] W. J. M. Rankine, *A Manual of the Steam Engine and other Prime Movers*, London and Glasgow, 1859, p. 479. See also M. Blaug, 'The Productivity of Capital in the Lancashire Cotton Industry during the Nineteenth Century', *Econ. Hist. Rev.* 2nd Ser. xiii, 13, 1960–1, pp. 380–1.
[37] Dickinson, p. 88; Boulton & Watt Catalogue of Engines.
[38] J. R. Harris, 'The Employment of Steam Power in the Eighteenth Century', *History*, lii, no. 175, June 1967.

another; and from casual evidence he infers that such occasions were hardly uncommon. Thus the number actually built must have been less than 1330 (if the latter is correct as the unadjusted total); this in turn will affect the calculation of backward linkages in Chapter 5. Moreover, the figures would greatly overstate the number of engines actually in use at any one time, which is what is required for the social savings calculation in Chapter 6. A few Newcomen engines were probably exported as well as the 12 Watts, e.g. the engine designed by Smeaton and sent from the Carron Company to Cronstadt (Russia) in 1777—though these may not have appeared in Harris's total anyway. It is improbable that any engines were imported in the eighteenth century. More importantly, the number of engines actually at work in the British Isles at any one time was likely to be well below the number installed. For instance only 57 of the 99 engines surveyed by Smeaton on the Northumberland coalfield in 1769 were at work,[39] whereas Harris's total includes them all. Matthew Boulton wrote to Watt, in about 1775, that '. . . The whole number [of pumping engines] in the County of Cornwall are exactly 40, but there are only 18 of 'em in work on account of ye high price of coals.'[40] The list of working Cornish engines given by Lord indicates that this was not uncommon, and Boulton & Watt engines were later to suffer equally on occasions.[41] Finally, of course, engines fell into disuse and were dismantled. Boulton recorded that there was only one Newcomen engine left working in Cornwall by 1783.[42]

Harris has not been prepared to go beyond a count of engines and to guess the total horsepower *in situ*. Dickinson put the aggregate power of Watt engines built by 1800 at 7500 horsepower, by averaging 500 engines at 15 horsepower apiece. But if he slightly over-estimated the number of engines Dickinson much more seriously underestimated their average size, probably through being misled by Lord's erroneous work. For instance, Lord allowed for 21 engines erected in Cornwall,

---

[39] For a list of the 99 engines, together with the cylinder size of many of them, see e.g. R. L. Galloway, *Annals of Coal Mining and the Coal Trade*, vol. i, London, 1898, p. 262.
[40] Roll, *An Early Experiment*, p. 67.   [41] Lord, pp. 155–8.
[42] Roll, *An Early Experiment*, p. 77.

and set down an impressionistic 20 horsepower per engine to get a total of 420 horsepower in Watt engines in the Duchy.[43] My figures in Chapter 6 show over 3500 horsepower in Cornwall by 1800, eight times as much, although the total number of engines counted is 53 only. An average of 15 to 16 horsepower an engine, as used by Lord and Dickinson, is reasonable for Watt's rotative engines, but his reciprocating (pumping) engines were much larger. Working up figures provided in the Boulton & Watt archives, I have pushed the total in Watt engines up to 12 750 horsepower.

Non-Watt engines were of lower power than Watt's, since an atmospheric engine had to be much larger in size to achieve the same effect as a Watt, sometimes impracticably so. Where Watt engines were installed at collieries, it was usually to provide greater power than one Newcomen engine could. True, an overwhelming proportion of atmospheric engines were found in mining, where power demands tended to be relatively large, but even there many of them were small engines, e.g. for winding up coal. An average of 15 to 20 horsepower for non-Watt engines is probably generous. This would give a range of total horsepower in engines of all types built by 1800 of from 25 200 to 29 350—say 25 000 to 30 000. How much of this still survived in 1800 is unknown; much less is it known how much was actually at work in or around 1800.

In the nineteenth century the quality of the data barely improves before the late censal years of the century. The only available 'guesstimate' is that by Mulhall.[44] He estimated 350 000 horsepower as stationary steam-power in 1840, rising

---

[43] Lord, pp. 167–75.
[44] M. Mulhall, *Dictionary of Statistics*, London, 1884. For other figures mainly based on Mulhall, see Matschoss, op. cit., pp. 135–6. At the time of going to press, an article by A. E. Musson appeared, covering much the same ground as the rest of this chapter ("Industrial Motive Power in the United Kingdom, 1800–70", *Econ. Hist. Rev.*, 2nd Ser., vol. XXIX, August 1976). The interpretations placed on the material seem to me similar to mine below, and I leave my judgements as originally expressed for the interested reader to compare. Three slight reservations about Musson's results may be worthy of note: (a) Musson uses Dickinson's figure of 7500 horse-power in 1800, that I have been discrediting above. (b) Musson seems to me to understate the steam horse-power in mining, and hence to overstate that in textiles. (c) The year 1850 that he uses as a base in the later parts of his article (op. cit., pp. 434–6) is a particularly bad one because of the rapid changes in ratios of nominal to indicated horse-power about that time, as indeed earlier sections of his article (p. 422) imply (see especially Chapter 7, below).

to 500 000 in 1850 and 700 000 in 1860 (2 000 000 in 1880). As with many of Mulhall's figures, the origins of these estimates are not known or revealed. It is my impression that the figures are too high for the early years of the century and hence underestimate the growth rate of stationary steam-power installed over the time with which I am concerned. To prove this would of course require the provision of an entire alternative set of estimates, and this book does no more than begin in one corner (textiles). It is, however, quite an important corner, or at least so Alfred Marshall believed: 'The conclusion that in 1839, the textile industries were the chief users (mining again being left out) of steam machinery is suggested by the fact that even in 1851 they owned ¾ of all the factories which employed more than 350 persons.'[45] Marshall noted that in 1907 textiles used one-quarter of industrial horsepower other than in mining, and clearly regarded this as a major reduction since the first half of the nineteenth century.

Now Mulhall's figures may come from extrapolating the figures of textile power to industry as a whole, using as a base the proportion estimated for 1890; but Marshall's comments and other information indicate that it would be misleading to assume that this proportion was constant. Another possible source of Mulhall's series is perhaps the occasional wild guess of contemporaries as to steam-power in use. We may dismiss the gross exaggerations of frustrated French engineers such as Dupin, Arago, and Pecqueur.[46] Partington in 1822[48] estimated that there were not less than 10 000 engines in the country producing at least 200 000 horsepower. This would mesh well with Mulhall's figure for 1840 (Partington is probably including steamboats), but it does not seem likely that eight to ten times as many engines would have been built over the war-influenced years 1800–20 as from 1712 to 1800, unless they were very small engines, of much less than the required 20 horsepower each.

[45] A. Marshall, *Industry and Trade*, London, 1919, p. 59.
[46] C. Dupin, *Discours et leçons sur l'industrie, le commerce, la marine*, . . . Paris, 1825, estimated "at least 320,000 HP" (p. 149 n.); D. F. Arago, *Œuvres*, v, Paris, 1848, believed there to be 10 000 engines with a total force of 600 000 horses in 1819 (p. 207); C. Pecqueur, *Économie sociale*, made various extravagant claims in computing the social savings of engine and machinery in textiles to be £20 m. in 1824 (Paris, 1839, i, p. 6).
[47] C. F. Partington, *An Historical and Descriptive Account of the Steam Engine*, London, 1822.

## The Engine: Technology and Scale

Some better-established estimates are available for particular cities:

(i) *London*. Farey made a count of engines in each industry in London in 1805,[48] and found a total of 112 steam-engines then at work, exerting about 1355 horsepower. Of these engines Boulton & Watt had manufactured 46, totalling 648 horses, at Soho by 1800. By the time he was writing, which seems to have been 1825, Farey could note:

In London there are about 290 steam-engines for water-works, small manufactories, and steamboats; they amount to about 5460 horse-power, or about equal to the strength of 48 000 men working continually. In Manchester there are about 240 large steam-engines in manufactures, exerting about 4760 horse-power, or 42 000 men. In Leeds about 130 steam-engines in manufactures, exerting about 2330 horse-power, or 20 500 men. In Glasgow there are 80 or 90 steam-engines. At Birmingham, Sheffield and particularly in all the smaller towns of Lancashire and Yorkshire, steam-mills are very numerous, and many of them are on an extensive scale.[49]

It does not appear that Farey is here trying to underestimate steam-power in any way, though his greater conscientiousness than most of his rivals as a historian of the steam-engine probably leads him to make much smaller upward allowances for error.

(ii) *Manchester*. Farey goes on to note a total of 430 horsepower in Manchester in 1800, from about thirty-two engines.[50] It is not exactly clear whether he means patent (i.e. Boulton & Watt) engines only, or all including 'fire engines', but the former is more likely. The industrious research of Musson and Robinson has shown fairly convincingly that '. . . it is probable that they [Boulton & Watt] built not more than a third of the engines in the county [Lancs.] during that time [1775–1800]'.[51] Musson and Robinson state that engines built by rivals were 'mostly in the Manchester area', though evidently they have wide boundaries in mind—only 6 of the 16 Bateman & Sherratt engines they concentrate upon are noted as being erected in Manchester–Salford, and only one of the other atmospheric engines. Given that most of the non-Watt engines were

[48] Farey, i. 654.   [49] Ibid. i. 7 n.   [50] Ibid. i. 654.
[51] Musson and Robinson, *Science and Technology*, p. 426.

probably quite small (several of Bateman & Sherratt's went to calenderers, etc.), a total of 650–750 horsepower seems plausible for Manchester at this date.

Apart from Farey's estimate of 240 engines totalling 4760 horsepower in Manchester in 1825, quoted above, there is supporting evidence from Baines,[52] who reckons 212 engines totalling 4875 horsepower in September of that year. Baines's estimate for the whole of Lancashire at that time is 1541 engines of 30 835 horsepower altogether, from which one should deduct 67 engines in Stockport (1965 HP) and 79 steam-boat engines at Liverpool (3931 HP), leaving 1395 engines and 24 939 horsepower—or say 25 000 horsepower, since 500 engines are averaged at the lowish figure of 15 horses each. Thus Manchester has nearly one-fifth of the steam horsepower of the county by the mid-1820s.[53]

A committee of the highly active Manchester Statistical Society ascertained 'the amount of steam-power employed in

TABLE 2.2

*Horsepower in use in Manchester, by occupation, 1837*

| Industry | Manchester | Salford | Together | % of Total |
|---|---|---|---|---|
| Cotton-spinning and weaving | 5272 | 764 | 6036 | 60·8 |
| Bleaching, dyeing, printing, etc. | 756 | 521 | 1277 | 12·9 |
| Cotton thread and smallwares | 270 | 36 | 306 | 3·1 |
| Fustian-shearing | 46 | 34 | 80 | 0·8 |
| Silk-throwing and manufacture | 237½ | 104 | 341½ | 3·4 |
| Flax-spinning | — | 70 | 70 | 0·7 |
| Woollens | 36 | 22 | 58 | 0·6 |
| Total textiles | 6617½ | 1551 | 8168½ | 82·3 |
| Total, all purposes | 7926½ | 1998 | 9924½ | 100·0 |

[52] E. Baines, sr., *History, Directory, and Gazetteer of the County Palatine of Lancaster*, Liverpool, 1825, ii, p. 740.
[53] If one accepts Musson and Robinson's assertion of Boulton & Watt building one-third of the engines in Lancs in 1800, and assumes they were the same average size as the others, then total Lancastrian horsepower in 1800 comes to 2670, of which Manchester ought to be contributing at least one-quarter. However, even if Musson and Robinson are right about the ratio of Watt to non-Watt engines, it is unlikely that the horsepower of the latter would be on average as great as the former, so that Manchester's predominance in 1800 would be still more imposing, and its relative decline thereafter more pronounced.

## The Engine: Technology and Scale

the various branches of manufacture, in the parliamentary boroughs of Manchester and Salford, 1837–8', as shown in Table 2.2.[54]

(iii) *Bolton.* Baines put the steam-power in Bolton and vicinity at 1604 horsepower (83 engines) in 1825. In 1836–7 Henry Ashworth, of the well-known cotton family, estimated the number and power of engines and water wheels in the 'whole district' of Bolton as:[55]

|  | Steam engines | Horse-power | Water-wheels | Horse-power |
|---|---|---|---|---|
| For all purposes | 308 | 5251 | 60 | 1170 |
| Cotton | 90 | 2448 | 18 | 491 |

This 'whole district' must have ranged quite widely. According to the Factory Returns,[56] the parish of Bolton contained 39 engines totalling 1082 horsepower for textiles in July 1835, while the neighbouring parish of Dean added another 15 engines of 307 horsepower altogether. In water-power Bolton was listed as having 10 wheels of 261 horsepower and Dean 7 wheels of 74 horsepower. This figure of Ashworth's is useful as one of the few district totals including a then-important colliery region.

(iv) *Leeds.* Farey reckoned 270 horsepower from 20 engines in Leeds by 1800; these must presumably be Boulton & Watt engines, since their own MSS include a count of 276 horsepower as early as 1797 by Lawson (though this count goes as far afield as Birstal). The town produced the great mechanic, Matthew Murray, but it is doubtful whether many of his engines were operating in Leeds before the turn of the century (Murray & Wood engines of 30 HP for the flax-spinners, Marshall & Benyon, in Holbeck, and 40 HP at Castle Foregate, were both supplied in 1799[57]). Goodchild noted 53 engines in woollen- or flax-mills in the West Riding by 1798 (including 8 Boulton & Watts totalling 184 horsepower).[58] Several had

---

[54] *J. Stat. Soc.* i, 1838; also P.P. 1871: XVIII, p. 1009.
[55] P.P 1871, loc. cit., also Ashworth MS. 2108.
[56] P.P 1836: XLV.   [57] Boulton & Watt MSS.
[58] J. Goodchild, 'On the Introduction of Steam Power into the West Riding', *South Yorks. J.*, pt. III, May 1971, and other information from Cusworth Hall Museum, Doncaster.

already been pulled down or destroyed by 1800. Of the non-Watt engines remaining there were 6 counted in Leeds itself, 2 in Holbeck, and 1 in Hunslet, with a likely total of 100–120 horsepower. A reasonable allowance for Leeds in 1800 might therefore be about 450 horsepower.

As well as Farey's 120 engines with 2330 horsepower in 1825, there are outside estimates for Leeds at this later date. In March 1824 one William Lindley calculated the number of engines in Leeds and its immediate vicinity as 129 and their total power 2318 horsepower, listing all users and manufacturers individually.[59] The close correspondence with Farey's figure suggests that it may be Farey's source (Farey was working for Marshall's, the great flax-spinning firm, in 1823). Of the engines then in use Boulton & Watt had built only 7 (of 170 horses); the local firm of Fenton, Murray & Wood had put up 70 engines of 1493 horsepower that were still operating. For March 1830 the local historian Parsons[60] reckoned there were 225 engines in the parish in all uses, of 4048 horsepower altogether.

(v) *Birmingham.* R. J. Forbes stated that by 1820 there were some 60 engines of about 1000 horsepower in total in Birmingham.[61] Fifteen years later the Birmingham Philosophical Society drew up a list of engines according to the date of erection and purpose to which they were applied.[62] Their result was 169 engines of 2700 horsepower then at work, not counting another 8 or more engines totalling 162 horsepower that had been pulled down. A parallel list for the borough in December 1838 (obviously based on the same information but with quite a few changes for individual engines)[63] reports 240

---

[59] Brotherton MS. 18, Univ. of Leeds.
[60] E. Parsons, *Civil . . . History of Leeds etc.*, Leeds, 1834, vol. ii, p. 203. Robert Baker, whose statistical inquiries on health, etc. are valuable for this period, puts the total horsepower in Leeds in 1858 at 5590, of which textiles contributed 3950 ('On the Industrial and Sanitary Economy of the Borough of Leeds in 1858', *J. Roy. Stat. Soc.* xxi, 1858). This would give a very low growth rate for steam-power 1830–58, so presumably the data of Parsons, etc. extend well beyond borough boundaries.
[61] In C. J. Singer, E. J. Holmyard, A. R. Hall, and T. I. Williams (eds.), *A History of Technology*, vol. iv: *The Industrial Revolution*, Oxford, 1954–8, Ch. 5. The horsepower obtained from counting the ensuing two lists is 1262 for 1820, of which some may have already been pulled down.
[62] W. H. B. Court, *The Rise of the Midland Industries 1600–1838*, London, 1938, p. 197.
[63] *J. Stat. Soc.* ii, Jan. 1840, p. 440.

## The Engine: Technology and Scale 35

engines of 3595 horsepower by that time, including 159 horsepower 'void and removed'.

(vi) *Glasgow*. The local historian James Cleland estimated 45 steam-engines averaging nearly 16 horsepower each in 1817 (i.e. 720 horsepower).[64] By 1825 he was able to count 242 engines of 4480 horsepower (excluding steamboats);[65] and in 1831 the census listed 250 engines of about 4400 horsepower.[66] Although the author and the district he surveyed remained the same in 1825 as 1817, it seems unlikely that the 1817 figure can be accepted—though it is more compatible with Farey's estimate of '80 or 90 engines', also for 1825.

(vii) *Ireland*. A more comprehensive account is available for Ireland in 1838.[67] In that year Belfast had 50 engines totalling 1274 horsepower, Dublin 29 engines of 438 horsepower, and the rest of Ireland 72 engines comprising 1399 horsepower. As with the Birmingham estimates for that year, all engines are individually listed, with date of erection. From 1810 the computed growth rate is 12 per cent p.a. for those engines that are dated, although this is likely to be an over-estimate, since the erection date of (presumably) older engines is often unknown, and this gives an artificially small base.

With all these figures uncertainty about the geographical coverage severely hinders any attempt to aggregate them into an over-all view. Court notes that the Boulton & Watt factory, as close as Soho, is not included in the Birmingham count for 1838.[68] It is not at all clear, either, whether steam-power was on balance moving into or out of the older-established industrial areas over this period. Their position may be approximately recapitulated as in Table 2.3. It seems unlikely that these six centres could have accounted for more than about 50 000 horsepower in 1840; whereas Mulhall's total for the U.K. in that year, noted above, was 350 000. Two plausible conclusions are possible: (i) measured by horsepower, the bulk of early steam-power arose in mining and similar scattered activities; or (ii) Mulhall's is an over-estimate. In Chapter 5 I shall very roughly put the horsepower in textiles alone (59 800 in 1838)

---

[64] J. Cleland, *Description of the City of Glasgow*, 2nd edn., Glasgow, 1840, pp. 98–9.
[65] H. Hamilton, *The Industrial Revolution in Scotland*, Oxford, 1932.
[66] Ibid., p. 211.
[67] P.P. 1837–8: XXXV, pp. 856–7.   [68] Court, op. cit., p. 256.

TABLE 2.3

*Approximate horsepower in some major industrial towns, 1800–1840*[69]

|  | Early 1800s | Mid-1820s | Rate of growth (%) | Late 1830s | Rate of growth (%) |
|---|---|---|---|---|---|
| London | 1355 | 5460 | 7·3 | — | — |
| Manchester | c. 650–750 | 4760–4875 | 7·7–8·4 | 9925 | 6·3 |
| Bolton | — | 1604 | — | 5251 | 10·4 |
| Leeds | c. 450 | 2318–2330 | 6·8 | — | 7·6* |
| Birmingham | — | 1000–1262 | — | 3436 | 5·7 |
| Glasgow | — | 4480** | — | — | — |

\* = 1800–30 growth rate (but cf.n.60).
\*\* = Farey's count of engines suggests that Glasgow had about two-thirds as many as Leeds.

at one-quarter the whole amount in the British Isles. Mulhall's figure is tantamount to assuming an annual growth rate of installed horsepower since the turn of the century of between $6\frac{3}{4}$ and $7\frac{1}{4}$ per cent, compatible with the results shown in Table 2.3 from 1800 to the mid-1820s, but less so thereafter. My alternative would bring the growth rate of steam-power down by one percentage point. In either case there is little indication in the Table of support for the thesis that steam-power led to an early concentration of power-using industry in a few major urban centres; this seems to have been a nineteenth-century rather than an eighteenth-century phenomenon. Total horsepower in Cornwall alone appears to have been about three times as large as in London and six times that of Manchester in 1800.[70]

The sketchy nature of the available data for the years after 1800 accounts for my diffidence in taking the social-savings calculations (in Chapter 6) beyond 1800. It would clearly be desirable to press ahead with gathering more soundly based information for these years. In view of the widely dispersed pattern of engines and horsepower thrown up by the above

[69] Sources as already described.
[70] Very crudely estimated from the data in Chapters 4 and 9, assuming full employment at "economical working" rates (cf. n. 33 of this chapter). 1800 figures from Table 6.5.

material, however, I do not believe that final salvation lies in working more and more intensively over such scraps. There is much more to be learnt by taking the matter industry by industry, and working forwards to and backwards from data on industrial output and power productivity. This is the approach I have chosen for one industry—textiles—in Chapters 7 and 8; there are still many years of work in prospect before any fairly satisfactory estimates can be expected.

# 3. The Economic Issues

PROCEDURES THAT have become orthodox over recent years, such as the concept of 'social savings', are employed in Part II of this book. No extended justification for either quantification or 'counterfactualism' will be attempted; indeed, many of the arguments raised against these approaches seem sterile where not actually misconceived. But this is far from declaring that the serious problems in using this methodology have all been resolved. In this chapter I shall be surveying some of these difficulties as a background to the techniques of empirical estimation that are used in Parts II and III.

## THE 'SOCIAL SAVINGS'

Professor Fogel gauged the impact of a new technique by its 'social savings'. This denotes the quantum of scarce resources formerly tied up in production (or transportation, etc.) now released to seek employment in other productive enterprises.[1] Full employment of the resources is assumed, both before and after; in other words, competitive equilibrium prevails, and prices equate with marginal costs. If there is unemployment in the world with the older technology, the social savings are likely to be overstated (price of labour or capital will be higher than marginal cost), but the contribution of the innovation would better be seen in terms of employment-creating effects than cost-reducing ones.

In his study Fogel compared railroad freight rates on wheat and other commodities between Chicago and New York in 1890[2] with those for other means of transportation. In fact, a

---

[1] R. W. Fogel, *Railroads and American Economic Growth: Essays in Econometric History*, Baltimore, 1964. For a review of reviews of Fogel's book, see J. A. Dowie, 'As if or not As if: the Economic Historian as Hamlet', *Austral. Econ. Hist. Rev.* vii, 1967. For discussion of the concept of "social savings" see P. A. David, 'Transport Innovation and Economic Growth: Professor Fogel on and off the Rails', *Econ. Hist. Rev.* 2nd Ser. xxii, 1969: A. Fishlow, *American Railroads and the Transformation of the Ante-Bellum Economy*, Cambridge, Mass., 1965: G. Gunderson, 'The Nature of Social Saving', *Econ. Hist. Rev.* xxiii, 1970.

[2] Freight rates from St. Louis to New Orleans are used for meat.

combination of canals and waterways turned out to be cheaper per mile for shipping wheat than rail, but the quantitative allowances Fogel made for a variety of supplementary costs incurred by water transportation reversed that result. The net cost advantage of railways was multiplied by the gross quantities actually shipped by rail in 1890 to establish the 'social savings' on each commodity.

Perhaps the most serious drawback of this approach is that the mode of calculation biases the result towards zero.[3] This is because under the pressures of market competition waterways which are more costly for shippers than the equivalent railway lines will simply be ignored, and cease operating. In the sixty-two years from the opening of North America's first railroad up to 1890 many waterways eventually succumbed to rail competition and were deserted—indeed, some of Fogel's later calculations postulated the reopening of these very waterways.[4] On the other hand, those canals which survived to 1890 did so either because they went unchallenged by railroads for some reason or, more generally, because they were able to continue to attract some custom away from their railroad competitors—presumably by setting freight rate charges on a par with rail rates. Thus comparing a rail service with a parallel shipping service when the latter has been able to keep a foothold—as was the case with the Chicago to New York routes—is likely to give an artificially small result.

In equilibrium, then, the higher-cost suppliers of waterway services will have disappeared. The long-run marginal cost curve[5] may be approximated as in Fig. 3.1. Before the advent of the railroads, waterways are shown to be transporting the quantity of goods the requisite distance ($TM'$) at a charge of $OP'$ per ton-mile, so that total waterway revenue is $OP'Q'TM'$. The coming of the railways forces charges down to $OP$, and waterways whose own minimum marginal costs are higher than

---

[3] This was noted by S. Lebergott, 'U.S. Transport Advance and Externalities', *J. Econ. Hist.* xxvi, 1966, and P. D. McClelland, 'Railroads, American Growth and the New Economic History: A Critique', *J. Econ. Hist.* xxviii, 1968. Both studies are seriously flawed when they attempt to correct for the biases specified.
[4] Fogel, op. cit., pp. 92–107.
[5] The emphasis on long-run costs implies that we are uninterested in the fixed versus variable costs dichotomy underlying the shape of short-run marginal costs; this was essentially McClelland's confusion (cf. n. 3 of this chapter).

this will be squeezed out. Some waterways, however, will be more favourably endowed or will have other attractions for shippers, and they will continue to transport *TM* at a price *OP*.[6]

Now the social savings approach conjectures that the whole of the new technology, here railroads, is destroyed by a kind of

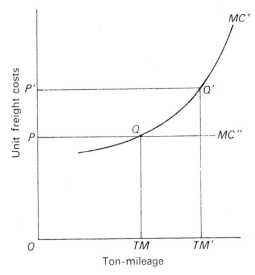

Fig. 3.1. Schematic cost curves for transportation with and without railways

strategic bombing, so carefully that, as a first approximation, all else is left intact. It is clear from the figure that taking water-freight rates as on currently functioning canals and waterways, such as the Chicago–New York route, is tantamount to assuming a long-run marginal cost curve that is horizontal, i.e. *MC"*, when it comes to foisting the rail traffic hypothetically on waterways. In historical fact the true marginal cost curve seems more like *MC'*, so that the calculation which uses *MC"* and the freight rate *OP* will, other things being equal, under-

---

[6] Even with competitive equilibrium fully established, the freight rates on each form of transportation need not be identical, i.e. at OP. Individual shippers may vary widely in what they require of a transportation service, so that in their demand functions relative price per ton-mile may be only one of n arguments. Fogel was of course broaching this by trying to evaluate the economic disadvantages associated with, e.g., slower speeds, but note that he was compelled to use market valuations to do so.

state the actual social saving. (Strictly speaking, Fogel does estimate some elements of that portion of $MC'$ lying above $Q$ by computing certain costs which were not paid as things were but would have had to be if greater strain had been placed on shipping capacity, e.g. the cost of winter freeze-ups. These help to explain why he eventually unearths a positive social saving. Nevertheless, at $TM'$ he still has a horizontal marginal cost curve.)

Professor Gary Hawke, in an equivalent study of English and Welsh railways for 1865, did not leave himself open to this objection, for he chose water-routing costs and charges as in the pre-rail era (mainly the mid-1820s). This, however, supposes zero technical progress in canals and waterways from 1825 to 1865—an assumption which, if historically incorrect, imparts a further upward bias to his results.[7]

The estimation of $OP'$ necessarily involves the unobservable. For example, the first calculation in Chapter 6 is the social saving from Watt's inventions. The method adopted is to 'freeze' the traditional technology at its state before the Watt engine took over (i.e. atmospheric engines, water-wheels, horse gins, etc.). The object is to obtain the cost structure, so that it will be possible under not too ferocious assumptions to guess what costs would have been in the year 1800, had Watt not gazed at the kettle boiling. That is, the same quantities of metal, wood, fuel, etc. per horsepower are required in counterfactual-1800 as in actual-1769, but priced at 1800 levels. In this fashion the point $Q'$ can be fixed.

Developments in the traditional technology that would have come about after 1769 but before 1800 without James Watt are admissible in theory (cf. Appendix 6.2). On the other hand, any improvements in the steam-engine which made use of the separate condenser (or parallel motion, etc.) in any way are not admissible. This cuts out, for instance, Hornblower's compound engine, Bull's inverted engine, the Trevithick high-pressure engines, etc. Hornblower and the rest may have been telling the truth when they claimed in their legal defence that they developed the separate condenser quite independently of Watt—it seems an obvious enough step, at least with hindsight.

---

[7] Hawke, *Railways and Econ. Growth*, esp. pp. 80–6.

But the question of whether the separate condenser would have been invented (and when) without Watt must remain no more than casual speculation within the methodology employed here. Naturally, if it could be proved conclusively that others would have developed the separate condenser, rotative motion, etc. independently before 1800, it would reduce the social savings contributed by James Watt at that date to zero.

There are further problems in envisaging a counterfactual world in which only the invention in which we are interested is bombed out of existence. It would be totally unrealistic to imagine that in the absence of the invention, all else would have in fact developed in the same way as it actually did. It is not difficult to imagine industries growing on different sites, and perhaps at much slower rates or not at all. However, under certain assumptions it can be taken that the adjustment of individual firms or whole industries to be deprived of the invention will be accommodative, and will go towards minimizing the costs of not having access to the railroad or steam-engine. If one can assume that the invention did not give rise to any irreversible innovations in related industries via its forward or backward linkages, then the computed social savings will constitute an upper bound on the 'true' social savings. Thus all coeval changes in the environment occurring in the hypothetical absence of the separate condenser (etc.) will cause the 'true' social savings to lie within the range calculated.

But the convenient assumption of there being no irreversible derived innovations in related industries is an extreme one, as historians such as Professor David have emphasized. For instance, John Lord, in his study of Boulton & Watt, stressed the unduly large concentration of their engines in the new industries of the late eighteenth century.[8] One might expect that the more significant the invention (in some sense as yet undefined), the greater these secondary and irreversible innovations might be. Thus some have argued that calculations of this kind should be limited to comparatively insignificant developments, taking place over a brief time span.[9]

[8] Lord, *Capital and Steam Power*, p. 176. Note also Mathias's references to the creative effects of steam-power (*The First Industrial Nation*, London, 1969, pp. 142–3, and *The Brewing Industry in England, 1700–1830*, Cambridge, 1959).
[9] J. D. Gould, 'Hypothetical History', *Econ. Hist. Rev.* 2nd Ser. xxii, 1969.

However, this advice evidently has its own limitations. By accepting it, we may be for ever finding that inventions are individually insignificant. Indeed the 'New Economic History' has been repeatedly attacked for narrow-mindedness and concentration on rather petty issues to the neglect of the sweeping generalizations of traditional economic historians.[10] Except in the few industries covered in Part III, I shall be making no great effort to paint across this broader canvas.

The social savings in Chapter 6 are computed for the year 1800. There is some economic justification for this, e.g. it is within two years of the date Rostow chose as ending the 'take-off'. The primary reason for choosing 1800, however, is not that, nor its chronometric simplicity, but because it marked the expiration of James Watt's patents and hence is the last date for many years for which there are, at present, any even remotely verifiable figures on the location and quantity of steam horsepower. As H. W. Dickinson remarks, so far as the technology of steam is concerned, one starts the nineteenth century with a clean slate.[11] Soon Trevithick is marketing his high-pressure engines, and the nature of best-practice technique is modified.

Luckily, 1800 or thereabouts is convenient for the kinds of assumptions detailed above. There is some indication that in the 1790s and early 1800s an inflationary gap was opening up. Resources were being strained in order to gear the economy to war. Inflation ran at high levels as aggregate demand outstripped aggregate supply. The drafting of half a million men into the Armed Forces substantially reduced the customary labour surplus. Even in the rural districts where under-employment was endemic, measures had to be taken to combat labour 'shortages' at peak times, e.g. by adopting the threshing machine, and by 'petticoat harvests' in Hampshire.[12] Since the chief alternative technologies 'adopted', water- and wind-power, were capital-intensive and labour-saving relative to

---

[10] To which the usual retort has been something like, 'We "may not be able to say very much, but at least the reader is aware of what has been said" ' (L. Davis 'Professor Fogel and the New Economic History', *Econ. Hist. Rev.* 2nd Ser. xix, 1966, 658).
[11] H. W. Dickinson, *A Short History of the Steam Engine*, Cambridge, 1939, p. 90.
[12] E. L. Jones, 'The Agricultural Labour Market in England 1793–1872', *Econ. Hist. Rev.* 2nd Ser. xvii, 1964–5.

steam, the demand for power would have to be even more elastic than that for labour not to reabsorb the capital actually employed in steam-engines. Accordingly, the full-employment assumption does not commit a grave injustice. It is unlikely that this would be so two or four decades later. On the other hand, under wartime exigencies it is unlikely that general equilibrium would hold, with prices equalling marginal costs wherever required.[13]

### BACKWARD AND FORWARD LINKAGES

A narrow version of the 'energy crisis' argument can be tested by the social savings, but as implied on page 42 a full assessment of the contribution of the steam-engine to British industrialization would trace the spin-offs generated through forward and backward linkages. The backward linkages are taken to be the effects on output and innovation in industries supplying the raw materials, i.e. the various metals used in constructing engines and fuels for keeping them at work. If the social savings are positive, the sum of backward linkages in output terms will by definition be less than from the older technology that is being replaced, because the social savings represent the bundle of resources freed. For this reason, the *direction* of backward linkages is critical. Steam-engines are visualized as important for backward linkages not because of the quantities involved but through their concentration in sectors that led Britain's industrialization. Iron, especially, had a high capacity for innovation and growth over these years. In a few cases, including that of John Wilkinson's famous boring mill, the innovation linkage from steam-engines can be directly isolated. An over-all indication can be gained from the importance of the demand for iron in steam-engines to total domestic iron output. Linkages to the coal industry can also be examined;

[13] Prices for coal in the major industrial towns, except London, are roughly 'normal' for their period in 1800 itself (App. 4.2); prices for pig iron in Birmingham and bar iron in Liverpool are also 'normal' (*Thesis*, Table I.8). Prices of agricultural products were, however, exceptionally high because of the disastrous harvests of 1799–1800, putting the competition to steam-engines from horse gins at a particular disadvantage. For a model of the British economy during the French Wars that takes general equilibrium to prevail, see G. Hueckel, 'War and the British Economy, 1793–1815: A General Equilibrium Analysis', *Expl. Econ. Hist.* 2nd Ser. x, 1973.

these have hitherto received less emphasis, partly because the coal industry is often taken to be well advanced before the 1760s, and partly because Watt's engine was designed to reduce coal consumption per horsepower.

Forward linkages are the effects of recruiting steam on industries undergoing mechanization. Compared to backward linkages there is little agreement in the literature about the form that derived innovations will take. The adoption of steam-power could conceivably have been sufficient to encourage attendant technical changes that hastened mechanization. For the industries studied in Chapters 7 and 8 I believe this to have been unimportant, though proof is scarcely possible. For instance, the great 'waves of gadgets' in these textile industries were developed for pre-steam forms of power supply. Forward linkages became truly noteworthy when the cost of supplying power fell and this happened to influence the nature, extent, and mode of employment of the machines driven by power.

The precise details of this type of linkage will be brought out in Chapter 7, but a few preliminary remarks may be found helpful. The economic problem has in fact already been raised, on page 10, where I asked what meaning could be given to a decline in the number of spindles driven by a unit of power in the first half of the nineteenth century. On further reflection, it will be seen that if the capabilities of the engine and its transmission do not change, spindles per horsepower can be increased only by driving them more slowly. Technically speaking, increases in spindleage per horsepower can be simply traded off against reductions in spindle speeds over quite a wide range of speeds. Equivalent technical situations, however, do not imply equivalent economic situations. Slower speeds will raise the average fixed costs of the mill. Why not, then, increase speeds to the technical maximum? Obviously because average variable costs will rise. Labour costs per unit of output might grow, though this is uncertain. The predominant cause of rising variable costs is that power costs per unit of output shoot up dramatically at the margin. The explanation lies in aerodynamic factors operating on machine and 'package' (cop, bobbin, etc.). The effect of air (or water) resistance in increasing fuel consumption has been set out by Fishlow for

American railroads and by Harley for British steamships.[14] It was, of course, an important reason for lowering speed limits on motor cars during the 'energy crisis' of the early 1970s.

A fall in costs of power supply therefore not only reduced average total costs but additionally increased the optimal spindle speed—and so encouraged a decline in spindleage per horsepower. In Chapter 10 it will be shown that international differences in power costs, working through these relationships, help to explain why countries chose different technologies (and products) for textile spinning.

Within Britain the historical problem associated with using this model is that of dating the fall in power costs. The backward and forward linkages, and especially the secondary derived innovations they gave rise to, can hardly be expected overnight. In his stage theory Rostow envisaged these linkages as carrying the economy on from 'take-off' into the 'drive to maturity', when the example set by a few progressive manufacturing industries (the 'leading sectors') spreads to the economy as a whole. As it happens, power costs did not fall substantially until the 1840s and 1850s.

[14] A. Fishlow, 'Productivity and Technical Change in the Railroad Sector, 1840–1910', National Bureau of Economic Research, in *Output, Employment, and Productivity in the United States after 1800* (Studies in Income and Wealth, vol. xxx), New York and London, 1966, pp. 636–7; C. K. Harley, 'The Shift from Sailing Ships to Steamships, 1850–1890: A Study in Technological Change and its Diffusion', in D. N. McCloskey (ed.), *Essays on a Mature Economy: Britain after 1840*, London, 1971, pp. 217 ff.

## PART II. THE EVALUATION

## 4. The Diffusion of Best-Practice Techniques

THIS CHAPTER explores some of the basic economic determinants underlying the decision that an entrepreneur of the Industrial Revolution period might have made in preferring one type of steam-engine to another. At first sight it may seem naïve to suppose that a businessman would have made up his mind on purely economic criteria. No doubt to some the charge will still stick after reading the chapter. It bears emphasizing, though, that what is at issue here is comprehending what decision the entrepreneur ought to have reached, in any specified set of economic circumstances. What he actually did do is, of course, a much more pertinent historical question, and, as ever, there is much less information readily available to answer it.

### FIXED COSTS

Cost figures for Savery and Newcomen engines are very scrappy, and awkward to compare in that they relate to widely differing times and places. By contrast, there is a veritable plethora of cost data, broken down as minutely as one could hope for, on the Boulton & Watt engine, both before and after 1800. For the last engine I shall be considering, the Cornish engine, accessible figures of costs are as rare as for the eighteenth-century engines.

Atmospheric engines—both Saverys and Newcomens—continued to be built until early in the nineteenth century. For technical reasons noted earlier, Savery engines were often chosen as an alternative to Newcomens when it was a matter of just a few horsepower. In such small sizes they were cheap to purchase. £50 was the standard price for a 1-horsepower engine early in the eighteenth century. Dr. Tann[1] notes that Savery engines late in the same century were usually insured at from £50 up to £200, while Musson and Robinson point to a 4-horsepower engine of the Savery type costing £200 *including* the water-wheel it drove, in 1790.[2] Farey reckoned that Savery engines cost only about one-quarter the price of Newcomens in these very small sizes; Newcomen engines were not very suitable anyway because of excessive friction.[3]

Above 4 horsepower or so, however, the balance tipped over towards the Newcomen engine, despite its somewhat ponderous appearance. To some degree the problems associated with little Newcomen engines were spread more thinly, but to a far greater degree the technical difficulties of the Savery pattern escalated. Table 4.1 has been concocted from a miscellany of sources, with emphasis on what people actually paid rather than on what practitioners said ought to be paid.

Apart from the acknowledged fact that Table 4.1 depends on a small number of observations, there are two problems that it highlights. The first is the question whether costs of delivering and erecting the engine are lumped in or not. Just by inspection of the Table it is obvious that it makes a great deal of difference. For instance, a 24½-inch engine at Wednesbury cost £419 all-up in 1727, including £32 for bricks and £60 for other aspects of erection. Engines of 6 horsepower (with 22-inch cylinders) were delivered for £295 at Hawkesbury Junction in 1725 and £350 completely erected at Newcastle at about the same date.[4]

Secondly, there was some tendency for costs to fall in the course of the eighteenth century. Early engines, often built on a 'one-off' basis, were comparatively expensive, such as the £1007 paid for a 12-horsepower engine at the Edmonstone

---

[1] J. Tann, *The Development of the Factory*, London, 1970.
[2] Farey, *Treatise* i. 107, 116, 125; Dickinson, *A Short History of the Steam Engine*, p. 25; Musson and Robinson, *Science and Technology*, p. 405.
[3] Desaguliers, *A Course of Experimental Philosophy*.
[4] C. T. G. Boucher, 'The Pumping Station at Hawkesbury Junction', *Trans. Newcomen Soc.* xxxv, 1962–3; R. A. Mott, 'The Newcomen Engine in the Eighteenth Century', in ibid.

## The Diffusion of Best-Practice Techniques

TABLE 4.1

Prices of certain Savery and Newcomen engines to customers[5]

| Horsepower | Type | Cylinder diameter (in) | Price (£) | Year (approx.) |
|---|---|---|---|---|
| 2 | Savery | — | 100 | 18th c. |
| 4 | Savery | — | 140 | 1790 |
| 6 | Newcomen | 22 | 270 | 1725 |
| 8 | Newcomen | 24 | 320 | 1727 |
| 16 | Newcomen | 33 | 730* | 1733 |
| 20 | Newcomen | 36 | 700 | 1736 |
| 28 | Newcomen | 42 | 1200* | 1740s |
| 28 | Newcomen | 42 | 907 | 1795 |

* includes erection

Colliery in 1727.[6] After the first third of the century prices fell; the chief reasons probably being the substitution of iron for brass, the cessation of the Savery–Newcomen patent in 1733, and perhaps advancing technical knowhow. In 1724 it was claimed that a 32-inch engine could not be erected on Tyneside for under £1200, unless it were a poor one; twenty years or so later this was the regular cost of an engine with a 42-inch cylinder.[7] The 42-inch engine at Elsecar, Yorkshire, which operated until fairly recently, cost £907 plus £69 erection in 1795.[8]

[5] Usher derived presumed power ratings as follows, from Beighton's calculations in 1717:

|  | 16 strokes/minute 6 ft stroke | 20 strokes/minute 8 ft stroke |
|---|---|---|
| 30 in Cylinder | 16 | 27 |
| 40 | 28 | 47 |
| 60 | 44 | 74 |

(A. P. Usher, *A History of Mechanical Inventions*, rev. edn., Cambridge, Mass., 1954, p. 352.)
[6] Farey, i. 230.
[7] A. Raistrick, 'The Steam Engine on Tyneside, 1715–1778', *Trans. Newcomen Soc.* xvii, 1936–7.
[8] A. K. Clayton, 'The Newcomen-Type Engine at Elsecar, West Riding', *Trans. Newcomen Soc.* xxxv, 1962–3.

Another way of seeing this is by looking at estimates rather than actual pay-outs, especially the estimates of the coal-viewer, John Curr, for erecting an atmospheric engine in the Sheffield area, whether for a colliery or for manufacturing. These are set out in Table 4.2.[9] Curr's figures are based on

TABLE 4.2

*Costs of installing atmospheric engines, c. 1797*

| Cylinder (in) | Engine, etc. (£) | Boiler (£) | Engine-house (£) | Labour (£) | Total (£) |
|---|---|---|---|---|---|
| 30 | 308 | 97 | 117 | 78 | 600 |
| 40 | 419 | 168 | 143 | 97 | 827 |
| 50 | 583 | 272 | 186 | 105 | 1146 |
| 60 | 775 | 362 | 241 | 178 | 1556 |
| 70 | 959 | 497 | 304 | 262 | 2022 |

what seem to be careful assessments of the relative proportions of various parts. They happen to agree fairly closely with the detailed breakdown for a 48-inch engine quoted by Farey as being purchased by a coal-mine twenty years earlier; they are also not too far out of line with the actual costs of the 66-inch engine built for the Cronstadt Dockyards, St. Petersburg, by the Carron Company in 1774.[10] However, even allowing for varying dates they seem a little modest alongside the actual outlays quoted in Table 4.1.

For separate-condenser engines of the kind made famous far and wide by James Watt there exists the voluminous and detailed cost information provided in the engine-books in the Boulton & Watt MSS. Observations on some 736 engines built by this company, the bulk of them between 1799 and 1815, were analysed in a variety of ways. More complex procedures were tried, but the figures in Table 4.3 are mainly obtained by simple averaging of engine costs according to horsepower for the years 1799 and 1800, and in fact concur closely with the results of regression analysis. The main exceptions are for

---

[9] J. Curr, *The Coal Viewer and Engine Builder's Practical Companion*, (1797) 2nd edn., London, 1970, pp. 94–5.
[10] R. H. Campbell, *Carron Company*, Edinburgh and London, 1961; App. 4.1.

boilers. It was common practice to purchase a boiler rated about 2 horsepower higher than the engine itself, partly to permit the purchaser to work the engine hard. Sometimes two or even more boilers were acquired for the one engine, and if so their combined horsepower was usually greater than that of the engine. For these and similar reasons, regression results obtained for boiler costs were unsatisfactory, and in some cases cost information as late as 1803 was borrowed. The sample size for Table 4.3 is small, and a number of the entries are

TABLE 4.3
*Costs of Boulton & Watt engines by horsepower, c. 1800*

| Horse-power | Cost of engine (£) | Cost of boiler (£) | Total cost (£) | Cost per horsepower (£) |
|---|---|---|---|---|
| 2 | 157 | 21 | 178 | 89·0 |
| 3 | 177 | 26 | 203 | 67·7 |
| 4 | 226 | 30 | 256 | 64·0 |
| 6 | 249 | 37 | 286 | 47·6 |
| 8 | 303 | 45 | 348 | 43·5 |
| 10 | 344 | 54 | 398 | 39·8 |
| 12 | 361 | 63 | 424 | 35·3 |
| 14 | 434 | 68 | 503 | 35·9 |
| 16 | 448 | 81 | 528 | 33·0 |
| 20 | 509 | 93 | 602 | 30·1 |
| 24 | 638 | 125 | 763 | 31·8 |
| 28 | 699 | 149 | 848 | 30·3 |
| 30 | 802 | 164 | 966 | 32·3 |

single observations, so that the figures are not completely reliable. Despite that, the heavy outlays per horsepower needed to construct small Watt engines come out clearly, both for engines and boilers. Above about 12 horsepower the capital cost per horsepower is more stable, although the particularly small number of observations for larger-sized engines precludes any very strong assertions on that point.

To translate these cost figures into prices that consumers paid it is necessary to distinguish the years before 1800, when Boulton & Watt's patents operated, from those after. Until 1800 Boulton & Watt claimed that they were not seeking to make any substantial direct profit out of the sale of their engines. Exercise of the patent rights meant that the Watt engine was a commodity whose supply was highly inelastic in the short run. The monopolistic control so provided (always bearing in mind the temptations it offered to rival 'pirates') encouraged Watt to charge for the quasi-rents that arose out of this scarcity value. These were captured by various kinds of premium. Watt indeed alleged that no charge over costs was made other than for the premium, though Roll has demonstrated that this magnanimous gesture was by no means strictly true.[11]

In the case of their reciprocating engines, then, Boulton & Watt introduced the policy of charging the users one-third of the savings in fuel costs that their engine effected over its predecessors; though as will be shown below the policy was more honoured in the breach than the observance. For rotative engines, the company capitalized the premium (as it was ultimately compelled to do, in fact, for most of the reciprocating engines, regardless of Watt's own objections[12]); setting it usually at £5 per horsepower per annum in the provinces and £6 in London. As 1800 and the expiry of the patent approached, the procedure of paying the royalties in a lump sum became increasingly common; moreover, this had always been practised for small engines (6 horsepower or less). Once the patent lapsed the costing was revised, and in 1801 a uniform mark-up of 25 per cent was imposed. Some of the material on prices relating to the period around 1800 is set out in Table 4.4.[13]

Boulton & Watt engines were however, not fully representative of all steam-engines then manufactured in Great Britain. Most commentators would agree that they were more professionally constructed to more exacting standards, and so likely

---

[11] Because they took some profit on the ironwork. Hence Boulton: "Dearman charges the Heating Cases 14/0 per hundred (weight), but I have charged them 16/0, and I think we should take a profit or commission upon everything we take the trouble to provide." Roll, *An Early Experiment*, pp. 56–7.
[12] e.g. Roll, op. cit., p. 117.
[13] For prices of earlier engines, see Hills, *Power in the Industrial Revolution*, p. 167. Sources for Table 4.4 are as follows—col. (i): Roll, p. 312, includes a premium charged for 3 years 10 months, as being the length of time presumably left to run (?). Col. (ii) and col. (iii): Roll, p. 241, from a table drawn up by M. R. Boulton. Col. (iv): From documents entitled "Calculations of the Costs and Prices of Small Engines October 9, 1801", and "Materials for New Estimates of Rotative Engines" (Boulton & Watt MSS., Box 7).

## TABLE 4.4
### Prices of certain Boulton & Watt rotative engines, c. 1800

| Horse-power | (i) 1795 (£) | (ii) 1798, London (£) | (iii) 1798, elsewhere (£) | (iv) 1801 (£) |
|---|---|---|---|---|
| 4  | —    | 354  | 338  | 350  |
| 6  | —    | 383  | 366  | 385  |
| 8  | 490  | 479  | 458  | 470  |
| 10 | 562  | 529  | 505  | 520  |
| 12 | 634  | 567  | 542  | 560  |
| 14 | 774  | —    | —    | 600  |
| 16 | 850  | 736  | 703  | 710  |
| 18 | 920  | —    | —    | —    |
| 20 | 988  | 810  | 774  | 820  |
| 24 | 1201 | 1053 | 1006 | 1020 |
| 28 | 1334 | —    | —    | —    |
| 30 | —    | 1134 | 1083 | 1160 |
| 32 | 1617 | 1170 | 1118 | 1200 |
| 36 | 1755 | —    | —    | 1261 |
| 40 | 1888 | —    | —    | —    |
| 45 | 2154 | —    | —    | 1526 |
| 50 | 2322 | —    | —    | —    |

to be heavier and dearer than most others. This implies that any calculations that might be based on the costs or inputs of their engines are likely to be upper bounds on the averages for the whole industry, and these biases will have to be borne in mind later. In Tables 4.5 and 4.6 I have put together accessible information on the engine prices of this and other companies for the first part of the nineteenth century, other than that already conveyed in Table 4.4. Table 4.6 applies only to some manufacturers of boat engines in the 1820s and 1830s. This does not reflect any sudden switching of my interest to steamboats, but that in practice boat engines were often used on land for industrial purposes over these years, and indeed were the arena for much of the technical development (e.g. towards direct-acting engines) at the time.

One might expect that price would be substantially affected by the price of the metal components. Pig and bar iron prices

TABLE 4.5
*Prices of steam-engines, various manufacturers*

| HP | (i) Boulton & Watt 1799–1804 (£) | (ii) Boulton & Watt 1840–1842 (£) | (iii) Fenton, Murray & Wood 1804 (£) | (iv) Fenton, Murray & Wood 1812 (£) | (v) Fenton, Murray & Wood 1830 (£) | (vi) Goodrich 1800 (£) |
|---|---|---|---|---|---|---|
| 2 |  |  | 175* |  | 234 |  |
| 3 |  | 300 |  |  | 270 |  |
| 4 | 338* |  |  |  | 324 | 327 |
| 6 | 373* | 412 | 380* | 400+40 | 396 |  |
| 8 | 460* | 470 |  | 468+60 | 450 |  |
| 10 | 491 | 517 | 510* | 557+70 | 520 | 491 |
| 12 | 525 | 617 |  | 644+82 | 600 |  |
| 14 | 584* | 720 |  | 716+100 | 690 |  |
| 16 | 880 | 785 |  | 770+110 | 755 |  |
| 18 |  | 975 |  | 827+130 | 800 |  |
| 20 | 1083 | 993 | 600 | 894+150 | 850 | 750 |
| 22 |  |  |  | 963+150 |  |  |
| 24 | 1276 |  |  | 1028+160 | 950 |  |
| 25 |  | 1088 | 715 |  |  |  |
| 26 |  |  |  | 1083+175 | 1030 |  |
| 28 |  |  |  | 1134+190 | 1080 |  |
| 30 | 1418 | 1285 | 830 | 1186+200 | 1140 | 1050 |
| 32 | 1830 |  |  | 1443 |  |  |
| 35 |  |  | 938 |  |  |  |
| 36 | 1800 |  |  | 1304+240 |  |  |
| 40 |  | 1435 | 1045 |  |  | 1440 |
| 45 | 1504* | 1550 |  |  |  |  |
| 50 |  | 2351 |  | 1487+? |  | 1727 |
| 60 |  | 2363 |  |  |  |  |
| 65 |  | 2975 |  |  |  |  |
| 80 |  | 2991 |  |  |  |  |
| 100 |  | 4275 |  |  |  |  |
| 110 |  | 4680 |  |  |  |  |
| 120 |  | 5225 |  |  |  |  |
| 125 |  | 5175 |  |  |  |  |
| 150 |  | 6325 |  |  |  |  |
| 185 |  | 7913 |  |  |  |  |

|   | (vii) Trevithick 1804 (£) | (viii) Trevithick 1825 (£) | (ix) Fixed condensing 1856 (£) | (x) Fixed non-condensing 1856 (£) | (xi) Portable 1856 (£) |
|---|---|---|---|---|---|
| 1 | 126 | | | | |
| 2 | 221 | | | | |
| 3 | 284 | | | | |
| 4 | 347 | | 120 | 100 | 160 |
| 5 | 399 | | 138 | 115 | 175 |
| 6 | 441 | | 156 | 130 | 190 |
| 7 | 473 | | 174 | 145 | 205 |
| 8 | 494 | | 190 | 160 | 215 |
| 10 | 536 | 420 | 200 | 175 | |
| 14 | 750* | 522 | | | |
| 18 | | 612 | | | |

*Sources:*

(i) Starred estimates from B. & W. MSS. ('Calculations of the cost of fitting and of the Prices of ye Small Engines, 9/10/1801', etc., 32 HP estimate from Marshall MSS. Other unstarred figures from Goodrich MSS. 18, 56, 444, etc. (The figures are lower than in Table 4.4 because rounding-up has not been included.)

(ii) 'List of Estimates rendered by B & W & Co between the 30th Sept. 1840 and 21 Jany. 1841' (B. & W. MSS.); available estimates averaged.

(iii) Fenton, Murray & Wood engine prices. Starred estimates taken direct from Goodrich MSS. Unstarred estimates from E. A. Forward, 'Simon Goodrich and his work as an Engineer', Part I, *Trans. Newcomen Soc.* iii, 1922–3. These figures exclude boilers and are thus not strictly comparable with the other columns.

(iv) Goodrich MS. 438—price of engine plus iron boiler from Fenton, Murray & Wood. Cf. also T. Turner, 'History of Fenton, Murray & Wood' (M. Tech. Sc. Thesis), U.M.I.S.T., May 1966. 26-, 32-, and 50-HP figures are from Marshall MS. 28.

(v) Goodrich MS. 1511; Turner, op.cit.

(vi) Forward op.cit. It is clear from inspection that various manufacturers are included.

(vii) Forward, op.cit., using Goodrich MSS. The price was said to increase by £21 for each HP over 8, though most such engines were probably small, since the Trevithick engines of this sort consumed more coals per unit output. Starred estimate is from F. Trevithick, *Life of Richard Trevithick*, London, 1872, vol. ii. Cf. also Raistrick, *Dynasty of ironfounders*, p. 205.

(viii) H. W. Dickinson and A. Lee, 'The Rastricks—Civil Engineers', *Trans. Newcomen Soc.* iv, 1923–4.

(ix)–(xi) W. Morshead, 'On the Relative Advantages' (1856). Lynch & Inglis exhibited portable engines at the Great Exhibition in 1851 for £120 (4 HP), £160 (6 HP), and £200 (8 HP). Cf. also *Artizan*, x, 1852, p. 164.

TABLE 4.6
*Prices of boat engines, various manufacturers*

| HP | (i) Brunel 1823 (£) | (ii) Boulton & Watt 1831 (£) | (iii) Maudslays 1831 (£) | (iv) Butterley Co. 1831 (£) |
|---|---|---|---|---|
| 6 |  | 500 |  |  |
| 10 |  | 800 |  |  |
| 16 |  | 1050 |  |  |
| 20 | 1200 | 1475 | 1400 | 1300 |
| 24 | 1300 |  |  |  |
| 30 | 1600 | 1850 | 1750 | 1750 |
| 40 | 2000 | 2300 | 2400 | 2175 |
| 50 | 2400 | 2650 | 2750 | 2575 |
| 60 | 2800 | 3150 | 3000 | 2950 |
| 70 |  | 3950 | 3500 | 3300 |
| 80 | 3200 | 4200 | 4000 | 3625 |
| 90 |  |  | 4500 | 3925 |
| 100 | 3500 |  | 5000 | 4200 |
| 110 |  |  | 5500 | 4500 |
| 120 | 3800 |  | 6000 | 4750 |

*Sources:*

(i) from a MS. letter from Sir Marc Brunel to J. Gamble, dated 6 Dec. 1823.

(ii)–(iv) Goodrich MS. 1629. Dickinson quotes £2500 for a 40-HP engine in 1826 ('The Lewin Diary: A Link with Rennie', *Trans. Newcomen Soc.* xix, 1938–9).

in fact fluctuated quite drastically from year to year.[14] Yet prices charged to customers for the individual components were kept virtually constant for long periods of time. Boulton & Watt in the late 1790s were charging practically the same amount per cwt as the Coalbrookdale Company in the 1730s.[15]

[14] See *Thesis*, esp. Fig. I.1 and Table I.8.
[15] Boulton & Watt data from MSS. ('Soho Foundry, July 3, 1828'); Coalbrookdale sold cylinder bottoms at 32s. 6d. per cwt. in 1736 (J. S. Allen, 'The 1712 and Other Newcomen Engines of the Earl of Dudley', *Trans. Newcomen Soc.* xxxvii, 1964–5), as compared with the 30s. that Boulton & Watt charged Sirhowy Ironworks (see App. 4.1) and many other clients in the late 1790s. In the 1750s and 1760s Coalbrookdale prices were 26s. 0d. for cylinders and 16s. 0d. for bottoms—in 1770–2 both were set at 30s. 0d. (see Raistrick, *Dynasty of Ironfounders*, App. IV and pp. 135 ff.).

## The Diffusion of Best-Practice Techniques 57

Indeed the great iron companies, such as the Carron Company, cartelized in 1762 in order to maintain prices of engine castings.[16] Charles Hyde's recent research seems to show that the prices of charcoal pig iron rose fairly sharply in the middle of the century, but this was offset by there being no very strong trend in prices for coke pig iron after 1760.[17] The enginebuilders thus appear to have been in a sufficiently entrenched oligopolistic position to be able to administer and maintain price stability. The fluctuations of pig iron prices (etc.) then in effect dictated the levels of their profits or losses, and it is likely that many of them actually took their cut in this way. Boulton & Watt, having maintained prices per cwt for cast iron, wrought iron, and brass at about the same level from the late 1790s up to the late 1820s, then reduced them around 1830 to levels that prevailed up to the mid-1850s.[18]

From these observations it ought to be possible to estimate the price of an engine erected at most times and places in the British Isles during the 1780–1840 period, without an extravagant amount of error. It is also worth extending this to the mid-nineteenth-century period. The rise of the high-pressure engine brought with it new cost patterns whose ramifications will be dwelt upon in Part III. In regard to fixed costs it seems that Boulton & Watt were building high-pressure engines even cheaper than their orthodox low-pressure types by the early 1840s. However, these were probably non-condensing versions of the high-pressure engine.

The high-pressure condensing engine, such as the Cornish engine, had long preserved a reputation for high fixed costs. As this engine combined elements of high- and low-pressure types, the conclusion seems hardly surprising. It is widely attested in the secondary literature.[19] Yet it is not particularly easy to verify from what primary sources there are. Sheer lack of data may account for the difficulty, but it is surely better to try to explain what there is rather than overlook it. Note that

---

[16] Campbell, op.cit., p. 216.
[17] C. K. Hyde, 'The Adoption of Coke-Smelting by the British Iron Industry, 1709–1790', *Expl. Econ. Hist.* N.S. x, 1973.
[18] Drawn from comparisons of weight and cost of parts in the engine-books.
[19] J. A. Phillips and J. Darlington, *Records of Mining and Metallurgy*, London, 1857; C. F. Partington, *A Course of Lectures on the Steam Engine*, London, 1826, p. 45; J. Bourne, *A Popular Description of the Steam Engine*, London, 1856.

the customary Cornish system of 'cost-book' accounting means that it is rare to find total fixed costs of engines in, say, individual mine-records. Data from the Cornish manufacturers of engines are more abundant. A very rough assessment of costs for the largest engine-builders in the late 1830s, Harveys of Hayle, is presented in Table 4.7. Other manufacturers were

TABLE 4.7
*Engine costs at Harveys, c. 1835–1838*[20]

| Cylinder diameter (in) | Cost (£) | Approx. horsepower |
|---|---|---|
| 8 | 175 | 2? |
| 12 | 225 | 3 |
| 18 | 350 | 5 |
| 20 | 440 | 6 |
| 22–4 | 540 | 7–8 |
| 26 | 580 | 10 |
| 30 | 630 | 12 |
| 36 | 700 | 17 |
| 40 | 760 | 21 |
| 45 | 810 | 25 |
| 50 | 1100 | 33 |
| 60 | 1600 | 51 |
| 80 | 2100–800 | 100 |

generally cheaper than Harveys, according to the morsels of information available.[21] Comparison with Table 4.5 suggests that they were hardly more expensive than those built elsewhere in the country, especially considering that the above power figures are for 'economical working', and much more work could be obtained safely enough from them. Their larger size per horsepower must be offset by a lower per unit charge for materials. Other considerations than simply the engine may have been responsible for the allegations of great cost. The Cornish engine required far more rigid foundations than a low-pressure one; whence the robust Cornish engine-house

[20] Adapted mainly from D. B. Barton, *The Cornish Beam Engine*, Truro, 1965, pp. 279–80. Horsepower figures from Table 4.12.
[21] Barton presents a number of cost figures from other manufacturers: ibid., pp. 73 n., 188.

## The Diffusion of Best-Practice Techniques

whose relics dot the countryside.[22] However, that aside, some of the allegations of expense are clearly directed at the two-cylinder engine such as Woolf's original models, and may not have been intended against the single-cylinder version.

Even so, it is difficult to see how the cylindrical boiler could not have been substantially more expensive than its counterpart elsewhere (Table 4.7 excludes boilers). Not only was it made of thicker cast or wrought iron, but its proportions involved nearly eight times as much boiler surface per nominal horse-power as Watt had allowed.[23] The same can be said of the masses of supporting bobs (or beams) required to give the Cornish engine some stability in operation.

### ANCILLARY FIXED COSTS

Fixed costs as seen by, say, the cotton-spinner or colliery-manager included more than simply the purchase price of engine and boiler, with or without premium. The major ancillary charges incurred were:

(i) Framework. Installation required a solid framing to hold the engine upright (with its overhead beam) and in many uses to minimize vibration. The cost of this framework depended on local access to cheap raw materials; wood was most commonly employed in the eighteenth century. Boulton & Watt quoted as much as £250 for a 46-inch engine and £200 for a 36-inch engine in 1790,[24] but this was for the Fenland, where wood costs were high. Oldknow was quoted £50 for a 4-horse and £77 for a 6-horse engine in 1791.[25] McConnel & Kennedy spent £96 on framing for the 16-horsepower Boulton & Watt engine they chose to purchase in 1797.[26] The timber framing of a large engine at the Portsmouth Dockyard cost £140.[27]

(ii) Erection. Apart from the materials such as wood, wages had to be paid for assembly and erection on the site. Boulton &

---

[22] Ibid., p. 170. A few costs of engine-houses are given by Barton on pp. 238–9.
[23] J. Parkes, 'On Steam Boilers and Steam Engines', *Trans. Inst. Civil Engineers*, iii, 1842.
[24] N. Mutton, 'Boulton & Watt and the Norfolk Marshland', *Norfolk Archaeology*, xxxiv, 1967.
[25] G. Unwin, A. Hulme, and G. Taylor, *Samuel Oldknow and the Arkwrights*, Manchester, 1924, p. 131.
[26] C. H. Lee, *A Cotton Enterprise, 1795–1840*, Manchester, 1972, p. 105.
[27] Goodrich MS., 101.

Watt added £50 to £80 to their estimate to allow for erecting medium-sized engines. Tann recorded just over £30 as labour costs of installing a 6-horsepower engine in the West Country;[28] while at the other extreme Simon Goodrich put workmen's wages for erection at Portsmouth as high as £320.[29] The Cornish engineman, John Budge, was paid £63 by Richard Trevithick senior for re-erecting the Carloose atmospheric engine at Dolcoath in 1775;[30] the entire engine as re-erected cost £2040 and was regarded as the best of its kind in Cornwall at that time. Naturally the spread of 'independent' and direct-acting engines curtailed expenses of these kinds.

(iii) Engine-house. Generally an engine-house had to be provided for, if the engine were not to be installed inside the factory; and sometimes a separate boiler-house as well. Farey puts the cost at £580 for a 48" × 7' pumping engine (atmospheric) to be installed at a colliery in 1775 (masonry and carpentry £420, brickwork £160).[31] Tann puts the item at an average of £250–300 for the West Country.[32] The 'House and all belongings' built for an early Boulton & Watt engine for the Birmingham Canal Company at Smethwick cost £452 for a 10-horsepower engine.[33] However, all these seem rather high. It is difficult to get any reasonable average, since Boulton & Watt did not usually include this item in their cost estimates, because of its great variability; while capital cost estimates for using firms often include the engine-house with the mill buildings. I shall put the cost of engine-house, etc. at 20 per cent of that of the engine and boiler. If anything this may be conservative for the later eighteenth century.

(iv) Steam-pipes. Cost of supplementary pipes depended in part on factory layout. Several estimates of this item are available; the difficulty is to sort out the piping that was required for motive power from that needed for heating the mill. Primary sources, relating mainly to large mills, give a

---

[28] J. Tann, 'The Employment of Power in the West-of-England Wool Textile Industry 1790–1840', in N. B. Harte and K. G. Ponting (eds.), *Textile History and Economic History*, Manchester, 1973, p. 217.
[29] Goodrich MS., 101.
[30] Trevithick, *Life of Richard Trevithick*, vol. ii, p. 21.
[31] Farey, pp. 232–3. On Tyneside costs seem to have been lower, cf. A. Raistrick (1936–7). This is the engine in App. 4.1(B).
[32] Tann, in Harte and Ponting, op. cit.  [33] Roll, p. 45.

range from £130 to about £200,³⁴ though a few very much larger estimates must include a large component for heating functions.

(v) Delivery. Any precise reckoning of standard delivery charges would evidently have to be based on information about mean delivery distances; about which very little is known. Simon Goodrich noted delivery charge on engines for the Portsmouth Dockyard (the main engine costing £3724 and the supplementary blast engine valued at £390) as £300:³⁵ this was from Birmingham by way of Liverpool in 1803. Matthew Murray, writing to Goodrich in the following year, quoted a neat 2-horsepower engine at £175 delivered in Leeds, and reckoned transport (presumably to London) '. . . conveyed very well by land carriage at about £8 per ton, or by water at £2'³⁶—for the 2-horsepower engine this would have come to £16 or less. Roll contains a full list of delivery charges on Boulton & Watt engines to Hull, according to size of engine.³⁷ These average between 4 and 5 per cent of the total cost of engine and boiler as in 1795, falling slightly as a proportion with horsepower. By the early 1840s Boulton & Watt seem to have charged as much as 8 to 10 per cent on some deliveries to London from Soho, though perhaps with additional on-site attendance. But by this date construction was very widespread. Glasgow had nine factories manufacturing steam-engines in 1840; even a place as unnoted for industrial advance as Dublin had four major producers in the late 1830s.³⁸

(vi) Time. In assessing the opportunity cost of engines one might also have to take account of the delays in delivery incurred in purchasing an engine, and time spent in actual erection, since these presumably might reflect some earnings foregone. It is widely acknowledged that Boulton & Watt were slow to fulfil their orders. Average delays in the 1790s seem to

---

[34] Especially Quarry Bank MSS. and other sources used in the cost estimates below. The huge mill built by William Fairbairn for Bailey at Staleybridge in the mid-1830s (220 horsepower and mill-buildings alone valued at £30 000) ranked steam and gas pipes as high as £2 400, but this surely must be largely for heating. See A. Ure, *The Cotton Manufacture of Great Britain*, 2nd edn., ed. P. L. Simmons, London, 1861, vol. i, pp. 314–16.

[35] Goodrich MS., 101.   [36] Ibid.   [37] Roll, p. 312.

[38] J. Cleland, *Description of the City of Glasgow*, p. 99; P.P. 1837–8: XXXV, p. 857.

have been in the range of seven to eleven months, plus another one to one and a half months for erection.[39] For the above-mentioned Portsmouth Dockyard engine in 1803 Goodrich averred 'The whole of the Machinery can be prepared ready to be delivered at Birmingham in 6 months after the receipt of the order; and the whole may be expected to be fixed up ready for use in 10 months after the order.'[40]

### VARIABLE COSTS: COAL

Quantitative information on lubricants is too scattered to help very much, but this does not seem to be a severe loss. The same cannot be said for fuel. The price of coal could easily dictate whether or not it was an economic proposition to resort to steam-power. As one cannot speak of any unique price for coal prevailing at any point of time the very diversity of coal prices across the country becomes highly significant in understanding diffusion patterns.

The quality of coal varied greatly according to its type (anthracite, bituminous, etc.), the region and particular mine from which it came, and even the different strata in the same mine. In some cases the fuel used was not coal at all. Peat was sometimes used in its place in the engines of Cornwall and Devon, as well as in Ireland. In these districts, of course, large supplies of cheap coal were not at hand.

Where coal was abundant and inexpensive, it was common practice to burn the inferior small coals, or 'slack', in steam-engines. Had the engines not consumed them, much would probably have gone to waste. Thomas Savery himself recognized this point, when in 1702 he boasted that 'The coals used in this engine are of as little value as the coals commonly burned on the mouths of the coal-pits are . . .'[41] If no use were found for them, disposal might even have cost money, if all they were doing was to pile up higher and higher the Alps of Wigan.

Apart from these disposal problems, the use of such slack can be shown to have been based on sound economic grounds. The Scottish engineer, Robertson Buchanan, wrote in 1810

---

[39] Hills, *Power in the Industrial Revolution*, p. 171; McConnel & Kennedy MS., 11 Sept. 1805.
[40] Goodrich MS. 101. This engine was built by Whitmore in 1803.
[41] In the *Miner's Friend*, London, 1702.

Small coal, or culm, is much used for steam engines. From repeated trials, made in the neighbourhood of Glasgow, it requires just double the weight of culm that it does of coal to produce the same heat. At the prices in 1808, the cost of culm was to that of coal, supplying the same work, as 12 is to 14, thus producing a saving of $\frac{1}{7}$ by the use of culm.[42]

Independent tests for Glasgow and elsewhere indicate that if anything he may have underrated the efficiency of slack; for example, John Smeaton found that the produce of slack was four-fifths that of Yorkshire coal.[43] Now the average price of large coal at the pit's mouth of a certain Staffordshire colliery between 1828 and 1836 was 8s. 3¼d. per ton; of intermediate coal ('lumps') 4s. 5d. a ton; and of slack 2s. 7¾d. a ton.[44] Taking Buchanan's ratio of effective work of 2:1 for large coals and slack, it follows that it is more economical to burn slack in a steam-engine at the pithead, since consumption per horsepower for a 24-hour day would cost 11s. 6d. in this situation, as compared with 18s. 1d. by burning large coals. However, haulage of coal by inland waterways to London (Paddington) cost 8s. 8¾d. per ton, including canal dues, etc. Thus in London the respective consumption costs would be 49s. 8d. a day for slack but only 37s. 2d. for large coals, so reversing the margin of advantage.

For present purposes, the question of the type of fuel is little more than a special case of the question of regional variations in prices of fuel; sometimes exacerbating the differences (as with slack), sometimes narrowing them (as with peat). As the last example shows, coal, being so bulky, was costly to

---

[42] R. Buchanan, *Essays on the Economy of Fuel and the Management of Heat*, Glasgow, 1810, p. 82; Cf. also R. Brunton, *A Compendium of Mechanics*, Glasgow, 1824, p. 109.
[43] e.g. for the York Waterworks Engine: 'According to the price of slack, when deposited in the yard [at York] at 11s. 8d. per waggon, a day's work in slack will amount to 4s. 6½d. But, according to the prices of coals, when deposited in the yard at 14s. 2d. the day's work of coals will amount to only 4s. 5½d., and if we could, in all cases, take off the water as fast as the engine can draw it, when self-worked, a day's work will be done with coals for 3s. 8½d.' Smeaton, *Reports*, vol. ii, p. 342. For an extended version of these comments, see J. Radley, 'York Waterworks, and other Waterworks in the North before 1800', *Trans. Newcomen Soc.* xxxix, 1966–7.
[44] P.P. 1837–8: XV, q. 2689, p. 151. Also T. Raybould, *The Economic Emergence of the Black Country*, Newton Abbot, 1973; J. Thomas, 'Josiah Wedgwood as a Pioneer of Steam Power in the Pottery Industry', *Trans. Newcomen Soc.* xvii, 1936–7.

transport. By the same token, the coming of a canal or a railway could alter the case in favour of adopting steam-power. Bridgewater's Sankey Cut after 1755 was supposed to have halved the price of coal in Manchester, from about 10s. 10d. a ton to 5s. 5d. Similarly, the price at Birmingham fell from 13s 0d. to 8s. 4d. (or 7s. 6d.) from 1770 as the direct result of being linked by canal to a coalfield.[45] Note also that the dominant portion of the social savings calculated by Dr. Hawke for railway freights in 1865 arose out of transporting coal.[46]

At the beginning of the nineteenth century slack in Leeds cost 1s. 3d. a ton at the pit (the Middleton Colliery), 1s. 6d. to 2s. 0d. at the staith in Leeds itself, 2s. 0d. in Wigan, and 3s 4d. in Manchester. At Saddleworth, high up between the Yorkshire and Lancashire coalfields, small coal cost 10s. 10d. and ordinary sizes about 15s. in the early nineteenth century.[47] Coal prices in some major industrial centres over certain spans of years from 1796 to 1855 are set out in Appendix 4.2.

Available data on coal prices in each region are summarized in Figure 4.1. For England and Wales, contours representing regions with equal coal prices (one for 10s. and one for 20s.) are sketched. The main source is a list entitled 'The Price of Coal in 480 Poor Law Unions, 1842–43'.[48] This was household coal, destined for the hearths of the poor, and not exactly what is wanted. But if the miscellaneous information that can be acquired on prices of slack, or whatever, is interpolated, one obtains a fairly comprehensive picture of engine-coal prices by district at the outset of the railway age. The areal coverage is actually weakest for regions where the coal was mined, partly because it was presumably less responsible for poverty in those places, partly because of a confusion of jargon relating to units of measurement in differing mining districts.

Access to coal at low cost has been used to help explain

[45] W. S. Jevons, *The Coal Question*, 3rd edn., ed. A. W. Flux, London, 1906, p. 122; G. C. Allen, *The Industrial Development of Birmingham and the Black Country, 1860–1927*, London, 1929; Court (1953) says 7s. 6d.
[46] Hawke, *Railways and Economic Growth*, Chs. 6–7. Of the total social savings on freight traffic in 1865 obtained in a preliminary estimate as £25 m. to £28 m., coal alone accounted for almost £20 m., or 70 per cent plus.
[47] Marshall MS. 57; J. de L. Mann, *The Cloth Industry in the West of England from 1640 to 1880*, Oxford, 1971, p. 155.
[48] P.P. 1843: XLV. See also PP.. 1871: XVIII, App. 17, for prices at English ports in 1785.

# The Diffusion of Best-Practice Techniques 65

4.1. Distributions of coal prices in England and Wales, 1842/3.

factory location from the time itself. John Marshall, for instance, stressed its importance in the success of his great flax-mill in Leeds at the turn of the century. In the same manner, John Foster elected to build his pioneering worsted-mill at Queensbury, halfway between Bradford and Halifax and high above them both, in order to capitalize upon the coalfield on top of the ridge at this point; despite the high costs of carting raw wool and finished goods to and from this mill, and despite having to construct a factory village around him.[49]

From the map and the Factory Returns for 1838 one can locate textile-mills according to three broad bands of coal prices, as in Table 4.8. In this Table the figures in the last three

TABLE 4.8

Location of textile-mills according to price of coal, 1838[50]

|  | Total steam horsepower | Coal less than 10s. per ton (%) | Coal 10s. and under 20s. (%) | Coal 20s. and over per ton (%) |
|---|---|---|---|---|
| Cotton | 40 783 | 96 | 4 | * |
| Worsted | 5863 | 86 | 12 | 2 |
| Woollen | 10 887 | 81 | 19 | 1 |
| Flax | 3134 | 69 | 29 | 2 |
| Silk | 2320 | 54 | 33 | 12 |

\* = less than 0·5 per cent.

columns are the percentage of total steam-horsepower in these textile industries located in the relevant price band. Cheap coal (under 10s. a ton) was confined to about 15 to 20 per cent of the land area of England and Wales, and medium-priced coals to another 35 to 40 per cent. There is therefore a marked concentration of the cotton and woollen industries, especially, in areas where the price of coal was low. Naturally this is far from saying that coal prices were paramount in *determining* the

[49] D. T. Jenkins, 'The West Riding Wool Textile Industry, 1780–1835', Univ. of York D.Phil. thesis, 1969. For Fosters, see E. M. Sigsworth, *Black Dyke Mills*, Liverpool, 1958. Best coals at Queensbury cost 5s. and average coals 4s. in 1850 (Black Dyke MSS.).
[50] Collated from P.P. 1839: XLII.

location of cotton- or worsted-mills in those areas. Yet the relationship is quite striking, particularly since received opinion is probably that coal was a much less important determinant for textile-mill location than, say, for blast furnaces.[51]

### COAL CONSUMPTION

J. G. B. Hills[52] has made perhaps the most careful recent appraisal of the thermodynamic efficiency of eighteenth-century engines. For average-quality industrial coals

It can be shown that, with coal of this quality, a 'duty' of n million lbs. of water raised one foot per bushel of coals (84 lbs.) corresponds to an overall thermal efficiency of $0.15$ n % and a fuel consumption of $170/n$ lb. per horsepower per hour. The ideal efficiencies of perfect heat engines, using steam as the working medium on the Rankine-Clausius cycle, were found to be about 7% for an engine with a loading of $9.7$ lbs. p.s.i. on the piston and about 5% with a loading of $7.5$ lbs. p.s.i.

| | Duty million | Efficiency per cent | Fuel Consumption lb/HP/hour | Source |
|---|---|---|---|---|
| 1725 Chester-le-Street | 3·75 | 0·56% | 45·3 | Nixon |
| 1769    75" | 4·59 | 0·69 | 37·2 | Smeaton |
| 1769    60" | 5·88 | 0·88 | 29 | Smeaton |
| 1769    ? | 7·44 | 1·12 | 22·8 | Rolt |
| 1772 Long Benton | 9·45 | 1·42 | 18 | Rolt |
| 1778 Ketley (atmospheric) | 5·77 | 0·86 | 29·5 | Mott |
| 1778 Ketley (Watt) | 14·6 | 2·18 | 11·7 | Mott |

A brief word should be inserted here about 'duty'. Horsepower is a measure of the output of energy in a certain amount of time. James Watt happened to decide upon 33 000 lb lifted 1 ft high in a minute (or 550 ft-lb per second). For duty, however, the divisor is not time but quantities of coal: as Hills states, the number of lb raised 1 ft high by consuming 84 lb of coal in the engine. Duty thus measures the productivity of the largest variable input directly, and with certain reservations can be used to assess the degree of technological attainment. By contrast, horsepower can be increased over quite a wide range by simply building larger and larger engines embodying much the same engineering knowhow. Horsepower can be derived from figures of duty if one knows the actual coal

[51] e.g. P. Mathias, *The First Industrial Nation*, p. 133.
[52] This succinct statement occurs in the course of a discussion of a paper by A. K. Clayton, 'The Newcomen-Type Engine at Elsecar, West Riding', *Trans. Newcomen Soc.* xxxv, 1962–3.

68    *The Evaluation*

consumption and the precise amount of time involved. To take an example: at the Poldice mine in Cornwall the two best atmospheric engines averaged 7·04 millions in duty in 1778, whilst a single Boulton & Watt engine not only did more work (horsepower) but also attained a duty of 26·71 millions in 1792.[53]

Achievements in Cornwall are rather misleading, however, for reasons mentioned below. In essence they tended to represent the performance of an engine under conditions that were much more favourable for getting an extremely high duty figure than would have been possible in the day-to-day working of, say, a textile-mill. In examining diffusion patterns later in the chapter I shall be using ideal performances for illustration, even though everyday running performances would obviously be more appropriate for the individual purchaser to compare. To compensate, the latter will be considered here.

Of the engines under consideration, the Savery engine was the most uneconomic in fuel consumption. One at the (London) City Gas Works consumed about 30 lb per hour, but its indicated horsepower would seem to have been little more than 0·5![54] Two Savery engines modified by Joshua Wrigley were stated to burn 30 to 31½ lb per horsepower per hour, though Farey thought they could have been worked better.[55]

Atmospheric engines of the Newcomen type were more efficient. The figures given by Hills in the extract quoted above are some indication, but they relate primarily to pumping engines. Data on engines used for manufacturing are less easy to come by. In Table 4.9 some examples are given; the first being from Smeaton's Reports, and the remainder all textile engines mentioned in the Marshalls MSS. operating in the 1790s.[56]

In the first half of the eighteenth century the general impression conveyed is that Newcomen engines consumed from 20 to 30 lb of coal an hour for each horsepower, or even more if they were to be found on or near coalfields. Later in the century

[53] D. Gilbert, 'On the Progressive Improvements made in the Efficiency of Steam Engines in Cornwall', *Phil. Trans.* 120, 1830, pt. I; W. J. Henwood, Presidential Address to the Spring Meeting of the Royal Institution of Cornwall, 1871.
[54] Partington (1822), p. 149.
[55] Farey, i. 125; Matschoss, *Die Entwicklung*, p. 303.
[56] Smeaton, vol. i, pp. 223 ff.; vol. ii, pp. 347–9, 401; and Marshall MS. 57.

TABLE 4.9
*Coal consumption of some atmospheric engines used in manufacturing*

|  | Location | Cylinder diameter (in) | Length of stroke (ft) | Coal burnt lb/HP/hr | Approx. horse-power |
|---|---|---|---|---|---|
| Old Mill (1776) | Hull | 25 | ? | 24·3 | 8 |
| Rogerson & Co. | Hunslet | 32½ | 8 | 16·0 | 18 |
| Holroyds | Sheepscar | 32½ | 7½ | 16·0 | 18 |
| New Constitution | Manchester | 16 | 5 | 16 | 3 |
| Simpson & Co. | Manchester | 64 | 7½ | 25·0 | 33 |
| Holdsworths | Wakefield | ? | ? | 16 | 4 |

efficiency improved markedly, for reasons (as noted in Chapter 3) that deserve to be laid chiefly at Smeaton's door. A 42-inch engine at the Long Benton Colliery, Derbyshire, was burning 26 tons a week in 1768; equivalent to 26·6 lb per horsepower per hour for a 78-hour week. After Smeaton's improvements, it averaged 17·63 lb.[57] Similarly, John Curr's engine at Attercliffe in 1790 burnt 17¾ lb.[58]

For the Boulton & Watt engine, data are once more liberally supplied. For the later years of their patent it seems reasonable to put average performance of their engine applied to industrial purposes at 12½ lb per horsepower per hour, as also does Dr. Tann.[59] For example, Farey himself measured a good Watt engine in a starch factory at Lambeth at 12·6 lb in 1805, and his other estimates cluster around this.[60]

The same writer believed that performances of Watt engines deteriorated after James Watt relinquished any interest in them. However, this does not stand out strikingly when the results are pooled. Thomas Wicksteed, writing in the *Civil Engineer and Architect's Journal* in 1840, used a range from 10 to 15 lb of coal per horsepower per hour for engines then employed in cotton-spinning—averaging 12, though he thought 12 lb

---

[57] F. Nixon, 'The Early Steam-Engine in Derbyshire', *Trans. Newcomen Soc.* xxxi, 1957–9.
[58] Curr, op. cit., p. 38.   [59] Tann, in Harte and Ponting.   [60] Farey, i. 492.

on the low side for Lancashire. The celebrated engineer, Sir William Fairbairn, accepted these estimates of Wicksteed.[61] Even the stoutest defender of all that was good in Lancastrian practice, Robert Armstrong, allowed $13\frac{1}{2}$ to 14 lb of good coal per horsepower per hour, in 1839.[62] The master spinner James Kennedy adopted the equivalent of $14\frac{1}{2}$ lb in a paper written in 1817, but believed rather more for smaller engines.[63] These last two figures include coal burnt in the engine-boiler but used for heating the mill.

One consequence of these figures is that the claims often made on behalf of the Watt engine, that it reduced coal requirements to one-half or even one-quarter of those of Newcomen engines, are wildly extravagant. It is perhaps not widely known that the supposedly selfless exaction of one-third of the fuel savings by way of a premium applied only to Newcomen engines unimproved by Smeaton, as Watt actually informed him.[64] In setting the ratio at one-third, Watt had in mind the engine as Newcomen had long since left it, rather than the rival his engine was actually competing with on the market.

Extensive discussion of the coal needs of the high-pressure engine is relegated to Chapter 9. It may be noted here, though, that the typical attainment of such engines when used in manufacturing was about 5 lb of coal per horsepower per hour, as for instance was achieved at Quarry Bank on their compound (McNaughted) engine in the early 1850s. This represents less than half the consumption of a Watt engine, but about twice as much as the best pumping engines in Cornwall.

ANNUAL COSTS

It will be convenient to bring the above information together in the guise of several estimates of annual costs. I have chosen years and dimensions to fit in with later discussion as far as possible. It is not worth covering every single permutation of circumstances that will arise in due course, however, and the

[61] Both in the *Civil Engineer and Architect's J.* iii, 1840.
[62] R. Armstrong, *An Essay on the Boilers of Steam Engines*, London, 1839; Marshalls MS. 39.
[63] Quoted by J. R. McCulloch, 'An Essay on the Rise, Progress, Present State, and Prospects of the Cotton Manufacture', *Edinburgh Rev.* xlvi, 1827.
[64] See the letter from Watt to Smeaton quoted by Farey (i. 329); also Roll, pp. 30–1.

present figures are thus no more than a rough guide. For one thing, the figures here are intended to apply to Manchester, and will therefore not suit any district with different coal (or other) prices.

The horsepower sizes considered are 10 and 30. The former is more appropriate for describing cotton textiles around 1800; the latter for cotton from about the 1830s on. This assertion rests on what is known about standard layouts for mills in each period, as established by industrial archaeologists.[65] On these grounds one would expect to find that some part of the increased efficiency that came about between the two periods arose out of economies of scale.

For industries other than cotton the use of Manchester costs makes the present estimates less valuable. However, the sizes considered can be reasonably helpful: in 1800 the average size of a Boulton & Watt rotative engine in Britain was about $15\frac{1}{2}$ horsepower, and of all their engines (i.e. including the reciprocating ones) 26·4 horsepower. Engines of other types were averaged at between 15 and 20 horsepower in the last chapter; though this was little better than a guess.

In Table 4.10 I have estimated capital costs in the most relevant sizes. Although not fully characteristic of its time, a 30-horse Watt engine for 1795 has been included for comparison, in order to bring out the effect of the premium charged on his patent.

The working day in textile-mills was abnormally long, even by the standards of the Industrial Revolution at large. The 12-hour working day was common by the end of the eighteenth century, and by 1816 had risen to 13 or even $13\frac{1}{2}$ hours in Lancashire.[66] In the 1830s the 69-hour week was usual. The 1834 inquiry allows greater precision; an average of 148 mills in Cheshire (which replied particularly extensively to this question) was a working year of 3568·2 hours, or $304\frac{1}{2}$ days of 11·72 hours each.[67] However, steam-engines were worked even longer than the bulk of the operatives, first to get up steam before the work commenced, then to keep it up during any

[65] See esp. S. D. Chapman, *The Early Factory Masters*, Newton Abbot, 1967.
[66] M. A. Bienefeld, *Working Hours in British Industry: An Economic History*, London, 1972, Ch. 2.
[67] P.P. 1834: XX.

## TABLE 4.10
### Capital costs of typical engines[68]

| Year:<br>Type:<br><br>Size: | (i)<br>1795<br>Watt<br>rotative<br>30 HP<br>(£) | (ii)<br>1800<br>Newcomen<br>pumping<br>10 HP<br>(£) | (iii)<br>1800<br>Watt<br>rotative<br>10 HP<br>(£) | (iv)<br>1800<br>Watt<br>rotative<br>30 HP<br>(£) | (v)<br>1835<br>Watt<br>rotative<br>30 HP<br>(£) |
|---|---|---|---|---|---|
| Engine | 805 | 326 | 452 | 955 | 1050 |
| Boiler | 140 | 68 | 68 | 205 | 200 |
| Premium | 750 | — | — | — | — |
| Engine-house | 190 | 104 | 104 | 232 | 250 |
| Framework | 96 | 80 | 80 | 116 | 125 |
| Erection | 80 | 50 | 50 | 90 | 107 |
| Steam-pipes | 80 | 50 | 50 | 80 | 80 |
| Total | 2141 | 578 | 704 | 1678 | 1812 |
| Cost p.a. | 224·2 | 85·0 | 102·0 | 218·7 | 211·5 |
| Cost per HP p.a. | 7·47 | 8·50 | 10·20 | 7·29 | 7·05 |

[68] These are mainly consensus figures. Col. (i): engine—Roll, p. 312; boiler—ibid.; premium—5 full years @ £5 per horsepower; engine-house—20 per cent of cost of engine and boiler; framework—from Lee, op.cit., p. 105, for 16-horse engine for McConnel & Kennedy's in 1797; erection—Boulton & Watt data; steam-pipes—see p. 61. Col. (ii): engine plus boiler—interpolated from Table 4.1; boiler—as for col. (iii); engine-house, framework, erection, steam-pipes—as for col. (i). Col. (iii): engine plus boiler—Table 4.4; boiler—Table 4.3 plus 25 per cent; others—as for col. (i). Col. (iv): engine and boiler—as for col. (iii); engine-house and framework—20 per cent and 10 per cent of cost of engine plus boiler; erection and steam-pipes—as for col. (i). Col. (v): interpolation from Boulton & Watt data for engine and boiler; others as for col. (iv). See also Hills, *Power in the Industrial Revolution*, p. 167; M. M. Edwards, *The Growth of the British Cotton Trade, 1780–1815*, Manchester, 1967, p. 206; Ashworth MSS. (Stock Book); P.P. 1833: XXI; Goodrich MS. 154; J. Butt (ed.), *Robert Owen: Prince of Cotton Spinners*, Newton Abbot, 1971. Depreciation rate set at 7½ per cent p.a. for engine, 12½ per cent for boiler, and 4 per cent for buildings (see *Thesis*). Interest rate set at 6 per cent for 1795 and 1800, and 5 per cent for 1835.

# The Diffusion of Best-Practice Techniques 73

breaks for meals, etc. One steam-engine, reporting week by week in 1815, worked 3996 hours.[69] The 'New Engine' at Gotts worked an average of 3928 hours a year between 1835 and 1837, and 3890 hours from 1838 to 1842.[70] I shall thus suppose a working year of 3900 hours for textile engines at the end of the eighteenth century, and 3800 hours in the 1830s. Note, however, that by this time textiles were notoriously working the longest hours of all industries, and even of all factory industries, so that the average for *all* industries using rotative engines would almost certainly be lower than this.

Hours of operation for reciprocating engines are more difficult to predict, in the sense that practice varied much more widely. In Cornwall the best engines were normally worked around the clock in the early nineteenth century, to prevent the mine from filling up overnight. On the other hand, many pumping engines were used only intermittently, as needs dictated. I shall make two suppositions for the Newcomen engine: one, that they averaged the same hours annually as rotative engines (i.e. 3900); and two, that they averaged 4500 hours a year—with the latter intended as an over-generous allowance for the engines that worked 24 hours a day.

The price of coal delivered in Manchester is taken to be 7s. 3d., 9s. 3d., and 9s. 0d., respectively, in 1795, 1800, and 1835.[71] These figures are obviously not to be trusted to the decimals of pounds, but nevertheless are probably sufficiently accurate for me to assert that there was little decline in the money costs of steam-power from the turn of the century until the late 1830s. Fixed costs of engine and boiler for Watt-type engines rose a little (see Table 4.10), though in annual expenses this was offset by a falling interest rate. Indeed, from the years in which sales of Watt engines first really boomed—the early 1790s—there was little change in the money outlays, since by this time the premiums for patent rights (and thus interest payments on them) were rapidly contracting.

Moreover, for the chosen site of Manchester the Newcomen engine was less than 10 per cent more costly to run than the

[69] P.P. 1816: III.
[70] Gott MS. (Brotherton 193) 211. Marshalls in Leeds worked their flax-spinning engines 3714½ hours in 1823 (MS. 39).
[71] See App. 4.2.

TABLE 4.11
*Annual costs of typical steam-engines, erected in Manchester*[72]

| Year:<br>Type:<br>Size: | 1795<br>Watt<br>rotative<br>30 HP<br>(£) | 1800<br>Newcomen<br>pumping<br>10 HP<br>(£) | 1800<br>Watt<br>rotative<br>10 HP<br>(£) | 1800<br>Watt<br>rotative<br>30 HP<br>(£) | 1835<br>Watt<br>rotative<br>30 HP<br>(£) |
|---|---|---|---|---|---|
| Capital costs | 224·2 | 85·0 | 102·0 | 218·7 | 211·5 |
| Material costs: | 255·1 | 160·0 | 118·3 | 337·2 | 326·0 |
| Labour costs: | 53·4 | 48·0 | 48·0 | 57·5 | 75·0 |
| Total costs | 532·7 | 293·0 | 268·3 | 613·4 | 612·5 |
| Costs per HP | 17·8 | 29·3 | 26·8 | 20·4 | 20·4 |
| Costs per HP per hour | 1·09d. | 1·80d.<br>(1·73d.)* | 1·65d. | 1·26d. | 1·26d. |

\* = averaged over 4500 hours.

Watt type. The implied effect of changing the site will be examined in the next section.

THE THRESHOLD FOR THE WATT ENGINE

Professor Paul David developed the concept of the 'threshold' to describe the cost configuration at which it would become economically rational to switch over from an old technique to a

[72] The only comprehensive estimate of the running costs of oil, tallow, stuffing (gasking), etc. I have encountered so far is that by W. Henshall in 1847, in which he puts the costs for a 100-horse engine at 27s. 0d. a week. Costs of these items for 10- and 30-horsepower engines are then extrapolated from that on the basis of proportional dimensions of their cylinders. The results agree satisfactorily with miscellaneous information from Quarry Bank records. 1800 and 1795 figures taken from this 1835 estimate by assuming similar price movements as for coal. Labour costs are put at 30s. 0d. weekly for 1835; the amount the Ashworths paid, for instance, on their 30-horse engine at New Eagley in the 1840s—other enginemen were paid much the same. Engineman's labour for 10 horsepower reckoned at 25s. a week in 1835. Other years extrapolated on Bowley-Wood series of ship-building and engineering wages, from B. R. Mitchell with P. Deane, *Abstract of British Historical Statistics*, Cambridge, 1962.

new one.[73] It is illuminating to use this concept to analyse the shift from atmospheric to Watt engines at the end of the eighteenth century.

By this date the figures quoted earlier suggest that a Savery engine cost little more than half a Watt engine of equal power to purchase, while Newcomens were roughly 75 to 95 per cent the price of Watts. Incidentally, it may come as a surprise that Newcomens should have been any cheaper than Watt engines at all, since Watt's inventions allowed his engines to employ much smaller cylinders, and sometimes smaller boilers, to produce the same nominal horsepower. (It has been estimated that in Smeaton's day a good Newcomen engine could develop about two-thirds of the power of a single-acting Watt engine with the same size cylinder.) Appendix 4.1, which compares costs for engines of each type that I trust are typical, indicates that the additional cast iron components required for Watt engines—steam-case, separate condenser, etc.—plus perhaps the extra fitting charges arising out of the greater precision needed for Watt cylinders, were enough to swamp the rest and tilt the balance in fixed costs over to atmospheric engines. The 'threshold' must weight these advantages against those in variable expenses favouring the Watt type. Initially the various thresholds are calculated for engines in optimal condition.

The threshold has to be computed separately for rotative and reciprocating engines, reflecting the different pricing systems Boulton & Watt tried to operate for the two. In making his own comparison of engine types, Farey deemed a Newcomen engine in excellent order to be capable of a coal consumption of 15·87 lb per horsepower per hour; this being recorded by the 52-inch engine that John Smeaton had improved at Long Benton. A Watt rotative engine could be worked at as little as 9·07 lb, though it would be in even better shape than the Newcomen to do so.[74] Now in the 1790s a Watt engine of 28

---

[73] P. A. David, *Technical Choice, Innovation and Economic Growth*, Cambridge, 1975, Chs. 4–5. An earlier hint of the threshold approach to the present issue is by Jennifer Tann: 'The advantage of a lower coal consumption and reduced possibility of failure of a Boulton & Watt engine were off-set by the disadvantages of a higher capital cost, an annual premium, and the fact that it was a more complicated engine to run.' ('Richard Arkwright and Technology', *History*, vol. 58, 1973).
[74] Farey, i. 368 n. In comparable condition to the quoted atmospheric engine, Watt's could achieve about 10½ lb/HP/hr (ibid., p. 488).

horsepower cost some £790 for engine plus boiler.[75] To this must be added the licence fee of £5 per horsepower p.a. in the provinces (£6 in London).

As a standard for Newcomen engines I take the long-lived 42-inch engine at Elsecar, which cost about £900 (Table 4.1), although Curr's figures in Table 4.2 would reduce this to about £700. Assuming in addition that the engine worked $12\frac{1}{2}$ hours daily and 300 days a year, then coal consumption annually would be 744 tons for the atmospheric engine. In these circumstances, an entrepreneur indifferent between the two engines on other grounds would select an atmospheric engine in preference to a Watt when coals cost less than 7s. 10d. per ton (or inside London when coals were less than 9s. 7d. which was never the case). If, however, the factory already had a Newcomen engine installed, it would be a different matter, since it would pay to scrap the atmospheric engine only when its variable costs rose above the *total* costs of the new Watt engine. If another £25 p.a. is allowed for extra repairs on the old (atmospheric) engine, then the threshold price of coal rises to 14s. 0d. a ton, under these conditions.

Watt's reciprocating engines were more economical on fuel than the rotative ones, partly because they could make some use of expansion of the steam in the cylinder. Engines in optimal condition could be working as low as 6·26 lb per horsepower per hour. For reciprocating engines, Watt usually tried to charge a premium amounting to one-third of the cost of fuel saved.[76] However, when the price of coal fell to 5s. or less, Boulton & Watt exacted one-half of the fuel saving. This figure would seem to have been chosen wisely, because the threshold price for coal in the second set of circumstances—replacing an atmospheric with a condensing engine, at a saving of £25 a year in repair bills—would be 5s 10d. if Boulton & Watt took one-third of the fuel economy.

As Watt's patent rights came towards expiry late in the 1790s the balance changed sharply. Taking £1040 as the all-in price

---

[75] Roll, p. 312.
[76] On my figures, one-third of the fuel saved was equivalent to charging £5 per horsepower when the price of coal was 18s. 8d. a ton; i.e., for any coal price less than 18s. 8d. the user was better off paying one-third of the fuel saving than the £5 per horsepower p.a.

## The Diffusion of Best-Practice Techniques

of 28-horse rotative Watt engines in 1799 and assuming (perhaps wrongly) that the price of atmospheric engines did not change, the threshold price for coal would have fallen to about 1s. 1d. a ton, comparing new engines and thus making no allowances for extra repairs. Even slack coal at the pithead was not valued as cheaply as this, so it is no surprise to find orders for new condensing engines pouring in in the late 1790s and early 1800s. Again, however, if the mill (or whatever) already had an atmospheric engine in operation, the threshold price would rise to 6s. 6d. a ton, and it is clear from earlier considerations of the regional patterns of coal prices that this would have prevailed for many manufacturing areas in 1800. Farey[77] noted that the last atmospheric engine in London was pulled down in 1813, but on some coalfields they persisted well into the nineteenth century and even occasionally into the twentieth.

Thus many colliery-owners preferred the Newcomen engine long after the Watt type became available. John Farey's father in his survey of Derbyshire in the first few years of the nineteenth century, remarked 'I met with no pumping engine on Boulton and Watt's principle at a coal-pit; the old atmospheric engines, well-contrived and executed, being thought to answer better in such situations.'[78] R. A. Mott summed up for the Tyneside collieries likewise:

... the greatest need in mining was for powerful pumping engines. This need was satisfied by Newcomen's engine and although Watt's improvements gave this a three-fold potential increase in power in the last decade of the 18th century, mining engineers largely ignored it and waited for the expiration of Watt's basic patent and for higher steam pressures before they applied extensively Watt's improvements of the single-acting atmospheric engine.[79]

From these results it is perfectly understandable why Watt jr. wrote to his father that 'The gross sum (including premium) which your engines cost at first startles all the lesser manufac-

---

[77] Farey, i. 253. There was a Savery engine working in St. Pancras in 1819 and for an unknown time thereafter; see G. Birkbeck and H. and J. Adcock, *The Steam Engine Theoretically and Practically Displayed*, London, 1827.
[78] Nixon, 'The Early Steam-Engine', quoting J. Farey, sr., *General View of the Agriculture and Minerals in Derbyshire*, vol. i, London, 1807.
[79] Mott, op. cit.

turers here', referring to Manchester.[80] From the numbers calculated it is apparent why potential customers felt aggrieved by the swingeing size of the premium, and why there was a standing temptation to 'pirates' to build their own separate condensers and dodge the payments.

These calculations are imperfect in a whole host of ways, some of which can be emphasized. In the first place, they assume that entrepreneurial expectations were 'myopic' to a degree unjustified by the reality. Anyone buying a Watt engine, or simply contemplating doing so, must have been aware that the premium payments were going to cease in 1800. To cope with this, it would be more logical here to compute the rate of return from converting from an atmospheric engine to one with a separate condenser. There would be a different result for each particular year of, say, the 1790s. Apart from the labour involved in carrying this exercise out, the reason for ducking it is the very range of other errors and approximations present.

Secondly, then, it must be acknowledged that Boulton & Watt often departed from the strict principles assumed for setting their premiums, as I have already implied in one or two respects. Watt himself was notoriously unwilling to comply with these lapses, but was generally overruled. In a number of minor cases the full rate had to be abated, especially where the power was lying partially unused.[81] Much more often the annual rate was capitalized into a lump sum, normally to the benefit of the user.

Thirdly, the calculations above were for best-practice engines. Average attainment in day-to-day industrial use was quite considerably poorer, as previous figures show. It is clearly illicit to suppose that the gap between average and best practice was proportionately the same for atmospheric and Watt engines. Most authors have supposed, as I did earlier, that Boulton & Watt's engines were built to more demanding and reliable standards than their predecessors. On the other hand, the complaints of many of their customers demonstrate that their engines were far from being trouble-free. Even if it could be proved that their competitors' products broke down

[80] Musson and Robinson, *Science and Technology*.   [81] Roll, p. 117.

## The Diffusion of Best-Practice Techniques

more often—probable, but not readily provable—one would have to contend with the view of several contemporaries that Newcomen engines were easier and cheaper to maintain and repair in the ordinary course of things. The very precision of the Watt engine made it prone to minor repair bills, e.g. with its nagging problem of keeping a satisfactory piston packing.

Fourthly, it cannot be assumed, as I have done here, that Newcomen and Watt engines were equally worthy, costs aside, for all industrial purposes. The crude construction of the former rendered its operation too erratic (in isolation) for the delicate work required in a textile-mill, for instance. Modifications that would have enabled a mill-owner to employ an atmospheric engine—of the kind to be discussed in Chapter 6—would lower the threshold price of coal.

### DIFFUSION OF THE HIGH-PRESSURE ENGINE—FACTORS

Why did the dissemination of high-pressure steam in British manufacturing enterprise take so long? For a start, some of the ingredients contributing to its precociously high level of efficiency in Cornwall—far in advance of contemporary scientific comprehension—also acted to limit its use for other ends. Very high duties were attained partly through running the engine at very slow speeds, with extensive pauses at the end of each stroke to perfect the vacuum, and with an especially slow return (pumping) stroke.[82] Such duties frequently occurred early in the life of an engine, when the pit was still comparatively shallow and the water load far from onerous. For instance, Austin's engine at Fowey Consols Mine, which performed at the extraordinary level of 125 millions in a special test in 1835, was regularly worked at only about 3 or 4 strokes a minute in

[82] Henwood's experiments on three engines in 1831 gave the following average durations of stroke (all times in seconds):

|  | Strokes per minute | Duration of working stroke | Duration of return stroke | Interval between strokes |
|---|---|---|---|---|
| Wheal Towan 80″ × 10′ | 5·35 | 1·6 | 4·8 | 4·8 |
| Binner Downs 70″ × 10′ | 7·49 | 1·34 | 4·23 | 2·4 |
| East Crinnis 76″ × 10¼′ | 3·5 | 1·7 | 4·17 | 11·2 |

('On the Expansive Action of Steam in some of the Pumping Engines on the Cornish Mines', *Trans. Inst. Civil Engineers*, ii, 1838.)

that year:[83] with a piston length of 10·3 ft its piston averaged under 35 ft per minute. Naturally the horsepower supplied by the engine suffered as a result. John Darlington in 1855 reckoned the power of Cornish engines as shown in Table 4.12, by rather generously averaging pressures at 40 lb p.s.i.[84]

TABLE 4.12
*Effective (indicated) horsepower of Cornish engines*

| Diameter of cylinder (in) | Strokes per minute | | Horsepower per stroke | |
|---|---|---|---|---|
| | Economical working | Safe working | Economical working | Safe working |
| 15 | 5 | 14 | 3 | $8\frac{1}{2}$ |
| 20 | $4\frac{1}{2}$ | 12 | 6 | 16 |
| 30 | 4 | 10 | 12 | 30 |
| 40 | 4 | 10 | 21 | 54 |
| 50 | 4 | 10 | 33 | 84 |
| 60 | 4 | $9\frac{1}{2}$ | 51 | 122 |
| 70 | 4 | 9 | 73 | 164 |
| 80 | 4 | $8\frac{1}{2}$ | 100 | 212 |
| 90 | 4 | 8 | 131 | 263 |
| 100 | 4 | 8 | 162 | 325 |

Barton notes that the customary speed of operation lay in the range of 4 to 6 strokes per minute, with the precise level depending, as might be expected, on local circumstances of pitwork, etc.—finding it involved more trial and error.[85] Only when there were exceptional unwatering problems would the large engines be likely to work at 10 to 12 strokes a minute, e.g. by Woolf's engines at Wheal Abraham in 1818.[86] Small engines, especially if double-acting or rotative, worked faster but at lower levels of duty.[87]

This slow operation of Cornish engines may help to explain why Peter Barlow found that their output, measured in horsepower, was only eleven-fourteenths of what he would expect from engines of similar size in the manufacturing districts such

---

[83] Farey, ii. 225 n.
[84] Phillips and Darlington, *Records* . . .; also Barton, *Cornish Beam Engine*, p. 100 n., quoting the *Mining Journal* for 1855.
[85] Barton, p. 101.  [86] Ibid., pp. 40, 102.  [87] Farey, ii. 254–60.

as Lancashire.⁸⁸ In fact they could have been more powerful, but a trade-off existed between maximum economy of fuel and maximum economy of time, and the solution in Cornwall was heavily biased towards the former. In Lancashire the lower coal prices in the 1830s (5*s*. to 9*s*. depending on exact location) dictated a different choice. The conventionally accepted velocity for the piston since the time of Watt had been 220 ft per minute, and although this was rarely achieved in practice, most tables of proportions of this period show a range from 190 to 210 ft per minute for engines of 8 horses and above.⁸⁹

The figures of piston speed are for double-acting engines in Lancashire, as against single-acting ones in Cornwall, and this reflects standard practice in the respective areas. As noted in Chapter 2, double-acting engines were never able to attain the level of fuel economy of the older single-acting type from the time they were developed by Watt. Offsetting that was an obvious gain in capital productivity, since about twice the power could be generated from the same engine when converted to double action.⁹⁰

The advantages of single-action lay not just in the rather greater fuel economy possible. Farey⁹¹ stressed that the operation of the Cornish engine owed much to the invention of the plunger pump by Watt's associate, William Murdoch, who had been working in Cornwall up to the turn of the century. This pump forces up the water when the pump rod *descends*, thus reversing the usual effect of lifting or sucking pumps.

Mr. Murdoch's application of plunger-pumps to the engines for draining deep mines in Cornwall, in lieu of the lifting-pumps formerly used there, conjoined with the increasing depth of the

---

⁸⁸ P. Barlow, 'A Treatise on the Manufactures and Machinery of Great Britain', *Encyclopaedia Metropolitana*, Mixed Sciences, vol. vi, London, 1836.
⁸⁹ Rees, *Cyclopaedia*; Barlow, op.cit., p. 191; J. Renwick, *Treatise on the Steam Engine*, New York, 1830.
⁹⁰ Some of the Watt engines used in Cornwall were employed in double-acting fashion, whether purchased that way (as with the great engine at Wheal Maid in 1787) or subsequently converted to meet the greater power requirements of greater depths of mine shafts; but the practice virtually died out with the gain from fuel economies registered in the early nineteenth century, until still greater depths reached in the 1840s caused the revival of interest in the double-acting engine.
⁹¹ Farey, ii. 146 ff., 270. Farey points out that Murdoch had quite different objectives in mind (148–9).

mines, offered every facility for the adoption of Mr. Woolf's system of high pressure steam operating with a great extent of expansive action, in single acting engines; because, with the plunger-pumps, the force of the steam is applied to lift the unbalanced weight of the pump-rod, instead of the columns of water; and then the variable force with which the high pressure steam operates during its expansion becomes unimportant circumstance, because it is the unbalanced weight alone which is subjected to the variable force, and the descent of that weight operates with uniformity, whilst it is producing the required effect of raising water in the pumps.[92]

In other words, there was a comparatively short, sharp indoor stroke, during which the steam entered the cylinder, then was cut off after about one-tenth (or more) of the stroke had been completed, following which the engine worked with steadily diminishing power as expansion was carried out. Thus the speed of the piston during the indoor stroke was rapid at first, tempered only by the need to overcome inertia: it is estimated that Taylor's famous 85-inch engine was being hit by about 100 tons on the piston each time the steam (at 50 lb p.s.i.) was let into the cylinder.[93] Moreover, once the cut-off had been applied, the rapid acceleration of the piston was reversed and output dropped away. However, as Farey states, none of this mattered for the typical Cornish engine because all that was happening during the indoor stroke was that the pump rods are being raised. Once the indoor stroke had been completed, and after a pause to improve the vacuum, the leisurely outdoor stroke proceeded by gravity[94]—only then was water actually pumped, and the work could be made very regular as a result.

All of this would have been quite unsuitable for turning delicate machinery: I have already pointed to the continuing problems over transmission even with the low-pressure Watt engine, Nevertheless, several possible solutions were already at hand.[95]

[92] Ibid. ii. 146.
[93] This is Barton's calculation—op.cit., p. 57. By my own computation it would need only 40 lb p.s.i. above atmospheric pressure to produce an impact of 101 tons.
[94] More precisely, by the excess of weight of the pump rods over that of the column of water being lifted. In a large engine J. S. Enys calculated in 1839 that the pump rods could weight 150 tons and the column of water 37 tons (*Trans. Inst. Civil Engineers*, 1839, p. 45).
[95] The 'returning engine' system of using steam to pump water to the top of a water-wheel was occasionally mooted, e.g. by Josiah Parkes ('On Steam Boilers', p. 67). It was more common to recommend extending the usual fly-wheel method, but this could scarcely cope with irregularities on such a scale (Farey, ii. 266–7).

## The Diffusion of Best-Practice Techniques 83

A preferred alternative was to adapt the system of using two half-size condensing engines arranged at right angles to each other and operated in such a way that through-cycle variations just cancelled each other out—to use Woolf engines in this manner.[96] Even with this arrangement it was found that the amount of expansion would have to be moderated to operate, say, a textile-mill. Permanent and satisfactory improvement awaited the development of a workable double-acting rotative version of the high-pressure expanding engine. Arthur Woolf in fact devised such an engine, and applied it to some small 'whims', but in adapting an important stamping engine at Wheal Vor he made the steam-valves too small, through over-rigid deductions from his totally false scientific views on the laws of expansion. According to Farey, this discredited the idea in Cornwall until 1835, when James Sims erected a high-pressure rotative stamping engine (single-acting)—with double-action employed from 1839.[97] There can be little doubt that many Northern manufacturers, even if they had heard of Cornish fuel economy, were under the impression that it was not possible to apply it either to double-action or to rotative propulsion.

Viewed in economic terms, all the alternatives suggested for moderating the fluctuations in power from a high-pressure engine reduce to (i) a loss in efficiency compared with that of pumping in Cornwall, or (ii) larger outlays on associated equipment, or (iii) both of these. The higher cost of the Cornish engine has been mentioned in an earlier section of this chapter; I concluded that the main reason was the higher costs of boilers and balance bobs.

Even despite the mounting attacks on the flimsiness of boilers of the Boulton & Watt type,[98] the full-sized Cornish-style cylindrical boiler did not rapidly assert itself in Lancashire.

Thus, it has been proved by direct experiment that the loss of fuel resulting from even a considerable diminution of heating surface, caused by cutting off 10, 20 or even 30 feet from the length of a

[96] Farey, ii. 309.
[97] Ibid. ii. 263. The explanation is not completely convincing, since by Farey's own admission the Woolf rotative engine, with more adequate steam-valves, continued to be built in London by Hall & Son from 1815, and used there and on the Continent.
[98] Parkes, pt. I, pp. 11–12.

long cylindrical boiler, has amounted to a very small percentage indeed upon the first cost of the extra boiler plate and brick work. This remark chiefly applies to cases where fuel is not very expensive, as in the manufacturing districts, and would require modification if applied to the case of boilers in Cornwall, where any saving of fuel, however effected, is perhaps to be preferred to every other consideration. It is found, however, in Lancashire that an effective heating surface of 10 to 12 square feet per HP is as much as can be economically allowed.[99]

Josiah Parkes noted that Lancashire engineers normally allowed 12 to 15 square feet of boiler to vaporize 1 ft$^3$ of water per hour, while the Cornish used 60 to 70, and this is roughly consistent with the above quotation.[100]

What I am arguing for is the existence of a trade-off between fixed and working costs. As already mentioned, the Lancastrian manufacturers saved capital by running their engines faster, but in addition they tended to economize on physical capacity by ignoring the huge boilers and unwieldy engines and enginehouses of the Cornish miners. The importance of high coal costs in spurring the Cornish to fuel savings is widely acknowledged, as in the quote above, and here taken as read. On my estimates, coal outlays for a 30-horsepower engine at Camborne would be substantially lower than for an equivalent engine in Manchester, in spite of the 25s. a ton paid for coal in the former in 1835. If coal requirements were cut by three-quarters as a result of installing a Cornish engine in Manchester, annual coal costs for this size of engine would have been reduced by about £185. On the other hand, if boiler costs rose in proportion to surface heating area (seven and a half times on above estimates), capital costs in Manchester would have risen nearly £230 a year, and so swamped the saving in fuel bills.

Thus it is not quite enough to explain Lancastrian tardiness by cheaper coal alone: if this were all, there would still be *some* advantage to going over to higher pressures in Lancashire. One has to recognize the attendant rises in capital costs. Lack of information precludes any more exact reckoning of threshold

---

[99] *Engineer & Machinist's Assistant*, p. 50.
[100] *Op.cit* Even before Fairbairn's patent in 1844 on the Lancashire boiler it had become common to build internal flues into northern waggon boilers, e.g. that of the Butterley Company (Fairbairn, *Mills and Millwork*, pp. 255-6).

## The Diffusion of Best-Practice Techniques

values than those just noted, but it is conceivable that capital was cheaper in Cornwall than in the newer manufacturing areas. Cornwall and Devon were favoured counties for early country banks, some of them with unusually strong links with local industry.[101] The difficulties encountered by mine-owners and manufacturers in other parts of the country in raising fixed capital may therefore have been less acute in the southwest. Secondly, the price of capital goods produced within Cornwall seems to have been a little lower, arising out of lower charges for iron and other metals. Thirdly, some writers maintained that the cost-book system of accounting encouraged capital intensity: 'I would observe that the Cornish system of mining accounts in which no reference is made to the capital expended, has afforded the mining engineers more liberty in the adoption of whatever proportions appeared advantageous in the boiler surfaces in the flues, or in the size of the cylinder for expansion, and in an increase of the strength of the pitwork . . .'[102] Certainly engineers had a high degree of independence from managers,[103] and their bonus system must have encouraged fuel savings at the expense of capital expenditure. However, Barton considers that minimum-cost constraints almost always curbed engineering improvements. As the *West Briton* newspaper wrote: '. . . All the coal saved above 70 million duty is paid for at too dear a price in the racking of the engine and pump-work and the increased liability to breakage.'[104]

I do not have direct evidence that textile mill-owners held off buying high-pressure engines solely for economic reasons. Analogous reasoning, however, was used by Simon Goodrich, in speaking of marine engines, whose boilers decomposed in two years through the effects of salt water: 'For a 40 horsepower [Boulton & Watt] engine the boiler will weigh about 24 tons, and every time a new one is required to replace the old it is done at a cost of 2400 or £100 per month for the wear of the boilers alone; a sum exceeding the whole expense of coals

[101] L. S. Pressnell, *Country Banking in the Industrial Revolution*, Oxford, 1956, Ch. 10.
[102] J. S. Enys, 'Remarks on the Duty of the Steam Engines employed in the Mines of Cornwall at Different Periods', *Trans. Inst. Civil Engineers*, iii, 1842, p. 458.
[103] E. Alban, *The High-Pressure Steam Engine Investigated*, trans. W. Pole, London, 1847–8, Intro., fnn. 3–5.
[104] Quoted by Barton, p. 59. No date given.

consumed by the present Cornish engines for the same quantity of work.'[105]

Northern manufacturers therefore attempted compromises between higher fuel economy and the heavy capital outlays that this might have involved. In Chapter 9 I show that even in Cornwall many of the gains could be made in the form of minor adaptations to existing plant. In Lancashire there was even less willingness to dismantle existing engines, and a frequent resort was to compounding by simply adding on a high-pressure cylinder.

A solution particularly common in the cotton area around Staleybridge was the use of 'Thrutchers'. These were horizontal, non-condensing engines that were not used in isolation but were attached usually to the main gearing or drive of existing engines. They employed steam at about 50 lb p.s.i., which after use they discharged into the main cylinder for condensation. Fairbairn,[106] the patentee of the Lancashire boiler, considered that they lost a goodly part of the possible economies in fuel consumption because of the excessive loss of heat entailed in conveying the steam around. However, he also noted that there was a considerable advantage from the point of view of capital costs from appending the Thrutcher rather than installing a completely new high-pressure engine.

Similar in intention, and rather more widely adopted in the manufacturing districts of the North of England, was the method of 'McNaughting', so named after its inventor. Again this consisted of appending a high-pressure cylinder to an existing low-pressure engine. The arrangement was rather neater than in the Thrutchers, because the new cylinder was attached to the working beam of the old, though at the far end to the old cylinder. The average saving of fuel was alleged to be 30 to 40 per cent if steam was used at 30 to 40 lb in the McNaught cylinder.[107]

There was nothing particularly original about these modifica-

[105] Goodrich MS. 1557.
[106] Fairbairn, op.cit., pp. 237–40. Note particularly his remarks: 'To accomplish the increase of pressure no change has taken place in the engine itself, beyond the strengthening of the parts, and the substitution of wrought-iron and steel for parts which were before considered sufficiently strong of cast iron' (p. 237).
[107] *Artizan*, viii, 1850, p. 257, and ix, p. 111, etc. See also 'Advertisements relating to the Great Exhibition' (British Museum).

tions—Pattison and Buddle had used a similar arrangement on the Newcastle coalfield by 1825,[108] and the same was true elsewhere; but McNaught's engines did get some recognition as being the safest of the alternatives.

Through the efforts of Woolf and others, London had a history of high-pressure engines dating back to the very early years of the nineteenth century, and this tradition was maintained by Hall & Son after 1815. However, they do not seem to have been able to make any notable inroads into major domestic markets, and indeed earned some opprobrium for rather clumsy engineering work.

Horrockses of Preston were said to be the first Lancashire textile-mill to erect a Cornish engine, in 1839[109]—although some internally flued boilers such as the Butterley Company's may have been found before that date. The patenting of the Lancashire boiler in 1844 gave greater facility for safe use of higher pressures, but it does not seem likely that much use was made of them until late in the 1840s. Between 1846 and 1851 there were said to be 63 engines McNaughted, with others in hand in 1851.[110] Black Dyke Mills were said to be the first in Yorkshire to compound, when they McNaughted their 45-horsepower engine in the Shed Mill after 1848.[111] Crump and Ghorbal mention that by the 1850s the old waggon boilers in the Huddersfield woollen industry were being pulled out and replaced by Cornish boilers.[112] The Gotts had their steam-engines and boilers at Armley Mills altered from 1853 to 1855, presumably for these reasons.[113]

In flax-spinning, the great firm of Marshalls recorded in their MSS. 'Notes on Processes' at their Shrewsbury Mill in 1829:

. . . Told Horsman to enquire if he could not get undressed coals to come in cheap—And to cover the cylinder of the 60 Horse with 2 thicknesses of hay band and then plaister neatly over it, and to note the saving of fuel . . .

[108] *Mechanic's Magazine*, iv, 1825, pp. 369–71. Indeed Woolf's first compound engine was of this type.
[109] Armstrong, *Essay*, p. 90.
[110] 'Advertisements relating to the Great Exhibition'.
[111] W. Cudworth, *Round About Bradford*, Bradford, 1876.
[112] W. B. Crump and S. G. Ghorbal, *History of the Huddersfield Woollen Industry*, Huddersfield, 1935, p. 120.
[113] Gott MS. 224.

Nov. 25th 1830 . . . Holmes says that the change in the setting of the valves of the 60 Horse to shut off at ⅜ the stroke is a great improvement; the engine takes less steam and works smoother. Horseman says the Cornwall engine cylinder casings are protected by lath and plaister leaving a space of 1″ between . . . and casing: putting the plaister on thick and making it neat and smooth and painting it outside—Ordered both engine cylinders, and steam pipes to be covered in this way . . .[114]

However, Marshalls did not compound their engines at Leeds until the years 1850 to 1866.[115] In the silk industry, Messrs. Penn of Greenwich supplied a high-pressure engine to a mill in 1841.[116] McNaughted engines were also apparently popular at an early stage in Glasgow.[117]

Factors other than economic rationality have sometimes been used to explain this slow diffusion of high-pressure steam in Britain. Safety is one. The loss of life from boiler explosions was by any industrial standards severe: in November 1850 the *Engineer and Machinist* reported that no less than 1600 people had been killed in the previous three years—a period in which the conversion to high pressures was gathering momentum. Many of the reported accidents occurred in fact through trying to exact high-pressure steam out of low-pressure boilers (often rather old and thin)—in the worst cases the Coroners' reports found that the safety-valve had been wedged down in utter disregard of human life.[118] However, a properly constructed high-pressure boiler solved the problem by using thicker boiler plates: ½-inch plates were the minimum used in Cornwall, compared to ¼-inch to $\frac{5}{16}$-inch for standard low-pressure boilers.[119] Fatal accidents from boilers occurred in Cornwall, of course, but most contemporaries agreed that their incidence was no greater than and perhaps less than elsewhere.[120] Farey

[114] Marshall MS. 37.
[115] Marshall MS. 26. Cf. also O. Ashmore, 'Low Moor, Clitheroe: A Nineteenth-Century Factory Community', *Trans. Lancs & Cheshire Antiq. Soc.* 73–4, 1966.
[116] Farey, ii. 314.   [117] *Artizan*, vi, 1848, p. 182.
[118] G. Dodd, *History and Explanatory Dissertation on Steam Engines and Steam Packets*, London, 1818; *Artizan*, iii, 1845, pp. 87–91.
[119] Parkes, op. cit.; Fairbairn, op. cit., p. 267. None of this is to deny that manufacturers elsewhere may have erroneously believed the high-pressure engine to be more dangerous and to have desisted from employing it accordingly; but I subsume this under the entrepreneurial arguments later. (e.g. E. Galloway, *History and Progress*, p. 43: '. . . there is still a mass of prejudice to contend with'').
[120] *Civil Engineer and Architect's J.* iii, 1840.

## The Diffusion of Best-Practice Techniques

clearly thought the Woolf water-tube boilers safest of all, and rather regretted their abandonment (until late in the nineteenth century, as it turned out) in favour of Trevithick cylindrical ones.[121]

Related is the argument that the heavy hand of James Watt and his deliberate eschewal of high pressures influenced English manufacturing practice for half a century. Undoubtedly it influenced his own firm for many years—and thus those who blindly relied on their quality, such as the Birmingham Waterworks—but Farey noted that no other firm was adhering steadfastly to low pressures by about 1840,[122] and even the Boulton & Watt Company was building high-pressure engines by 1842. High-pressure engines were being erected in the city of Birmingham from 1834;[123] over the years 1834-8 some 42 per cent of the horsepower erected was in high-pressure engines, and in the last of those years (1838) 57 per cent. It is doubtful whether their reputation would have been influential far beyond.

It is unlikely that the sheer technical (as distinct from the economic) problem associated with irregular power delivery accounts for too much of this lag behind Cornish practice, because their adoption in pumping operations such as waterworks did not come much earlier. In London, where coals were more expensive than in Cornwall, Thomas Wicksteed had to plead to try out a Cornish engine, and then was forced by his employers to wait until a second-hand one of suitable dimensions came on the market[124]—this in 1838. Harveys sent 60-inch engine to Carlisle also in 1838.[125] For other areas, such as North Wales, adoption often required the prior dispatch and recommendation of individual Cornish engineers, such as John Taylor. Overseas there was no such reticence about the Cornish engine, as will be amplified in Chapter 10.

For all this, one cannot dismiss entrepreneurial lags in diffusion summarily. Farey[126] spoke of a 'state of apathy' in regard to compounding in the late 1830s. It does not seem entirely explicable in terms of vintages of capital, partly

---

[121] Farey, ii. 192.    [122] Ibid. ii. 305.    [123] *J. Stat. Soc.* Jan. 1840, p. 440.
[124] T. Wicksteed, *An Experimental Inquiry concerning Cornish and Boulton & Watt Pumping Engines,* London, 1841.
[125] Barton, p. 263.    [126] Farey, ii. 307.

because there are some indications that replacement by better equipment was neglected:

> We have often been surprised at the tenacity with which some firms will stick to the old-fashioned low pressure steam, 4 or 5 lbs. p.s.i., even when the chance of putting in new boilers occurred . . . The government dock-yards are particularly notorious in this respect. Some thousands of pounds per annum are wasted by the high-pressure-phobia of the authorities. It must have cost them a severe pang to allow 15 or 16 lbs. p.s.i. in the fast mail-boats.[127]

What all this adds up to in quantitative terms is unclear. The failing may have been one of the inventors rather than the businessmen: inventors were unable to come up with a satisfactory high-pressure rotative engine until about the mid-1830s. Exactly what technical weaknesses the entrepreneurs saw in existing high-pressure rotative engines, especially Woolf's, is unknown. Maybe they simply had not heard about them. Certainly many of the engineers they may have consulted remained sceptical of Cornish attainments until the later 1830s.[128] Equally certainly, many of them, unaware that the high-pressure engine could be made just as safe as low-pressure ones by using thicker plates, considered them outlandishly hazardous. Yet there is also contradictory evidence at hand in some cases.[129] As the Marshalls did at Shrewsbury in 1829, they seem prepared to undertake the simpler improvements, such as lagging pipes and cylinders, without yet going the whole hog of enlarging and improving the boilers, providing a cylinder case, and so forth.

It therefore seems that in regions where coal was cheap, and the pay-off to fuel-saving innovation thus less impressive, the outlays on such capital equipment embodying new technology (for example, tubular boilers) long proved too high to be justified on economic grounds. If the figures used so far are

---

[127] *Artizan*, vii, 1849, p. 25. The editor of this journal was John Bourne.
[128] G. H. Palmer still flatly refused to believe the Leans's Reports as late as 1838 because they contradicted the caloric theory of heat ('On the Application of Steam as a Moving Power . . .', *Trans. Inst. Civil Engineers*, ii, 1838). The more widespread doubts persisted over (a) safety and (b) application for rotative purposes, as already described.
[129] Several of the London-based engineers appearing as witnesses before a Royal Commission in 1817 (P.P. 1817: VI) were well aware of Woolf's improvements in Cornwall in 1815.

*The Diffusion of Best-Practice Techniques* 91

anything like accurate—and it would be impolitic to insist that they are—coal prices above a threshold of about 12s. a ton would be required to make the Cornish engine worth while (disregarding all other considerations of course).

It is also quite clear that the high-pressure engine was often overlooked in areas that suffered through much higher coal prices, e.g. many of the engines going up in London. My argument probably narrows the geographical scope for sociological-entrepreneurial explanations of the dilatoriness of these manufacturers, but I would not for a moment claim to have laid them to rest.

SUMMARY

Two functions have been served by this chapter. In the first place, I have set out much of the basic data on costs (both fixed and working) that will act as reference points for the rest of the book. Secondly, in the course of the chapter I have utilized the cost information to derive approximations to the threshold price-vector. For instance, in the eighteenth century the Savery engine remained economic but only for very low horsepower ratings (probably under 6 horsepower). The reasons were essentially the high fixed costs per horsepower of the Newcomen engine for these small engines, and the inefficiency in fuel (coupled with other well-known limitations) of the Savery engine. Both Savery and Newcomen engines gave way to Watt's improved type towards the end of the century, at a rate depending on (i) the higher fixed costs of the Boulton & Watt engine, plus—most importantly—the length of time their patent had left to run; and (ii) the level of coal prices per ton in the region concerned. Similarly, the low-pressure Watt-type engine yielded to the high-pressure versions of the beam engine[130] at an earlier stage where coal prices were rather high, to compensate for the extra capital costs. Technological adaptation later favoured high-pressure engines of the compound kind even where coals were not so expensive.

The threshold calculations attempted were based on an undeservingly simple set of assumptions. Yet what they

[130] Economy of space often conflicted with economy of fuel, e.g. in marine engines of this era. Thus the introduction of direct-acting engines adds a further dimension to the story.

indicated seems to agree tolerably well with what is known about diffusion patterns for each successive improvement of the steam-engine. I would not, however, claim to have set aside all non-economic explanations of diffusion, by any means. Quite apart from the fact that the economic assessments depend in turn upon technical relationships that are not always fully understandable (as with the high-pressure rotative engine), there appear to be lags unexplained by my simple economic formulation; especially in regard to the high-pressure engine. Non-economic factors cannot yet be consigned to the scrapheap of historical Aunt Sallys.

# APPENDIX 4.1

(A) *Weight of 20-horsepower engine for Sirhowy Ironworks, 28 August 1799, by component and material, ordered from Boulton & Watt*

(I) Cast iron

| | | | |
|---|---|---|---|
| Cylinder | 3563 lb @ | 30s. 0d. per cwt = | £47 14s. 3d. |
| Piston | 429 | 30s. 0d. | 5 14s. 9d. |
| Air-pump | 1506 | 30s. 0d. | 20 3s. 4d. |
| Hot-water pump | 127 | 30s. 0d. | 1 14s. 0d. |
| Cold-water pump | 474 | 30s. 0d. | 6 6s. 11d. |
| Steam-case | 1296 | 18s. 8d. | 10 16s. 0d. |
| Miscellaneous | 340 | 18s. 8d. | 2 16s. 8d. |
| Main gudgeon | 517 | 18s. 0d. | 4 3s. 1d. |
| Perpendicular steam-pipe | 677 | 18s. 0d. | 5 8s. 9d. |
| Eduction pipe | 476 | 18s. 0d. | 3 16s. 6d. |
| Condenser | 1136 | 18s. 0d. | 9 2s. 7d. |
| Safety-pipe, etc. | 166 | 18s. 0d. | 1 6s. 8d. |
| Boiler steam-pipe | 616 | 18s. 0d. | 4 19s. 0d. |
| Miscellaneous | 932 | 18s. 0d. | 7 9s. 9d. |
| Stand plate for cylinder | 326 | 16s. 0d. | 2 6s. 7d. |
| Total | 14 188 | 23s. 2d. | 146 17s. 7d. |

(II) Wrought iron

| | | | |
|---|---|---|---|
| Cylinder | 247½ lb | | |
| Steam-case | 24½ | | |
| Piston | 254¼ | | |
| Air-pump | 294¾ | | |
| Parallel motion | 949½ | | |
| Hot-water pump | 25½ | | |
| Cold-water pump | 40¾ | | |
| Safety-pipe | 15 | | |
| Miscellaneous nuts, washers, etc. | 960½ | | |
| Total | 2904 lb @ | 96s. 7d. per cwt = | 125 4s. 4d. |
| Boiler | 4900 lb @ | 42s. 0d. | 92 18s. 6d.* |

(III) Brass

| | | | |
|---|---|---|---|
| Cylinder | 12 lb @ | 18d. per lb | 18s. 0d. |
| Piston | 16 | 18d. | 1 4s. 0d. |
| Air-pump | 68 | 18d. | 5 2s. 0d. |
| Parallel motion | 107 | 18d. | 8 0s. 6d. |
| Hot-water pump | 4 | 18d. | 6s. 0d. |
| Cold-water pump | 22 | 18d. | 1 13s. 0d. |
| Safety-pipe | 4 | 18d. | 6s. 0d. |
| Miscellaneous | 115 | 18d. | 8 12s. 6d. |
| Miscellaneous | 45 | 28d. | 31 8s. 10d. |

\* Includes carriage.

## (IV) Copper

| | | | |
|---|---|---|---|
| Pipes | 122 | 13 2s. | 5d. |
| Miscellaneous | 13 | 3 3s. | 9d. |
| Total brass and copper | 530 | 47 15s. | 0d. |

(V) Miscellaneous (leather, lead, cement, wood, patterns, etc.)    17 3s. 7d.
    Carriage    3 10s. 0d.

                                              £430 9s. 0d.
Fitting excluded.

(B) *Cost of materials for a complete atmospheric engine for draining mines, 1775. Source: Farey, op. cit., pp. 232–3.*

Cylinder 48″ × 7′; capable of 12 strokes per minute but usually worked at 9. Piston loaded to 7½ lb p.s.i. Engine therefore 36 nominal horsepower.

### (I) Cast iron

| | | | | |
|---|---|---|---|---|
| Cylinder and piston | | 80 cwt @ | 30s. 0d. per cwt = £120 | 0s. 0d. |
| Furnace doors, etc. | | 7½ | 28s. 0d. | 10 10s. 0d. |
| Furnace grate bars | | 29 | 12s. 0d. | 17 8s. 0d. |
| Injection pump | | | | 7 0s. 0d. |
| | Total | | | 154 18s. 0d. |

### (II) Wrought iron

| | | | | |
|---|---|---|---|---|
| Cylinder | | | | £2 10s. 0d. |
| Piston | | 1½ cwt @ | 37s. 4d. | 2 16s. 0d. |
| Great beam, axis, etc. | | 8¼ | | 15 8s. 0d. |
| Great chains | | 32½ | 42s. 0d. | 68 5s. 0d. |
| Miscellaneous | | | | 12 10s. 0d. |
| | Total | | | 101 9s. 0d. |
| 2 boilers | | 200 cwt @ | 23s. 0d. | 230 0s. 0d. |

### (III) Brass

| | | | | |
|---|---|---|---|---|
| Eduction valve flap | | 40 lb @ | 24d. per lb = | £ 4 0s. 0d. |
| Steam-passage regulator | | 168 | 18d. | 12 12s. 0d. |
| Great beam | | | | 6 0s. 0d. |
| Miscellaneous | | | | 22 8s. 0d. |
| | Total | | | 45 0s. 0d. |

### (IV) Copper

| | | | | |
|---|---|---|---|---|
| Eduction pipe | | 124 lbs @ | 24d. per lb = | £ 12 8s. 0d. |
| Sundry pipes, etc. | | 20 | 24d. | 2 0s. 0d. |
| | Total | 144 | 24d. | 14 8s. 0d. |

### (V) Lead

| | | | | |
|---|---|---|---|---|
| Injection and snifting pipes, etc. | | 30 cwt @ | 18s. 0d. per cwt = | £ 27 0s. 0d. |
| Hot well and cistern | | 49 | 15s. 0d. | 36 15s. 0d. |
| | Total | 79 | | 63 15s. 0d. |

## The Diffusion of Best-Practice Techniques

(VI) Timber

| | | | |
|---|---|---|---|
| Great beam, etc. | 35 | 0s. | 0d. |
| Hot-well cisterns, etc. | 14 | 0s. | 0d. |
| Framing | 53 | 10s. | 0d. |
| Injection pump and waste-pipes 60 fathoms | 30 | 0s. | 0d. |
| Miscellaneous | 3 | 10s. | 0d. |
| Total | 136 | 0s. | 0d. |
| Whole total of materials | 745 | 10s. | 0d. |

(VII) Engine-house

| | | | |
|---|---|---|---|
| Masonry and carpentry (walls, roof, doors, windows, etc.) | £420 | 0s. | 0d. |
| Brickwork for 2 furnaces and setting the boilers | 160 | 0s. | 0d. |
| Total for engine-house | 580 | 0s. | 0d. |

# APPENDIX 4.2

*Price of coal in some industrial areas, 1796–1855*

Shillings per ton

|  | London | | | Birmingham | Liverpool | Manchester | Leeds | Edinburgh |
|---|---|---|---|---|---|---|---|---|
|  | (i) | (ii) | (iii) | (i) | (i) | (i) |  |  |
| 1796 | — | — | 33·9 | 6 | 8·5 | — | 5·9 | — |
| 1797 | 33·0 | 26·1 | 32·0 | 8 | 8·5 | — | 5·9 | — |
| 1798 | — | 28·9 | 33·0 | 8 | 9 | — | 5·9 | — |
| 1799 | 38·0 | 31·2 | 38·7 | 7·5 | 9·5 | — | 6·0 | — |
| 1800 | 45·6 | 38·7 | 49·4 | 8·7 | 11·3 | — | 5·9 | — |
| 1801 | — | — | — | 8·7 | 12·1 | — | 5·9 | — |
| 1802 | — | — | — | 8·7 | 12·5 | — | 6·0 | — |
| 1803 | — | — | — | 10·0 | 15 | 16·3 | 6·8 | — |
| 1804 | — | — | — | 10·8 | 16·5 | — | 7·1 | — |
| 1805 | 33·2 | — | — | 10·8 | — | — | — | — |
| 1806 | 32·9 | — | — | 10·8 | — | — | — | — |
| 1807 | 33·9 | — | — | 10·8 | — | — | — | — |
| 1808 | 36·5 | — | — | 11·0 | — | 14·2 | — | — |
| 1809 | 40·3 | — | — | 11·6 | — | — | — | — |
| 1810 | 38·2 | — | — | 12·1 | — | — | — | — |
| 1811 | 35·3 | — | — | 12·3 | — | — | 7·4 | — |
| 1812 | 33·2 | (iv) | — | 12·3 | — | — | 7·4 | — |
| 1813 | 38·8 | 42·0 | — | 12·3 | — | 13·3 | 7·1 | — |
| 1814 | 43·7 | 44·7 | — | 12·6 | — | — | 7·0 | — |
| 1815 | 34·6 | 39·0 | — | 13·3 | — | — | — | — |
| 1816 | 30·9 | 34·7 | — | 12·1 | — | — | 7·2 | — |
| 1817 | 29·8 | 31·8 | — | 10·8 | — | — | 6·8 | — |
| 1818 | 29·5 | 32·8 | — | 12·8 | — | 10·8 | 6·6 | — |
| 1819 | 30·9 | 30·6 | — | 13·3 | — | — | 6·3 | — |
| 1820 | 30·9 | 30·6 | — | 12·8 | — | — | 6·2 | — |
| 1821 | 31·5 | 32·5 | — | 12·1 | — | — | 6·3 | — |
| 1822 | 30·9 | 30·3 | (v) | 10·8 | — | — | 6·1 | — |
| 1823 | 33·4 | 33·3 | 36·7 | 11·8 | — | 10·4 | 5·8 | — |
| 1824 | 30·0 | 33·3 | 32·8 | 12·3 | — | — | 5·9 | — |
| 1825 | 29·5 | 23·5 | 32·3 | 15·3 | — | — | 6·2 | — |
| 1826 | 26·8 | 25·3 | 29·1 | 15·3 | — | — | 6·2 | — |
| 1827 | 27·6 | 28·3 | 30·1 | 13·3 | — | — | 6·4 | — |
| 1828 | 27·4 | 27·5 | 30·0 | 13·8 | — | 10·4 | 6·0 | — |
| 1829 | 25·1 | 23·5 | 27·5 | — | — | — | 6·0 | — |
| 1830 | 26·1 | 27·5 | 28·4 | — | — | — | 5·8 | 6 |
| 1831 | 23·9 | 22·8 | 26·0 | (ii) | — | — | 5·5 | — |
| 1832 | 20·8 | 19·5 | 22·3 | 5·5 | — | — | — | — |
| 1833 | 17·2 | 15·8 | 18·3 | 6·5 | — | 9·6 | — | — |
| 1834 | 19·4 | 19·0 | 20·3 | 7·5 | — | — | — | 5 |
| 1835 | 20·8 | 20·3 | 21·7 | 7·4 | — | — | 5·0 | 5·5 |
| 1836 | 21·8 | 21·8 | 22·8 | 10·3 | — | — | 5·4 | 6·5 |
| 1837 | 22·9 | 22·8 | 23·8 | 9·8 | — | — | 5·7 | 7·4 |
| 1838 | 23·4 | 22·8 | — | 8·5 | — | — | 5·7 | 6·6 |
| 1839 | 22·6 | 23·0 | — | 8·5 | (ii) | — | 5·4 | 5·7 |
| 1840 | 22·5 | 22·8 | — | 8·3 | 13 | (ii) | — | 5·5 |
| 1841 | 21·3 | 20·3 | — | 7·9 | 10·5 | 6·7 | — | 6·0 |
| 1842 | 20·1 | 20·5 | — | — | 10 | 6·3 | — | 5·6 |
| 1843 | 19·1 | 20·3 | — | — | 9 | 6·4 | — | 5·1 |
| 1844 | 21·8 | 24·5 | — | — | 9 | 7·3 | — | 5·4 |
| 1845 | 18·1 | 17·3 | (vi) | — | — | 7·7 | — | 7·0 |
| 1846 | 16·8 | 14·5 | 18·0 | — | — | — | — | 7·6 |
| 1847 | 19·8 | 17·0 | 20·5 | — | — | 6·8 | — | 7·2 |
| 1848 | 17·1 | 14·5 | 18·0 | — | — | 6·2 | — | 5·8 |
| 1849 | 16·6 | 15·5 | 17·5 | — | — | — | — | 5·3 |
| 1850 | 16·0 | 13·5 | 17·0 | 4·8 | — | 6·3 | — | 5·7 |
| 1851 | 15·0 | — | 16·0 | — | — | 6·3 | — | 5·6 |
| 1852 | 15·4 | — | 16·5 | — | — | 6·3 | — | 5·5 |
| 1853 | 20·1 | — | 22·0 | — | — | 7·4 | — | 7·0 |
| 1854 | 22·7 | — | 23·5 | — | — | — | — | 8·1 |
| 1855 | 20·8 | — | 22·0 | — | — | 7·1 | — | 8·3 |

## The Diffusion of Best-Practice Techniques

*Notes and Sources:*
(1) London. (i): Best Newcastle coals at the ship's side in the Pool, from Mitchell and Deane, *Abstract.* p. 482. (ii): Prices in Pool, including ingrain, from P.P. 1800 (1st Ser. x, Coal Trade), App. 16 (iii): Average monthly price of coals delivered at the Foundling Hospital, ibid., App. 14. (iv): First-Quality Newcastle and Sunderland Coals at beginning of June, from G. R. Porter, *The Progress of the Nation*, 1851 edn., pp. 277–8. (v): Average monthly prices of Stewart's Wallsend (Sunderland) coals, from P.P. 1836: XI. (vi): Price of Wallsend and Hetton, from Mitchell and Deane, p. 483.
(2) Birmingham. (i): Boulton & Watt Estimates, 1 Oct.–30 Sept. (ii) Prices paid for coal received at Soho Foundry, Boulton & Watt MSS.
(3) Liverpool. (i): J. Langton, 'The Geography of the South-West Lancashire Mining Industry, 1590–1799' (Univ. of Wales Ph.D. thesis 1970). (ii): P.P. 1845: XXX (Contractors' Prices for supply of Coal to Navy).
(4) Manchester. (i) P.P. 1833: XX (evidence of H. Hoole). (ii): P.P. 1871: XVIII (selling price at Bridgewater Trustees' Wharf at Castlefield, Manchester), p. 1294.
(5) Leeds. W. G. Rimmer, 'Middleton Colliery near Leeds, 1779–1830', *Yorks. Bull.* VII, 1955, App. III. Price at Colliery. Price at Leeds Staith 1803 was 7s. (statutory). Delivery to households in Leeds from the Staith was 0·4 to 0·6s. per ton in 1778, but up to 2·4 for Armley or Headingley. Aikin in 1795 claimed that coal was delivered at the Staith in Leeds for 13s. a waggon, i.e. 4s. 4d a ton (*Description of the County*, p. 576).
(6) Edinburgh. P.P. 1873: X. Price at Elphinstone Colliery, 10 miles by rail from Leith. N.B.: prices paid in the Firth of Forth for English and Scottish coals in 1791:

|  | Newcastle | Forth (St. Davids) |
|---|---|---|
| Best Quality | 13·1 | 9·0 |
| 2nd Quality | 12·4 | 8·3 |
| 3rd Quality | 12·0 | 8·2 |
| Slack | 10·9 | 6·0 |

(Source: P.P. 1871: XVIII, App. 23, App. 54.)

## 5. The Backward Linkages

A HYPOTHESIS that has received wide and practically unquestioned support in the literature on eighteenth-century British economic history is the idea that the iron industry and steam-engines were two growth points for the nascent industrial economy, and, moreover, that their growth was mutually supportive. Hence the dramatic improvements in size and efficiency of the engine in mid-century have been seen as critical to the emergence of an iron industry based on mineral fuels, through its intensive demands on energy. These would be classed here as the direct and forward linkages of steam to iron. In addition, there were backward linkages from steam to iron; namely the iron components of the engine (contrasting with the largely wooden water-wheel, etc.). Hence the roughly synchronous appearance of Boulton & Watt's engine, the coke-smelting of iron by Darby's method, and Cort's patents for puddling and rolling, is frequently taken to have been no accident.

I have not made any serious attempt in this book to investigate the forward linkages from steam to iron. Some of the necessary research has been undertaken by C. K. Hyde, in his recent thesis on the British iron industry. His summary statement on steam-power and the smelting branch may be quoted in full:

> Other historians have also argued that the development of the Boulton and Watt steam engine was crucial to the spread of coke-smelting. Their argument is refuted by the fact that a large and viable coke-smelting sector based on the cost-superiority of coke-smelting was established well before the Boulton and Watt engine had been developed. Had the iron industry been forced to rely upon water power and the Newcomen engine, its growth to about 1790 would not have been significantly slower than the actual performance.[1]

This disposes of any immediate need to reassess the impact on the smelting branch. On the other hand, Hyde's work on the

---

[1] C. K. Hyde, 'Technological Change and the Development of the British Iron Industry, 1700–1870' (Univ. of Wisconsin Ph.D. thesis 1971), p. 4.

## The Backward Linkages

refining branch raises fresh problems. Hyde has shown that the traditional emphasis on the expansion of wrought iron from the early 1780s ought to be attributed to the rise of the 'potting' technique. Cort's comprehensive advances spread only from the mid-1790s.[2] Yet the latter date is in closer agreement with what I have been claiming for the Watt engine in this book. Clearly, this should be left as an open question.[3]

In this chapter I am concerned only with the backward linkages, treating them as one half of the interdependence. The other important metallic inputs into engines, brass and copper, are also looked into. It will emerge, though, that far more important than any of the backward linkages to the various metals are those to coal. A few authors have appreciated the importance of this connection, without venturing beyond casual observation. By and large, however, it has been played down in the literature. The most likely explanation for this partial oversight is that the mining of coal—as distinct from its application to a new range of industrial uses—is not normally taken to be one of the chief springs of British industrialization during the Industrial Revolution. Through the research and influence of writers such as J. U. Nef,[4] it has long been accepted by most economic historians that coal had achieved self-sustaining levels and rates of growth long before the classic take-off period. Its continued expansion over those take-off years is generally assumed to be a rather passive response to the widening fuel needs of autonomously expanding industrial sectors—satisfying the demands of steam-engines and the like, but not much more than that.

In formally testing the backward linkages of the steam-engine, I shall be focusing mainly on the year 1800, paying special attention to the Boulton & Watt engine. This reflects conditions at the climax of the kind of take-off period envisaged by Rostow. Nevertheless, it does not do full justice to the backward-linkage concept. One would expect such linkages to be at their height not so much during the years of especially rapid growth of the leading sector(s) as during its maturity thereafter, when its size would be greater even if its

---

[2] C. K. Hyde, 'Technological Change in the British Wrought Iron Industry'.
[3] For Hyde's views, see his thesis (cf. n. 1), pp. 89 ff.
[4] J. U. Nef, *The Rise of the British Coal Industry*, London, 1932.

growth rate were slower. As a result, I have attempted like calculations for later years. These are based on much more fragmentary or restricted data, and are thus intended to be purely indicative.

CONSUMPTION OF METALS IN 1800

The primary source for this year is again the Boulton & Watt collection housed in the Birmingham Reference Library. The extent of information on weight of components is more limited than on costs. For this and related reasons I have recruited an estimating procedure which fits a pattern to all the ascertainable (but scattered) data, rather than putting too much faith on individual cases.

The difficulty arises because the total inputs of each kind (cast iron, wrought iron, and brass and copper) do not rise proportionately with horsepower. In other words, it is invalid (except by way of first approximation) to multiply the total number of Boulton & Watt engines by their average size (27·1 horsepower) and to relate the answer to the weight of metal needed for this engine size. Such a procedure would take no account of the *distribution* of engines by horsepower.

In Table 5.1 are set out the results of a sample taken from Boulton & Watt engines built over the years 1797 to 1800.[5] The columns show the hundredweight (cwt) of each metal per horsepower for the standard sizes of small engines.

By inspection it seems sensible to fit a second-degree polynomial of the form:

$$X = a + b \cdot HP + c \cdot HP^2 + u$$

where X is the weight of each metal input in cwt, HP the amount of horsepower, and u the stochastic error term. The equations thus estimated read:

$$\text{Cast iron} = 23 \cdot 1 + 13 \cdot 096 \, HP - 0 \cdot 1061 \, HP^2 \quad (5.1)$$
$$\text{Wrought iron} = 2 \cdot 11 + 1 \cdot 507 \, HP - 0 \cdot 0116 \, HP^2 \quad (5.2)$$
$$\text{Boiler} = 5 \cdot 71 + 2 \cdot 065 \, HP + 0 \cdot 0085 \, HP^2 \quad (5.3)$$
$$\text{Brass} = 1 \cdot 73 + 0 \cdot 194 \, HP + 0 \cdot 0002 \, HP^2 \quad (5.4)$$

[5] The sample size is fairly small, so that a number of the entries are based on only one observation.

The Backward Linkages 101

TABLE 5.1
Weight (in cwt) of each input per horsepower for Boulton
& Watt engines, c. 1800

| Horsepower | Cast iron per HP | Wrought iron per HP | Boiler per HP | Brass per HP |
|---|---|---|---|---|
| 2  | 25·0 | 1·43 | —    | 0·90 |
| 4  | 20·7 | 2·50 | 3·42 | 0·67 |
| 6  | 15·6 | 1·87 | 3·17 | 0·46 |
| 8  | 15·6 | 1·81 | 2·63 | 0·44 |
| 10 | 14·0 | 1·65 | 2·86 | 0·40 |
| 12 | 12·2 | 1·41 | 2·71 | 0·34 |
| 16 | 12·9 | 1·41 | 2·63 | 0·35 |
| 20 | 11·7 | 1·30 | 2·24 | 0·28 |
| 22 | 13·1 | 1·51 | 2·68 | 0·28 |
| 28 | 12·1 | 1·33 | 2·51 | 0·28 |
| 30 | 9·5  | 1·16 | —    | 0·25 |

Equations 5.1 and 5.2 provide coefficients with the expected signs, i.e. positive on horsepower and negative on horsepower squared, reflecting the importance of cylindrical shapes and thus what are sometimes called "volumetric" economies of scale. Because of data difficulties they have been estimated only on smaller engines (up to 30 horsepower in the case of cast iron and 45 horsepower for wrought iron). This leads to problems in extrapolating to larger engines—the coefficients on horsepower squared ($c$) are great enough to induce absolute reductions in the consumption of metals in very large engines, and in both cases the estimated input of metal in engines of over 100 horsepower actually comes out a minus quantity. Thus in the calculation rather arbitrary values for iron consumption per horsepower are inserted for these large engines. It is unlikely that consumption is much underestimated by doing this; if anything, the opposite.

Boilers were kept separate in Table 5.1 and eqn 5.3, since different clients ordered them in different materials. The coefficient $c$ is positive for boilers; but this may be quite plausible since it was common to buy larger boilers than were strictly necessary for large engines. For the Boulton & Watt engines constructed up to 1800, the total consumption of cast and wrought iron is estimated in Table 5.2.

TABLE 5.2

*Estimated consumption of cast and wrought iron in Boulton & Watt engines to 1800*

| Range of horsepower | Number of engines | Total horsepower in range | Cast iron consumed (cwt) | Wrought iron consumed (cwt) | Boiler metal consumed (cwt) |
|---|---|---|---|---|---|
| 1 and under 4 | 15 | 40 | 860 | 93 | 169 |
| 4 and under 6 | 31 | 128 | 2337 | 252 | 445 |
| 6 and under 8 | 26 | 161 | 2603 | 287 | 489 |
| 8 and under 10 | 47 | 385 | 5792 | 643 | 1090 |
| 10 and under 12 | 41 | 413 | 5913 | 661 | 1123 |
| 12 and under 14 | 37 | 450 | 6166 | 693 | 1188 |
| 14 and under 16 | 22 | 316 | 4351 | 492 | 851 |
| 16 and under 18 | 28 | 449 | 5764 | 651 | 1149 |
| 18 and under 20 | 11 | 211 | 2588 | 293 | 574 |
| 20 and under 22 | 59 | 1193 | 14 427 | 1646 | 3006 |
| 22 and under 24 | 6 | 135 | 1584 | 181 | 339 |
| 24 and under 26 | 14 | 339 | 3892 | 444 | 851 |
| 26 and under 30 | 14 | 382 | 4221 | 485 | 959 |
| 30 and under 32 | 16 | 481 | 5134 | 592 | 1207 |
| 32 and under 36 | 13 | 430 | 4423 | 512 | 1084 |
| 36 and under 40 | 6 | 221 | 2169 | 259* | 559 |
| 40 and under 45 | 13 | 548 | 5026 | 625* | 1403 |
| 45 and under 50 | 15 | 706 | 6067 | 784* | 1829 |
| 50 and under 55 | 7 | 367 | 2926 | 400* | 962 |
| 55 and under 60 | 12 | 685 | 5144* | 733* | 1815 |
| 60 and under 70 | 7 | 469 | 3117* | 488* | 1276 |
| 70 and under 80 | 17 | 1261 | 7818* | 1286* | 3480 |
| 80 and under 90 | 13 | 1096 | 6142* | 1096* | 3124 |
| 90 and under 100 | 8 | 767 | 3865* | 759* | 2255 |
| 100 and above | 12 | 1657 | 6210* | 1558* | 5435 |
| Total | 490 | 13 288 | | | |

\* = based on extrapolated figures of consumption per horsepower.

If for the moment it is supposed that all boilers were built of wrought iron, then the aggregate consumption of all Watt engines built before the patent expired comes to some 6237 tons of cast iron and 2627 tons of wrought iron, 1831 tons of the latter in the form of boilers. To set these figures in context I

## The Backward Linkages

have put together some of the estimates of British iron production, obtained from secondary sources alone, in Table 5.3.

TABLE 5.3
*Output of pig and bar iron, 1770–1810*[6]

| Year | Pig iron produced | Bar iron produced |
|---|---|---|
| 1770 | 32 000 tons | — |
| 1788 | 68 300 tons | 32 000 tons |
| 1796 | 125 000 tons | — |
| 1800 | 156 000 tons | 125 000 tons |
| 1805 | — | 100 000 tons |
| 1806 | 258 000 tons | — |
| 1810 | — | 130 000 tons |

If the figure quoted for British pig iron production in 1800 is correct—and it looks perfectly reasonable—then the cumulated demands from building *all* the Boulton & Watt engines over the thirty-one years in which their patents operated represented about 6·2 per cent of the output of pig iron from domestic sources obtained in just the one year, 1800. To get this I have multiplied the wrought iron figures by conventional ratios to convert them to pig iron equivalents.

Of the 490 engines, 113 rated at 5145 horsepower were built in the final six years from 1795 to 1800. Assuming steady growth of pig iron output over those years, the cast iron required for Boulton & Watt's steam-engines would amount to just over $\frac{1}{4}$ of 1 per cent of annual cast iron production in Britain. The figure would decline if any imported iron was used in manufacturing steam engines: imports were nearly one-third of total British supply in 1800.

The wrought iron figures noted in Table 5.3 form a less consistent pattern, and it seems probable that the figure quoted for 1800, obtained by R. A. Mott[7] by supposing only

---

[6] Pig iron figures are from A. Birch, *The Economic History of the British Iron and Steel Industry, 1784–1879*, London, 1967, pp. 18, 45. Wrought iron figures for 1788 and 1800 from ibid., pp. 44–5 (cf. also p. 25). Figures for 1805 and 1810 from Hyde, 'British Wrought Iron Industry', Table 6.

[7] R. Mott, 'The Coalbrookdale Group Horsehay Works', pt. II, *Trans. Newcomen Soc.* xxxii, 1959–60, p. 44.

20 000 tons of direct castings in that year, is over the mark. On the basis of the ratio of bar to pig in 1788 and 1810, 70 000 to 80 000 tons would seem more reasonable, leaving direct castings at 52 000 to 65 000 tons. Thus wrought iron consumed in all Watt engines would amount to from 3·3 to 3·8 per cent of production in the single year 1800. Annual consumption of British wrought iron in their steam-engines would come to about $\frac{1}{5}$ of 1 per cent of domestic production, taken over the years 1795–1800.

The lower proportion of wrought iron even than cast consumed by their engines is especially interesting, because in the late years of the eighteenth century (once Darby's technique of coke smelting had become widely disseminated in the 1750s and 1760s) it was to the wrought iron branch and finishing processes that most inventive effort was directed. In other words, not only was the annual demand of the Boulton & Watt firm a tiny proportion of the annual production of iron in Britain, but the bulk of that demand was directed to the sector —cast iron—which was considerably less valuable in terms of value added and by then relatively stagnant in its technology.

There are no exact figures of the total output of copper in 1800. The best indication available is the sales at 'ticketings' (i.e. public auctions) in Cornwall.[8] These figures are too low for total output (a) because not all Cornish copper was sold through the 'ticketings', and (b) because Cornwall produced only about 80 to 90 per cent of total British copper output in the late eighteenth century. Sales at the 'ticketings' in 1800 were 56 000 tons. Thus total British output in 1800 must have been at least 62 000 to 70 000 tons, not counting sales outside the 'ticketings' in Cornwall.

Equation 5.4 returns a positive coefficient to the term in horsepower squared, but it is not significantly different from zero at acceptable levels of confidence. Accordingly, I have equated it to zero and estimated on the basis of average horsepower of all engines. The result comes to just 175 tons, barely $\frac{1}{4}$ of 1 per cent of output in 1800 alone.

The calculation of the consumption of brass and copper is complicated by the uncertainty about the number of copper boilers the firm supplied (remember that in the above calcula-

[8] Mitchell and Deane, *Abstract*, pp. 151, 157.

tion I supposed that all boilers were of wrought iron and the estimate for wrought iron was exaggerated on that account). Early in the eighteenth century virtually all boilers were of copper, but as the techniques of hammering and rolling wrought iron plates improved through the century, these usurped the use of copper because of their much lower cost. In the 1790s the price of copper rose sharply through wartime demands, and copper boilers were rarely constructed. A generous allowance would be to suppose one-quarter of all boilers for Watt engines to be of copper, so that the total consumption of brass and copper would rise to almost 600 tons. This is still less than 1 per cent of 1800 output alone, and of course the wrought iron figures should be reduced *pari passu*.

To these computations for Boulton & Watt's engines one ought to add the inputs into those engines built by rival manufacturers who pirated the various Watt patents. The total horsepower of these was estimated in *Thesis* at about 2500. It is indeed something of an exaggeration to treat them for present purposes as Watt engines at all, since most of them probably pirated only the separate condenser, to the neglect of steam case and other ancillary items Watt thought necessary for perfection.

For both pirated engines and atmospheric engines I have to assume in the absence of other information that the distribution of engines by horsepower paralleled that for Watt engines, except around a lower mean.

Estimation of the inputs into atmospheric engines must needs be based on very flimsy evidence. In Appendix 4.1 the weight and cost of components for an allegedly typical fire-engine with which Smeaton was involved were given, along with similar data for a 20-horsepower Watt engine built twenty-four years later. These figures are arranged in summary in Table 5.4. With due allowances for economies of scale associated with the larger atmospheric engine, it would appear that the Newcomen engine consumed about the same relative amount of brass and copper, fractionally more wrought iron, much more lead, but little more than half the cast iron of a Watt engine. The main structural differences between the engines, viz. the steam-case, condenser, and air-pump, were all principally of cast iron; whence the slightly lower fixed cost of the Newcomen engine.

TABLE 5.4

*Weight and cost per horsepower of medium-sized Watt (W) and Newcomen (N) engines*

| Material | Units | Weight per horsepower | | Cost per horsepower | | Cost per unit of materials | |
|---|---|---|---|---|---|---|---|
| | | W | N | W (£) | N (£) | W (£) | N (£) |
| Cast iron | cwt | 6·33 | 3·37 | 7·34 | 4·30 | 1·16 | 1·28 |
| Wrought iron | cwt | 1·30 | 1·38 | 6·26 | 2·82 | 4·83 | 2·05 |
| Brass | lb | 19·8 | 16·3 | 1·57 | 1·25 | 0·08 | 0·08 |
| Copper | lb | 6·75 | 4·00 | 0·82 | 0·40 | 0·12 | 0·10 |
| Lead | cwt | 0·01 | 2·19 | 0·01 | 1·77 | 1·16 | 0·81 |
| Boiler | cwt | 2·19 | 5·56 | 4·65 | 6·39 | 2·12 | 1·15 |

To be reasonably safe, I have conjectured that the atmospheric engine required the same amount of brass and copper, 10 per cent more wrought iron, and 40 per cent less cast iron, as a Watt engine of the same power, for all engines. This will overstate the proportionate needs of the small Savery engines, but by continuing the patterns of size distribution as for Watt engines, these will not figure heavily in the computation.

The aggregate for each class of engine is given in Table 5.5. Figures are rounded to the nearest thousand tons, and even this amount of rounding probably exaggerates the accuracy of the results. Once again, the figures in these columns represent the total for Britain in the eighteenth century, and for comparison I include British production of the same metals in just the one year, 1800, in the final column.

TABLE 5.5

*Metal consumed in all engines built by 1800, in thousands of tons*

| Metal | Watts | Pirates | Newcomens | Total | 1800 output |
|---|---|---|---|---|---|
| Cast iron | 6·2 | 1·2 | 4·0 | 11·4 | 52–65* |
| Wrought iron | 2·2 | 0·4 | 2·6 | 5·2 | 70–80 |
| Brass etc. | 0·6 | 0·1 | 0·6 | 1·3 | 62–70 |

\* = castings only.

## CONSUMPTION OF METALS IN THE 1830s

The somewhat makeshift analysis of trends in horsepower carried out in Chapter 3, as well as the conventional wisdom of scholars such as Clapham, gives reason to think that the impact of the engine may well have been greater around the second quarter of the nineteenth century than at its beginning. At the same time, the crudeness of the aggregate figures on horsepower installed over those years, which caused the earlier discussion to be so makeshift, evidently complicates any pretence here at a calculation. On the other hand, these drawbacks are probably outweighed if only through providing a comparison, however rough and ready, for the above estimate for 1800.

Regression analysis was again employed to read a pattern into the raw data. In the equations 5.1 to 5.4 I took a cross-section of engines built at more or less the same point in time. In this case, I have brought in *all* the observations available from the Boulton & Watt engine-books, applying to a span of almost fifty years. This gives something over 700 cases to work with. Production functions of a engineering kind were fitted for each type of metal. The estimating equations used to construct these functions contained up to ten variables at a time; many of them dummies, but the bulk of them with some kind of connotation about time period.

The assumption that Boulton & Watt engines were, if anything, bulkier than those of their competitors is retained—with the important exception of their boilers—and this permits me to assert that the estimates for individual engines tend to be too high.

The production functions so estimated confirm that there was a non-linear relationship between horsepower and metal consumption, i.e. that generally inputs of metal rose less than proportionately with the power of the engine. What this means is that to apply the results of the regressions adequately one has to know the horsepower of every single engine in the country at the particular date wanted. It hardly needs pointing out that this cannot be done for *any* year of the nineteenth century. All is not lost, because it can in fact be done on a more limited scale, by carrying out just such a calculation for the textile industries alone in 1838. By good fortune, the Factory Returns for that year quote the horsepower of all textile engines in the U.K. according to their size range. Hence total consumption of metals in all textile engines can be obtained by the laborious process of substituting each engine into the equations estimated for each metal. This gave the results shown in Table 5.6.

TABLE 5.6

*Metal consumed in all textile engines in 1838, in thousands of tons*

|  | Cast iron | Wrought iron | Brass and Copper |
|---|---|---|---|
| Cotton | 47·7 | 3·0 | 0·7 |
| Woollen | 12·2 | 0·8 | 0·2 |
| Worsted | 6·2 | 0·4 | 0·1 |
| Flax | 7·6 | 0·5 | 0·1 |
| Silk | 2·6 | 0·2 | * |
| Total | 76·3 | 4·7 | 1·1 |

* = less than 500 tons.

Before putting these in context, I shall add in boiler inputs, derived by a quite different method. Boulton & Watt's boilers were sometimes used, but most critics thought them too flimsy in construction except perhaps for the conservatively low pressures recommended for their own engines. Yet in 1839 Robert Armstrong[9] reckoned that about 80 per cent of the boilers in the Manchester area were of the waggon type (of which Boulton & Watt's was a version), and that until about 1830 the waggon boiler had been all but universal in that district. Accordingly, the following assumptions are made: (i) all boilers were of the waggon type, built to the norms set out by Armstrong; (ii) all boilers were 2 horsepower greater than the nominal capacity of the engines they supplied; (iii) all boilers were of wrought iron. These suppositions would bring total consumption of wrought iron in engines and boilers up to 16 000 tons at most.

[9] R. Armstrong (1839), p. 17. A boiler for engines of up to 25 horsepower of this kind weighed 17 to 18 lb per sq. ft of surface, and Armstrong believed that some 17 to 18 sq. ft of surface were sufficient for 1 horsepower. These data are used in the calculations.

## The Backward Linkages

Now the total output of the iron industry in 1836 was 1·2 million tons of pig iron, according to R. C. Taylor; for 1839 Mushet quotes the figure of 1 248 781 tons.[10] The kind of comparison between the stock of engines and the flow of iron output made for 1800 can be repeated here. If all the engines operating in the textile industries had suddenly been swallowed up by the ground in the middle of 1838, and all blast furnace capacity in the country had then been set to work to smelt the iron required to rebuild them, it would have taken under a month to complete the task.

It seems more sensible, though, to compare the flow of engines with the flow of iron production. One can try to assess the proportion of annual iron output going into engine- and boiler-building. The Factory Returns for 1835 can be used to estimate the net investment in textile steam-engines between 1835 and 1838, and thus the net consumption of iron over those years. The 1835 Returns were not completed for all regions, so that the figures appearing in Table 5.7 are extrapolations from northern and north-west England. To the extent that the regional censuses are underestimates in 1835, as often alleged, the investment figures will be overstated. On the assumption

TABLE 5.7

*Net consumption of iron in engines, 1835–1838, thousands of tons, textile industry only*[11]

|  | Cast iron | Wrought iron | Brass and Copper |
| --- | --- | --- | --- |
| Cotton | 15·3 | 0·9 | 2·0 |
| Woollen | 2·3 | 0·2 | 0·4 |
| Worsted | 2·1 | 0·1 | 0·3 |
| Flax | 1·8 | 0·1 | 0·2 |
| Silk | 0·5 | * | 0·1 |
| Total | 22·0 | 1·3 | 3·0 |

\* = less than 500 tons.

that this amount of investment was equally divided among the three years (recall, for instance, that the pig iron production

---

[10] Birch, *British Iron and Steel Industry*, pp. 124–6.
[11] See *Thesis*, Table I.25 for fuller details.

estimated for 1839 is virtually the same as for 1836), then the sum of the three columns in Table 5.7 comes to about 0·6 per cent of the annual output of pig iron. Contemporaries were agreed that these few years represented an extraordinary surge of engine installation in textiles,[12] so that the ratio for an average year in the second quarter of the century was almost certainly lower than this. The main reservation concerning these findings is that it may be more desirable to have figures on gross consumption rather than net consumption of iron. Obviously the Factory Returns cannot help very much in trying to gauge the extent of replacement investment. On the other hand, any use of the broken-up engines for scrap ought then to be allowed for.

STEAM-POWER AND COAL CONSUMPTION

The case for a powerful backward linkage from steam-engines to iron production therefore cannot be said to stand up. Coal is another matter altogether. However, for reasons touched on in the introduction to this chapter, most authors have paid it scant regard. The impact of steam-power on coal can be assessed in the same way as for iron, though more simply and crudely.

A rotative Boulton & Watt engine in average working condition would have consumed nearly 22 tons of coal per horsepower a year, using the values estimated in the previous chapter.[13] Their reciprocating engines should have done a shade better than this even without expansion; and with expansion[14] their normal consumption would probably have been about 15 tons per annum for each horsepower. On the other hand, these estimates all assume a working year for the engine of 3800 hours, which is intended to apply to the rotative rather than the pumping engine. The latter probably worked longer hours on average, so that it should not be too far wrong to take 22 tons annual coal consumption for every operating horsepower of engines of the Boulton & Watt type. The earlier results showed that the consumption of atmospheric engines

---

[12] Faster than the rate of growth of output, as the ratio of horsepower to spindleage rose by about one-quarter (cf. Ch. 7).
[13] i.e. $12\frac{1}{2}$ lb per horsepower per hour.
[14] Watt experimented with expansion at Soho in 1776 before first trying it in practice with the Shadwell Waterworks engine in 1778 (Farey, *Treatise* i. 339, 341).

could vary substantially; an average of half the efficiency of Boulton & Watt engines might not be hopelessly astray[15] (equivalent to 25 lb of coal burnt per horsepower per hour).

To get the aggregate coal consumption in 1800 requires knowledge of the amount of working horsepower in the U.K. The figures generated in Chapter 3, however, refer to the total amount ever constructed. This is too large for present purposes: (i) it includes engines that were exported, e.g. some 537 horsepower in the form of Boulton & Watt engines alone; (ii) it includes engines that had subsequently been burnt down or otherwise destroyed; (iii) it assumes full-capacity working. Leave aside the third of these reservations for the moment. If just the known cases of Boulton & Watt engines exported[16] or pulled down are deducted, then there remain 12 300 horsepower in such engines installed in Britain and Ireland. So if all this power was being worked at full capacity at average rates of coal consumption, then coal requirements would come to some 270 000 tons for Boulton & Watt engines alone in 1800.

Now total output of coal was reckoned by the Royal Commission on Coal Supply in 1871 to have exceeded 10 million tons. J. U. Nef believed this figure was probably too low—he calculated an output of 10·3 million tons as early as the decade of the 1780s (cf. 7 millions by the 1871 Commission), although not all of the components of Nef's estimates seem well substantiated, and where there was any doubt he tended to take the upper limit.[17] But taking the conservative 1871 Commission figure gives a hypothetical consumption in Watt and pirated engines of some $3\frac{1}{4}$ per cent of annual output. The percentage, though, rises particularly steeply once the atmospheric engines are included. All engines together, if worked under full loading for an average of 3800 hours a year, would have brought consumption up to almost 10 per cent of 1800 coal output. Some will no doubt still find that kind of proportion scarcely worthy of attention. My own view is that the possibility that

[15] Cf. also Tann (1970).
[16] I am assuming that engines erected abroad did not require imported coal from Britain.
[17] Nef, op. cit., vol. i, p. 20; vol. ii, pp. 353 ff. The Royal Commission figure was extrapolated from shipments into London, giving 6·89 m. tons in 1785, 7·62 m. in 1790, 10·68 m. in 1795, and 10·08 m. in 1800 (P.P. 1871: XVIII, p. 852).

as much as one-tenth of annual coal output was going into fuelling steam-engines is quite noteworthy.[18] The second-round effects, i.e. coal used in producing iron to produce steam-engines, etc., ought to be added to the result, though the previous calculations indicate that these would be slight.

There are some serious upward biases to qualify the kinds of percentages computed, however, and I shall take them in increasing order of importance. Firstly, the use of Nef's coal-production data would lower the ratios a little. As implied above, Nef's figures are probably based on more extensive research than the 1871 Commission's (the latter extrapolated from London shipments), but are likely to be over-generous, so that this reservation need not be very restrictive.

Secondly, I have almost certainly exaggerated the number of horsepower-hours put in by all engines in 1800. There is no way of knowing exactly how many engines of those still standing were not at work for all or part of the year 1800, were being worked for shorter spells or with the piston less than fully loaded, etc.

Thirdly, the estimates are all so far in volume terms. Yet it has been established—in Chapter 4 of this book as well as by other authors—that many engines, especially Newcomen engines, were located on coalfields and quite rationally burnt otherwise barely saleable slack. If, therefore, I were to assess the contribution instead to the *value* of coal output, by adjusting the above percentages for differences in quality, I would expect a lower set of results.

The fourth reservation goes wider, and will be deferred until I have made a roughly comparable estimate of the effects on the coal industry for the middle of the century.

By 1856 those engines in the textile industries that had been compounded were capable of working with a coal consumption of around $4\frac{1}{2}$ to 5 lb per horsepower per hour. If it is supposed (partly on the basis of evidence to be provided in Part III) that mean coal consumption in stationary engines was about 9 lb an hour, then each horsepower would take about 12 tons a year (the year, of course, being shorter). Aggregate consumption by textile engines in the British Isles then comes to

[18] Especially as the bulk of coal sold must still have been for domestic purposes (two-thirds??).

## The Backward Linkages

just under 2 million tons[19] in the unlikely event that all were worked at full capacity throughout the year. For all engines the consumption might then be put at 8 to 10 million tons. Thus the first-round demands of stationary steam-engines could have accounted for as much as one-eighth to one-sixth of the over-all increase of 56 million tons of U.K. coal output between 1800 and 1856.

Few writers have placed as much weight on the backward linkages from engines to coal as on the forward linkages from coal to steam-power. An exception are the authors of the *Short History of Technology*, who write:

> By 1800 output [of coal] had risen to about 10 million tons, and by 1850 it was 60 million tons, the increase being mainly due to the demands of steam-engines.[20]

Even if locomotive and marine engines are included (using data from Hawke *et al.*), their statement is evidently exaggerated, but not unperceptive.

The year 1850 probably marks a high-water mark of the impact of steam needs on coal production. Thereafter increasing economy in the use of fuels caused the rate of growth of fuel requirements to taper off, at least in comparison with the growth rate of coal. This in turn raises a methodological reservation that has already been given brief notice in Chapter 2.

These calculations have hitherto been carried out in a most orthodox fashion—relating actual consumption from one particular source to actual output of the product is common to both new and old economic historians in computing the backward linkages. In turning to what I have called the direct linkages, however, some of the new economic historians have been wont to construct states of the world that never actually existed to answer their queries. This procedure will be followed in the next chapter. Now if there is any attempt to cumulate the various linkages (forward, backward, direct, lateral, or other), it seems only consistent to apply the same approach in each case, so far as is humanly possible.

---

[19] Total nominal horsepower installed in all major branches of the textile industry in 1856 was about 159 000.
[20] T. K. Derry and T. I. Williams, *A Short History of Technology from the Earliest Times to A.D. 1900*, Oxford, 1960, p. 312.

In that sense, the backward linkages established in this chapter have been for what Hawke calls the 'investment effect' of steam-engines, i.e. the linkages as historically experienced. I suppose, as others have done, that had the engines not been built, nothing else would have been forthcoming to supply any of the 'missing' power. In technical terms, I have made the assumption of a perfectly inelastic supply curve. In everyday terms, I have made an excessively strong assumption. In relaxing it, some substitution of other forms of power for the steam-engine ought to be allowed for. These alternative power sources will naturally make their own demands on metals and other inputs. The chief difficulty is that the *degree* of substitution that might have been expected in practice depends on the relevant supply and demand elasticities, and these are singularly problematical to estimate. To establish or even guess at the net outcome would be hazardous in the extreme.

An alternative procedure that is marginally more enlightening is to do what I do in Chapter 6. There I first posit the hypothetical replacement of all Boulton & Watt engines by some other technology. It can be safely be asserted right away that the hypothesized alternative will require as much as or more of nearly all iron and coal inputs, as well as of other inputs that need less consideration here. For instance, the 270 000 tons of coal burnt by Watt engines alone in 1800 would rise to 600 000 tons on the basis of the average requirements utilized earlier in this chapter.[21] At this extreme then, the net 'innovational' impact of James Watt on the iron industry must be negative, if attention is limited to the backward linkages. Of course, the elasticity assumptions made in Chapter 6 may be fully as implausible as those above, and possibly much more so. But if the other limitations of the analysis are ignored for the moment, the net backward 'innovational' impact should lie between these two bounds.

The second calculation attempted in Chapter 6 replaces all steam-engines by other forms of power: water-wheels, windmills, etc. By any standards, the engine would have consumed more iron and coal than even the same amount of power supplied by such other means, so that the freak effect for Watt

[21] This reflects consumption by Newcomen engines, plus the increment needed to turn water-wheels in lieu of Watt rotative engines: see Ch. 6.

## The Backward Linkages

engines alone mentioned in the last paragraph will not eventuate. Moreover, the major inputs required for water-wheels, windmills, and so forth, on the eighteenth-century plan, were by this stage subject to diminishing returns—most obviously, timber (though this itself may have induced quicker substitution of iron for wooden wheels[22] in my hypothetical world).

### SUMMARY

In this chapter I have carried out the conventional appraisal of the backward linkages, by estimating how important the engine was to the industries which supplied its inputs. The traditional emphasis on the relationship with the iron industry is shown to be somewhat misplaced: the backward linkages from steam to iron are distinctly unimpressive at any time up to the mid-nineteenth century, and in addition are least directed at the particular branches of metallurgy that were then undergoing rapid technological progress.[23] The linkages back to coal were far stronger proportionately, varying between one-tenth and one-fifth of annual output on present assumptions—which however, probably err on the high side. Furthermore, on these conventional assumptions one is led to the position that the more inefficient and wasteful the engine, the more powerful (again, proportionately) were the backward linkages to coal. In absolute amounts, of course, this could be offset by a suitably elastic demand schedule for power.

---
[22] Cf. App. 6.2.
[23] Note that the Watt engine is a worse offender than the Newcomen in this respect.

# 6. The Social Savings

THE INFORMATION acquired in Chapter 4 on fixed and variable costs can now be put to another use, and that is to establish the cost savings attributable to the steam-engine until the cessation of James Watt's patent rights in 1800. In fact, two calculations will be attempted: first, the social savings attributable to the Watt engine alone; and second, those arising out of all types of steam-engines put together. The second calculation is clearly more general, but to be so rests on shakier foundations. The former estimate has its own appeal. Most authors who have drawn some connection between steam-power and industrialization in Britain have been thinking of the 'improved' steam-engine. Moreover, the extent of derived and irreversible changes in related industries—if any—must be greater for all engines than just for the Watt. To revert to what I said in Chapter 2, this means that one's confidence in the results as being upper bounds is likely to be greater with the calculation for Watt alone.

This narrower calculation is in some ways the mirror image of the estimates of threshold prices of coal obtained in the later sections of Chapter 4. Instead of deriving the point at which one or other engine becomes the more economic proposition, I am here totting up the cumulative effect of the hundreds of instances in which the Watt engine was, in practice, more economical. However, the estimate attained in this chapter is not strictly comparable with the earlier values obtained for thresholds, because I shall be introducing some of the practical difficulties that were skipped over in Chapter 4. These problems are enough to require some consideration of the costs of forms of power other than the steam-engine.

When in turn it comes to estimating the social savings on all engines, elaborate detail on the costs of other forms of power is essential. I shall examine the costs of horses, windmills, and water-wheels, as possible eighteenth-century alternatives to steam-power. Since mechanics (and entrepreneurs) of the time viewed these as feasible alternatives in certain circumstances,

The Social Savings

they tried to establish rule-of-thumb equivalences among them. One such, published in 1832, is reprinted as Table 6.1.[1]

TABLE 6.1
'Relative Effects of Machines and First Movers'

| Number of Horses | Number of Men | Newcomen cylinder inches | Watt cylinder inches | Radius of Dutch sails, ft | Radius of Smeaton sails, ft | Overshot wheel, 10′ dia., gals. |
|---|---|---|---|---|---|---|
| 1 | 5 | 8·0 | 6·12 | 21·25 | 15·65 | 230 |
| 2 | 10 | 9·5 | 7·8 | 30·4 | 22·13 | 390 |
| 4 | 20 | 11·5 | 8·8 | 42·48 | 31·3 | 660 |
| 6 | 30 | 14·0 | 10·55 | 52·3 | 38·34 | 970 |
| 8 | 40 | 16·8 | 12·8 | 60·9 | 46·27 | 1350 |
| 10 | 50 | 18·5 | 14·2 | 67·17 | 49·5 | 1584 |
| 12 | 60 | 20·2 | 15·2 | 73·54 | 54·22 | 1900 |
| 14 | 70 | 22 | 17 | 79·49 | 58·57 | 2300 |
| 16 | 80 | 23·9 | 18·3 | 84·97 | 62·61 | 2686 |
| 18 | 90 | 25·5 | 19·6 | 90·13 | 67·41 | 3055 |
| 20 | 100 | 27 | 20·7 | 95 | 70 | 3420 |
| 22 | 110 | 29·8 | 23 | 104·86 | 76·68 | 4000 |
| 25 | 125 | 33·6 | 25·5 | 116·35 | 85·5 | 5250 |

ANIMALS

Recent literature has emphasized how common it was to employ animals for motive power in factories in late eighteenth-century Britain. The horse-mill especially was to be encountered in many collieries, carding-mills, breweries, and the like. Tann refers to eleven mills in Glasgow and Paisley in 1797 that employed animals to turn anywhere from 14 to 24 cotton mules per mill.[2] There were 18 mills in Oldham in 1791 (25 in the whole parish three years before), most of which R. L. Hills considers were driven by horses at this time.[3] Animals other than horses, such as donkeys, were also used on occasions. It is important to stress that many of the early textile inventions

[1] W. Grier, *The Mechanic's Calculator* . . ., Glasgow, 1832.
[2] Tann (1970), p. 49. She names the eleven mills.
[3] Hills, *Power in the Industrial Revolution*, pp. 91–2. Also G. H. Tupling, *The Economic History of Rossendale*, Chetham Soc. N.S. lxxxvi, Manchester, 1927; H. Brunner and J. K. Major, 'Water Raising by Animal Power', *Industrial Archaeology*, ix, May 1972; F. Atkinson, 'The Horse as a Source of Rotary Power', *Trans. Newcomen Soc.* xxxiii, 1960–1; A. Gray, *A Treatise on Spinning Machinery*, Edinburgh, 1819, p. 52 (cattle for flax-mills).

were developed for animal-powered mills, if not for man-power itself. Hargreaves' jenny and Crompton's mule fell in the latter category; while Cartwright's mill at Doncaster incorporating the first power-looms was driven by oxen; Paul and Wyatt used asses for their rudimentary roller spinning frames; and Arkwright first used horses for his improved frame.

It is probably no accident at all that Watt denoted his unit of measurement the *horse*power. He clearly believed that the replacement of animals by his engines would be one of his firm's most abundant markets and that the cost savings here would be particularly satisfactory.[4] He defined nominal horsepower as 33 000 lb raised 1 foot high per minute; in doing so he deliberately overestimated the average maintained power of horses in order not to disappoint customers. Many contemporaries thought that this was as much as 50 per cent greater than an ordinary mill-pony could keep up throughout its shift. It must also be remembered that the poor blind mare characteristically found working gins was a far cry from the stately draught horse—Rankine's data for the latter indicate they were some 42 per cent more powerful.[5]

Nevertheless, there appears to have been an economic advantage in employing such poor specimens, since they rarely cost more than £4 to £8 each to purchase.[6] They could not be made to work 'efficiently' for more than 8 hours; if they were to power a mill for a 12-hour day, relays were automatically necessary. The 10-horsepower mill driving Arkwright frames, so typical of the industry *c*. 1800, could thus have been powered by about 22 horses. Now the life-expectancy figure most often quoted for such horses is 10 years (Mathias says 10 to 12 years). With straight-line depreciation as for steam-engines, the capital

---

[4] Hills shows that several of the early textile engines were replacing horses, e.g. at McConnel & Kennedy's (*Power in the Industrial Revolution*, pp. 161–2). This was also common in breweries, e.g. Farey, *Treatise* i. 437. Most of the scattered contemporary calculations of the social savings took the horse as the only alternative, e.g. T. G. Cumming, *Rail and Tram Roads and Steam Carriages or Locomotive Engines*, Denbigh, 1826; G. Dodd, *Steam Engines and Steam Packets;* and the French authors quoted in Ch. 3.
[5] Rankine, *Manual of the Steam Engine*, p. 88.
[6] This price range comes from Lee, *A Cotton Enterprise;* the horses were sold dead for 25*s*. Mathias notes that Truman valued his entire set of five horses at only £7. 10*s*. in 1766 and £5 more the following year (*The Brewing Industry in England*, pp. 79–80).

The Social Savings 119

costs of the animals (including interest) for such a mill range from £15 to £25 or so p.a. In addition there was the cost of the gin: in the late eighteenth century horse-wheels were usually insured for £50 or less.[7] Two gins would be needed for the Arkwright mill, so that over-all capital costs amount to, let us say, £25 to £35 a year.

Current costs for horses would be vastly higher. Smeaton estimated that the cost of gin-drivers for a 9-horse team in 1776 would be £20 (2 drivers), with a horse-keeper adding another £18. 5s. 0d.[8] A fair proportion might be 5 drivers for 22 horses, although there are indications that this could have been reduced later in the century. Smeaton thought repair costs greater for horses than engines (farriers, collar-makers, etc.).

The sum total of these is in fact to leave little difference in costs between horses and steam-engines, once labour and capital items are aggregated. Any difference in total costs must thus be explained in terms of materials. Smeaton in the 1770s reckoned cost of upkeep at £25 per horse per year. The sum of £40 or so recurs in a number of estimates towards the end of the century, when grain was more costly.[9] The figure £50 is encountered on occasions, but is almost certainly too high for mill-ponies.[10] As late as 1836 Barlow calculated power costs for eight hours as follows:

Now the value of 80 lbs, or about a bushel of coals [for the engines], we can scarcely estimate at more than 1s., the expense of 1½ horse-

---

[7] Tann (1973) quotes £50 to £150 for horse gins in the West Country. Prices were reckoned at from £35 to £150 in Yorkshire, but the higher figures (at least) seem to include the cost of the threshing-mill being driven; see A. and J. K. Harrison, 'The Horse Wheel in North Yorkshire', *Industrial Archaeology*, x, Aug. 1973. Note also D. T. Jenkins, 'Early Factory Development in the West Riding of Yorkshire, 1770–1800', in Harte and Ponting, *Textile History and Economic History*, p. 253.
[8] Smeaton, *Reports*, vol. ii, p. 375.
[9] e.g. by Joseph Delafield for Whitbread's in 1786, quoted by Mathias, *Brewing Industry*, p. 93; also £39 mentioned in a court action, representing the purported losses incurred when deprived of water rights, derived from Joyce H. M. Bankes, 'Records of Mining in Winstanley and Orrell, near Wigan', *Trans. Lancs & Cheshire Antiq. Soc.* liv, 1939, p. 48.
[10] Costs of feeding and maintenance at Goodwyn's brewery, London, in 1784–5 came to about £50, but this is mainly for the more powerful drays. Similarly, Blenkinsop claimed that his pioneering patent locomotive would save the labour of 40 horses costing £50 each in first cost and £55 a year to maintain. Obviously this is either gross exaggeration or a quite different breed of horse. But Wood quoted 27s. to 29s. a week in a Cumberland colliery in 1803 (O. Wood, 'A Cumberland Colliery during the Napoleonic War', *Economica*, N.S. xxi, 1954).

keep will perhaps be, at a medium about 3s. 6d., while the expense of 7½ men, at the medium rate of wages for good English labourers, will be 21s. This, however, is not a fair view of the subject, because in the latter case the whole expense is included, whereas in the horse we must consider the first purchase, the expense of harness, stabling, shoeing, grooming, &c.; this perhaps may raise the daily expense of the horse to 4s. 6d.; and in the steam-power we have also, besides the expense of the fuel, to include that of oil, tallow, the engineer, stoker, the first purchase, erection of building; and this will probably in a medium-sized engine, amount to another shilling per day.[11]

The proportions here are in accord with mine above; the cost of maintenance comes to £35 per animal per year. Morshead[12] put the figure at £30 in 1856, though for an agricultural district. If feed and upkeep are taken at £25 as for Smeaton's assessment, the total annual cost of the 22 beasts will be about £650; if at £40, as for instance by Joseph Delafield (London, 1786), then around £980. The latter would be twice the yearly cost of a Watt engine of 10 horsepower in London in 1795, when the patent had five years left to run. Thomas Young noted in 1807: 'On the authority of Mr. Boulton a bushel of coals is equivalent to the daily labour of 8⅓ men or perhaps more; the value of this quantity of coal is seldom more than that of a single labourer for a day, but the expense of machinery generally renders a steam engine somewhat more than half as expensive as the number of horses for which it is substituted.'[13]

H. W. Dickinson[14] summarizes a 1752 estimate on Tyneside of the comparative cost as representing a saving of 30 to 40 per cent in substituting an atmospheric engine for horses, and, allowing for different feed prices in mid-century, for additional capital costs of both engines and horses, and for the substantially greater fuel bill of Newcomen than Watt engines, this result is probably a little less favourable towards steam than was Matthew Boulton above.

---

[11] 'A Treatise' 91. Andrew Ure claimed that the cost of horse feed was not less than 1s. 2d. a day in 1835, i.e. £21 p.a. (*The Philosophy of Manufactures*, new edn., London, 1967). Note that many manufacturers would have had stables for cartage anyway, even if using water- or steam-power (e.g. the Ashworths).
[12] Morshead (1856).
[13] T. Young, *Lectures on Natural Philosophy*, vol. i, London, 1807, quoted in Singer *et al.*, *History of Technology*, Book IV.
[14] Dickinson (1939), pp. 56–7.

## The Social Savings

At the Middleton Colliery near Leeds one of the first Blenkinsop locomotive engines was installed in the early years of the nineteenth century, with the hope of great monetary savings in prospect.[15] In fact when the engines needed replacement in 1831 the Colliery reverted to horses, despite having its own cheap coal, because the price of fodder had dropped sufficiently since the war years. In line with this situation I have graphed the ratio of oats to coal prices in Figure 6.1.[16]

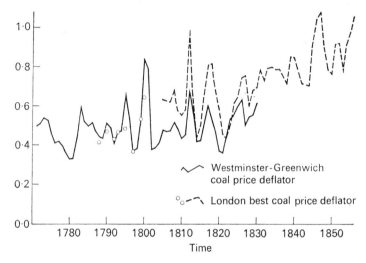

Fig. 6.1. Price relatives: oats and coal, 1771–1856
(On the vertical axis the price of oats per quarter is expressed as a fraction of the price of coal per chaldron)

Allowing for the difficulties of obtaining a suitable coal-price deflator, it would seem that in the second quarter of the nineteenth century oats were normally 50 to 100 per cent more expensive in terms of coal than in the last quarter of the eighteenth century. Hence even without the increase in steam-horsepower that actually took place in the early nineteenth

---

[15] Rimmer (1955), p. 45. Rimmer notes that the introduction of bran cheapened feed costs after 1800.
[16] Oats prices from Mitchell and Deane, *Abstract*, p. 488; coal prices in London from Ch. 4 above. This is not to say, of course, that rational decisions rested on this one price relative. See Tann (1973), p. 33, for other considerations temporarily favouring horses in the mid-eighteenth century.

century, the social savings of steam over animals would have risen on this account.

### WINDMILLS

The usual function of windmills was, of course, for milling corn. Steam-power was utilized for this same purpose on several occasions, most notably the Boulton & Watt fiasco at the Albion Mills, London, from 1785. However, the other less important functions of windmills, such as sawing timber, grinding logwood for dye, etc., constituting perhaps a fifth of the total number of windmills then in existence, were rather more important uses of steam-power. The most common overlap was in the pumping of water, e.g. in the Fenlands.

The foremost student of windmill technology, Rex Wailes, stated that '. . . In their hey-day, there may have been as many as 10 000 windmills at work in England at one time . . .',[17] of which over 2000 were sited in the Fenlands.

The corn windmills were naturally most plentiful where land was fairly level and water power lacking; that is, broadly speaking, east of a line drawn from Newcastle to Portsmouth. A fair number were to be found scattered over the remaining Midlands and Southern Counties, and they were fairly numerous also in Somerset, Wirral, the Lancashire Fylde and Anglesey, the granary of Wales . . .[18]

Probably the most careful count to date, though now perhaps needing further revision, was that taken by W. C. Finch in 1933 for Kent.[19] He counted 410 altogether, of which 51 were recorded on a map in 1736, 95 on another in 1769, and as many as 239 occurred in the first half of the nineteenth century. One J. A. Clarke mentioned that there had probably been more than 700 windmills at work on the Fens between Cambridge and Lincoln at one time, though by the time he wrote (1852) there were only about 220.[20] Most authorities seem to agree that the peak came towards the end of the eighteenth century, though it is rather doubtful whether there was any absolute decline

---

[17] R. Wailes, *Windmills in England*, London, 1948, p. 1.   [18] Ibid.
[19] W. C. Finch, *Watermills and Windmills*, London, 1933. I am grateful to Professor N. Rosenberg for this reference.
[20] In a book called *Fen Sketches*, quoted by R. Wailes, *The English Windmill*, London, 1954, pp. 79–80. Cf. also H. C. Darby, *The Draining of the Fens*, Cambridge, 1940, pp. 221, 225.

## The Social Savings

until about the second quarter of the nineteenth century. It would seem plausible that as many as 5000 windmills were operating when wind conditions were favourable c. 1800, and if so could develop altogether at least half as much total power as the steam-engines of all hues then installed; it is by no means inconceivable that their aggregate power was still in 1800 greater than that of steam-engines.

The lower estimate is based on an average of 3 horsepower per windmill, which seems reasonable for the Fens but perhaps too low for other areas. C. F. Partington wrote that,

A windmill with four sails, measuring 70 feet from the extremity of one sail, to that of the opposite one, each being $6\frac{1}{2}$ feet in width, is capable of raising 926 lbs., 232 feet in a minute [i.e. $6\frac{1}{2}$ horsepower]; and of working on an average 8 hours per day. This is equivalent to the work of 34 men; 25 square feet of canvas performing the average work of a day labourer. A mill of this magnitude seldom requires the attention of more than two men; and it will thus be seen that making allowance for its irregularity, wind possesses a decided superiority over every species of animal labour.[21]

This windmill is no doubt larger than the average in practice, but it is by no means out of the ordinary. R. L. Hills states that '. . . the best fen windmills were producing about 40 horsepower when there was sufficient wind'.[22]

Little information survives in readily accessible form about the costs of purchasing and operating windmills, though this partly reflects the lack of attention of economic historians to the problem. Certainly they could be costly to erect. The Union Mill at Cranbrook—said to be the most powerful ever built in Britain, and still standing—was claimed to have cost £3500 in 1814.[23] An average smock-mill was considered to cost £1000 in the later nineteenth century in Kent;[24] smock-mills were perhaps more expensive than other leading types (post-mills or tower-mills), but were most easily adaptable to containing

---

[21] Partington (1822), p. x.
[22] R. L. Hills, *Machines, Mills, and Uncountable Costly Necessities*, Norwich, 1967, p. 32. Derry and Williams (*Short History of Technology*), however, argue that on theoretical considerations the traditional windmill could not supply more than 30 horsepower and that about 5 horsepower was the likely output from 24-foot sails.
[23] C. P. Skilton, *British Windmills and Watermills*, London, 1947.
[24] Finch, op. cit., p. 49.

machinery. The largest windmills on the Fens cost some £1200 to £1400 in the early nineteenth century, while a large mill to drain 100 acres was reckoned at £1000 in the same area.[25] However, small windmills could be had for £80 apiece.[26] In Scotland, Donnachie & Stewart put the cost of erecting a windmill at £200 to £300 in 1797.[27]

As to working costs, only Hills's work is at all illuminating. Attendance and repairs on a windmill were said to vary from £150 down to £5 in the Fens in 1820. The usually expressed average was £80 for large mills and £25 for small.[28] A millwright in Hawkhurst, Kent, reckoned the useful life of a postmill as 200 years and of a smock-mill as 100 years.[29] However, casual observation, taking into account such factors as the vulnerability of windmills to fire, indicates that these figures are too high. Since repairs have here been counted separately, a depreciation rate of 2 per cent p.a. seems not inappropriate. Annual costs thus run from £6 to £10 per horsepower for large engines and £14 per horse for small. These are cheaper than a steam-engine, especially in areas such as the Fens where coals were so expensive.[30] The difference, of course, arises out of the irregularity of wind. According to recent measurements at Mildenhall, the percentage of hours for which the wind speed is greater than about 18 m.p.h. is 5·2 per cent, but it is only in this range that a windmill will begin to turn at all well.[31] It can work at low power with a wind speed of over about 12 m.p.h., which occurs another 20 per cent of the time. In other words, for three-quarters of the time a windmill was pretty ineffective in this district.

All the same, where time was no object—or, more precisely, where it could be regarded as cheap in terms of inventory accumulation, etc., as may well have been true in many branches of manufacturing to which it was applied—Partington's statement quoted above about its 'decided superiority' over animal power deserves bearing in mind. The windmill was drastically improved in the course of the eighteenth century,

[25] Hills (1967), pp. 31, 125.   [26] Ibid., p. 36.
[27] I. L. Donnachie and N. K. Stewart, 'Scottish Windmills', *Proc. Soc. Antiq. of Scotland*, xlviii, 1966, p. 285.
[28] Hills (1967), pp. 75, 124–5.   [29] Finch, p. 56.
[30] Hills (1967), pp. 123–4, states 15s. to 20s. a ton at Deeping Fen, 1828–31.
[31] Measurements are for 1950–9; ibid., pp. 32–3.

e.g. by Edmund Lee's invention of the fantail (patented 1745 and improved by Meikle in the 1750s), to keep the sails pointing towards the wind; by the 'spring sail' and 'roller reefing sail' (1772 and 1789 respectively; to adjust the sail area to wind speeds—made automatic by William Cubitt's patent sail of 1807); and by the use of cast iron parts (promulgated by Smeaton in 1754; afterwards he undertook extensive empirical inquiries into the design of sails and mills). Smeaton indeed thought that wind might well be cheaper than a Watt engine, though with reservations about availability of the requisite wind.[32]

### WATER

Much more serious as a rival to steam-power than either animals or windmills was water-power, for most industries and regions. By 1800 water-power was being utilized in most fields of manufacturing. Although much has been made in the literature of the locational flexibility allowed by steam, in most cases industries tended to concentrate in or nearby areas chosen originally for their water resources. Moreover, in some manufacturing processes the water *per se* was of considerable importance, the establishment of textile-weaving in the upland parts of Lancashire and Yorkshire being particularly well known. Many water-powered plants coexisted with steam long into the steam era.

Few writers have attempted the painstaking task of estimating total horsepower installed in water-mills, in, say, 1800. There were 5624 water-mills for corn-milling listed in the Domesday Book.[33] It seems beyond question that the total power installed in 1800 was vastly higher in water-wheels than in steam-engines, and additionally that it was still on the increase.

Water-wheels before the advent of steam-power could be vertical or horizontal. In each category it is conventional to classify them according to the point at which the water was laid on. Horizontal wheels were said to be numerous on the

---
[32] Campbell, *Carron Company*. Glynn noted that in the Fenlands, when there was most rain there was least wind ('Draining Land by Steam Power, *Trans. Soc. Arts & Manufactures*, ii, Derby, 1836). This has not been my own experience of the Fens.
[33] P. N. Wilson, 'Water Power and the Industrial Revolution', *Water Power*, vi, no. 8, Aug. 1954; J. Reynolds, *Windmills and Watermills*, London, 1970.

Isle of Man, in Ireland, in Continental Europe, etc.,[34] but were evidently less frequent within Britain. Joseph Glynn[35] believed there to be some resurgence in horizontal wheels about the time he was writing (1853), but these probably embodied technical improvements that will be discussed in App. 6.2. Costs will be established initially for vertical wheels, i.e. (i) overshot, (ii) breast, and (iii) undershot wheels.

In the eighteenth century and earlier the last of these, the undershot wheel, was easily the most common. The undershot was perhaps the most natural way of harnessing water: the water was often simply allowed to flow through unfettered, driving the wheel largely by the strength of the current as it did so. In a series of celebrated experiments in the 1750s Smeaton showed that the convenience of the undershot wheel was offset by its inefficiency.[36] According to his experiments on a model, the laying-on of water by means of a conduit over the head of the wheel to make it into an overshot wheel roughly doubled its efficiency. For reasons of hydrodynamics, the impulse provided by a free-flowing current was restricted by interference with the floats of the wheel, and in the existing state of knowledge it was therefore better to rely on gravity, i.e. the weight of water in the buckets as in the overshot wheel.

Despite this widely disseminated advice, the overshot wheel was not adopted all that often even in the 'power-hungry' take-off period. This was not due to lack of information, because the true overshot wheel seems to have been most widely adopted in agriculture, where one supposes information flows are weakest.[37] In favourable rural areas there could be a plentiful supply of water with a moderately high fall, and such localities were ideal for the overshot wheel. The problem of hindrances to navigation would not arise there, and the high fall could be tapped readily and cheaply.

However, in most manufacturing areas the intermediate 'breast' or 'breech' wheel was more common.[38] The breast

[34] Reynolds, op.cit.; J. Banks, *On the Power of Machines*, Kendal, 1803.
[35] J. Glynn, *Rudimentary Treatise on the Power of Water*, London 1853.
[36] J. Smeaton, *Experimental Enquiry concerning the Natural Powers of Water and Wind*, London, 1794.
[37] *Engineer and Machinist's Assistant*, 1850; Morshead, op. cit.
[38] Fairbairn, *Mills and Millwork;* J. Millington, *Elementary Principles of Natural and Experimental Philosophy*, vol. i, London, 1823; D. T. Jenkins (1969).

wheel would evidently have to be larger than an overshot one to cope with the same 'fall', but might even so cost less, because it was likely to involve a less complex structure of conduits to feed water to the wheel.

... the breast wheel has taken precedence of the overshot wheel, probably from the increased facilities which a wheel of this description affords for the reception of the water under a varying head. It is in most cases more convenient to apply the water of high falls on the breast at an elevation of about 30° from the vertical diameter, as the support of the pentrough is much less expensive and difficult than when it has to be carried over the top of the wheel. In cases of a variable head, when it is desirable to work down the supply of water, it cannot be accomplished without a sacrifice of power on an overshot wheel; but when applied at the breast, the water in all states of the river is received upon the wheel at the highest level of its head at the time, and no waste is incurred. On most rivers this is important, as it gives the manufacturer the privilege of drawing down the reservoir 3 or 4 feet before stopping time in the evening, in order to fill again during the night; or to keep the mill at work in dry seasons until the regular supply reaches it from the mills higher up the river . . .

Another advantage of the increased diameter of the breast wheel is the ease with which it overcomes the obstruction of back water. The breast wheel is not only less injured by floods, but the retarding force is overcome with greater ease, and the wheel works in a greater depth of back water.[39]

The customary procedure for assessing the efficiency of water-wheels was by adapting Parent's method of calculating the power-effect ratio.[40] Smeaton showed that an overshot wheel could be made to achieve about two-thirds of the potentially attainable power.[41] Fairbairn reckoned his best high-breast wheels could effect as much as 75 per cent of the available power, and deemed this to be a maximum for a water machine.[42] (Morshead[43] thought breast wheels only about two-thirds as efficient as overshot ones in rural districts.) By contrast, Smeaton computed the efficiency of *undershot* wheels at 33 per cent, from a model. Nevertheless, undershot wheels

[39] Fairbairn, p. 113.
[40] See Cardwell, *From Watt to Clausius*, who shows that Parent's further speculations on this subject were unsatisfactory.
[41] *Experimental Enquiry* . . ., pp. 28 ff.    [42] Fairbairn, p. 123.    [43] Op. cit.

remained popular for streams with very little fall, partly because laying on a sufficient head for a more efficient (overshot) wheel would require miles of costly leats, and partly because where economy of water was unimportant they could be built very cheaply. In 1850 it was considered that undershot wheels were generally preferred for falls of $4\frac{1}{2}$ feet or less.[44]

If one assumes Fairbairn's very high ratio of 75 per cent efficiency, then nearly 12 cubic feet of water falling 1 foot per second would be required to provide 1 horsepower. At the Ashworths's mills on the Eagley Brook, near Bolton, 1 foot of fall was indeed regarded as equivalent to 1 horsepower.[45] Obviously more would be obtained from a larger stream—Fairbairn[46] estimated 3·3 horsepower per foot of fall on the River Bann in Ulster (presumably 40 cu. ft per second).

Calculation of the power of wheels is thus easy once the flow of water is known. At Arkwright's mill at Shudehill, Manchester, 6000 gallons per minute falling 30 feet drove 4000 cotton spindles.[47] If one assumes 100 of Arkwright's frame spindles per horsepower the water-power would be 73 per cent efficient, since it would be providing $54\frac{1}{2}$ horsepower at 100 per cent efficiency. The two huge wheels at the Catrine Cotton Works, Ayrshire, each $50' \times 11'$, consumed 70 tons of water a minute from the River Ayr,[48] which reduces to a similar rate of efficiency as the wheels were rated together at 200 horsepower (these were Fairbairn's wheels). However, short of an exhaustive microcosmic study of economic geography, the data on rates of flow of water over much of the country are not readily obtainable.

Fortunately it is possible to devise indirect procedures for measuring costs of water-power which, though necessarily more approximate, partly evade this data problem. First, it is necessary to establish that wheels of the size considered could sustain the power required. Table 6.2 shows the dimensions of wheels built by William Fairbairn in Manchester in the early nineteenth century.

[44] *Engineer and Machinist's Assistant*, p. 243.
[45] House of Lords Record Office, MS. HC 1843.3:B2, evidence of J. F. Bateman and William Fairbairn.
[46] p. 77.    [47] Tann (1970).    [48] P.P. 1834: XX.

The Social Savings 129

TABLE 6.2
Proportions of Fairbairn's wheels, c. 1850[49]

| Diameter | Breadth within buckets | Depth of buckets | Revolutions per minute | Speed of periphery per second | Fall of water | Cu. ft. of water per sec. | Estimated Horse-power |
|---|---|---|---|---|---|---|---|
| 65 ft | 6 ft | 1' | — | — | — | — | — |
| 36 | 16' 7" | 1' 4" | 1·95 | 3' 7" | 33' 2" | — | — |
| 28 | 13' 0" | 1' 10" | — | — | 26' 6" | — | — |
| 20 | 17' 0" | 1' 8" | — | 3·82' | 16' 6" | 46 | 60 |
| 18 | 21' 0" | 1' 8" | 4·78 | 4' 6" | 13' 3" | — | — |
| 18 | 20' 0" | 1' 6" | — | — | 16' 0" | 36 | 52 |
| 18 | 18' 0" | 1' 10" | 6·15 | 5' 8" | 10' 0" | — | — |
| 18 | 12' 0" | 1' 5" | — | — | 16' 0" | 20 | 30 |
| 16 | 21' 0" | 2' 0" | — | — | 9' 0" | 116 | 70 |
| 16 | 20' 0" | 1' 9" | 7·8 | 6' 5" | 7' 10" | — | — |
| 16 | 18' 0" | 1' 8" | — | 5' 6" | 9' 6" | 45 | — |
| 16 | 16' 0" | 2' 0" | — | — | — | — | — |
| 16 | 14' 9½" | 1' 9" | — | 6·3' | 8' 0" | — | — |
| 15' 6" | 17' 6" | 1' 8" | — | 5·54' | 8' 0" | — | — |
| 15 | 6' 0" | 10" | — | — | 14' 6" | 8 | 12 |

As a rule of thumb, suppose some rough proportionality between power and the multiplicative function, length × breadth. Marshall[50] noted that Joshua Wrigley recommended a wheel of approximately 16' × 9' to drive 2000 spindles; in practice Garforth & Sedgwicks claimed to run 2000 spindles off a wheel of 28' × 5', and Thackray & Whiteheads off one of 12' × 5' (this last is small), while Sedgwick & Co. of Bingley drove 3000 spindles from an 18' × 12' wheel. It may not be too far wrong to allow a wheel such as that at Bingley (with a cross-product of 216 sq. ft) as suited to driving 30 horsepower in the late eighteenth century, wherever an appropriate water supply could be found.[51]

Dr. Tann notes that small (often undershot) wheels were usually insured for £30 to £60, while larger wooden wheels were rated at about £100 for insurance purposes.[52] These figures are extended by S. D. Chapman,[53] using the Sun Fire insurance records for the North-West in the 1790s—for cotton-mills there seem to be rather more valuations in the range of

[49] *Engineer and Machinist's Assistant*, p. 244. Cf. also the figures for brake horsepower of large wheels estimated by P. N. Wilson (1954).
[50] Marshall MS. 57.
[51] Hills gives slightly lower figures, viz. 15' × 12', by implication (*Power in the Industrial Revolution*, p. 101), but the wheel was not yet working at full power.
[52] She states £30 to £50 for West Country undershot wheels in her 1973 article (p. 200); other figures are from the 1970 book, p. 65.
[53] S. D. Chapman, 'The Cost of Power in the Industrial Revolution', *Midland Hist.* i, 1970, pp. 21–3.

£100 to £500 (and occasionally more) than Tann hinted at. Chapman considers that there was little variation between one district and another so far as 'millwright's work' was concerned, although this is not entirely borne out by such horsepower figures as he quotes. Anywhere from £10 to £30 per horse is found in a wider examination of mill valuations as the worth of the wheel itself. The quoted insurance valuations cluster at the lower end of this range. Direct evaluations seem higher, partly because subsidiary costs were included, e.g. gearing. Thus Quarry Bank rated its 90-horsepower wheel at £2400 in 1831, of which the wheel alone represented £1410 at the time of construction in 1818.[54] The iron wheel Hewes (or Hughes) built for the Arkwrights at Bakewell, probably of 70 to 80 horsepower, was valued at £1600 in 1830.[55] At the Ashworths's mill the 45-horse wheel erected by Fairbairn and Lillie (40' diameter) after 1822 cost £880; while the huge 62' wheel of 110–40 horsepower at their Egerton mill, also completed by Fairbairn & Lillie, cost £4800 including erection in 1829.[56]

On these data there can be no doubt that water-wheels cost far less per horsepower for purchase and erection than did steam-engines. Morshead stated that 15' × 3' wheels in 1856 (i.e. roughly 4 horsepower) cost £40 if mainly of iron, plus another £20 for erection. Eight-horse wheels cost around £120 plus erection of £50. Thus in these small sizes wheels cost only about one-fifth as much as beam engines; and little over half the cheapest portable engines exhibited at the Great Exhibition. For large wheels the cost was around half that of steam-engines of equal power.

Of course the story did not end here. Wheels needed foundations and erecting as did engines, but their foundations (i.e. framework) were substantially more expensive than for engines. The Gregs at Styal valued the foundations of their wheel at £1000 in 1831, compared to the £2400 for wheel and gearing.[57] Fairbairn reckoned on as much as 65 man-weeks of erection time for his 36-foot wheel (as noted in Table 6.2);

---

[54] Quarry Bank MS, C5/8/16/2; W. Lazenby, 'The Social and Economic History of Styal', Univ. of Manchester M.A. thesis 1949.
[55] Chapman, op. cit., p. 17.
[56] R. Boyson, *The Ashworth Cotton Enterprise*, Oxford, 1970, pp. 15, 21.
[57] Quarry Bank MS. C5/1/1/2, cf. C5/8/16/2.

## The Social Savings

46 man-weeks for the 18-foot wheels; and 36 man-weeks for the 16-foot wheels.[58] For small wheels, Morshead's information suggests erection came to around 50 per cent of the cost of the wheel alone, compared to about 25 per cent for small engines.

Moreover, the water to supply the wheel might have to be dammed up in order to secure enough—the accumulation of water in dams overnight effectively doubled the supply available for a 12-hour day. From there the water may have been channelled in through an elaborate system of leats and weirs. The Gregs spent £2005 just on local labour to improve the tail-race conducting the water away from the wheel in 1817.[59] Overshot wheels, albeit more efficient than undershot, raised greater problems with the supplying conduits. As Fairbairn noted,

... these conduits are not infrequently as difficult of construction and as expensive as the weir. In several large works with which I have been connected, the cost of conduits has extended to many thousands of pounds, as at the Catrine works in Ayrshire, or the Deanston in Perthshire. In the former case a large tunnel, with retaining walls and embankments several hundred yards in extent, had to be constructed, and at the latter a wide and spacious canal, nearly a mile long, before the water reached the mills ...

The extreme case of the Catrine wheels illustrates the point:

| | |
|---|---:|
| Water privileges and land | £4000 |
| Cost of weir | 1000 |
| Headrace, tunnel, and canal | 3000 |
| Archways, cisterns, sluices, &c | 1000 |
| Wheelhouse and foundations | 1500 |
| Tailrace | 1500 |
| Water-wheels and erection | 4500 |
| Contingencies | 1500 |
| Total | £18 000 |

The cost of power independent of mill-work equivalent to an annual rental for interest of capital, repairs and wear and tear, at 7%, amounting to £1260.[60]

Thus the capital costs of Catrine (including contingencies) work out to no less than £90 per nominal horsepower. Smeaton

[58] *Engineer and Machinist's Assistant*, p. 244.
[59] Lazenby, op. cit.    [60] Fairbairn, pp. 88–9.

took the average cost of a reservoir 3 ft deep to be 6d. per sq. yd, i.e. £120 per acre.[61] At Quarry Bank it was thought in 1856 that 'To excavate 5 acres 2 feet deep would probably cost £480 . . . but I think 2′ deep would not be enough.'[62]—i.e. £144 per acre 3 ft deep. At the New Tarbet Mills in 1771 it was estimated that building a dam and sluice cost about 5s. 6d. per cu. yd, plus another 1s. 6d. a cu. yd for rubble filling. In addition, an aqueduct 3 ft deep cost 6d. per yd to cut, while embanking cost 4d. per yd. However, through hard rock Smeaton estimated that a 2-mile leat would cost £400, i.e. 2s. 4d per yd.[63] At Papplewick, Notts, Chapman[64] puts the wheel etc. at £1950 (for 63 horsepower from one 42-foot and one 30-foot wheel), with application involving 31 acres of reservoir and 1100 yards of aqueduct costing another £4000, i.e. £94 per horsepower. He states this to be a high upper bound on costs per horse. The developments at Styal Mill in 1818–20 thus come to £77 per horsepower.

Water-mills tended to be very long-lived. Several of the large mills listed by P. N. Wilson continued to near the middle of the present century. Wooden wheels might be expected to last twenty years before replacement;[65] iron much longer. The repairs on the Catrine wheel were said to be minimal for thirty years.[66] At Egerton Mill the Ashworths rated their wheel at 7 per cent for depreciation—the same as for their auxiliary engines—and 4 per cent for weirs, etc.[67] Working costs for some scattered cases are given in the costings in App. 6.1. Wheels required comparatively little labour in operation and of course no coal.

Even so a water-wheel did require 'fuel', and the fuel was water. Water was not chargeable as a variable cost in the same way as tons of coal could be. Instead, water rights as the property of landowners who owned the streams as well as the land surrounding them were leased out to mill-owners. Thus water was not a free good.[68] Water rights were competed for by

---

[61] Smeaton, *Reports*, vol. ii, pp. 392–3.   [62] Quarry Bank MS. C5/3/2.
[63] Tann (1970, p. 65) quoting Smeaton's Reports.
[64] Chapman, p. 6.   [65] Hills, *Power in the Industrial Revolution*, p. 113.
[66] Fairbairn, p. 126.   [67] Ashworth MS. 1201.
[68] Cf. the statement of Hills, ". . . water was free while coal cost money" (*Power in the Industrial Revolution*, p. 93). The effects on pushing up land values are noted in P.P. 1846: VI (I).

aspiring entrepreneurs. There is a host of evidence for the immense amount of litigation that was entered into on this account, dating back many centuries. The rapidly expanding demands imposed by the Industrial Revolution accentuated this competition for the scarce supplies of water in favourably sited areas. The entrepreneur was faced with a choice—either he could go farther afield to where competition abated or he could remain in the best-situated area and pay more for his water rights as competition forced up the 'price' of water. The limitation on the first alternative was, of course, that by moving away from the highly favoured sites that were coming to be so heavily in demand he incurred extra costs. By standard Ricardian rent theory one would expect that the charges he paid for renting his water supply would diminish in proportion to the rise in the additional costs that ensued (additional transportation of raw materials and finished products, additional housing costs and other diseconomies, perhaps additional labour costs to attract skilled labour—though unskilled would normally be cheaper in the more remote region). The internal margin and the external margin would be equalized by rents being forced upwards at the internal margin and 'cultivation' (of water) extending outwards at the external margin. At the external margin rents on water would, by definition, be zero.[69]

For most purposes this saves my having to specify the whole variety of cost conditions over different parts of the country. It will be enough to establish total costs for the firm making normal profits, as these will be the same for every such firm, however much the components may vary from one district to the next. But the result can be generalized further than this. If equilibrium in the industry is assumed, that is to say, no tendency for a net rise or fall in the number or size of its firms (individual water-mills), then the costs of water-power will be the same as the cost of steam-power, after all attendant charges on both sides are taken into account. If water-power were more expensive than steam, marginal firms using water would be

---

[69] A detailed geographical analysis of the spread of water-mills for fulling and scribbling in the West Riding of Yorkshire by M. T. Wild reads like Ricardo's *Essay on the Profits of Stock*, once 'most favoured water-power sites' is substituted for 'most favoured corn lands'. See J. G. Jenkins (ed.), *The Wool Textile Industry in Great Britain*, London, 1972, Ch. 13.

making losses and in due course be forced out of business. The external margin would contract, and hence the internal margin in proportion, i.e. rents would fall by similar amounts on each grade of water supply. The decline would continue until water-power again became just competitive at both margins with steam. In other words, under the assumption of competitiveness any inequality between the costs of the two sources of power reflects a state of disequilibrium in the industry/economy and will be corrected by such disequilibrium being squeezed out.

Several authors have investigated the matter of comparative power costs, but by failing to appreciate how rents are determined they have answered the question already. S. D. Chapman, in a major study in this field, notes, 'Though the evidence in this paper is very slight, it points to the tentative conclusion that water was capable of competing with steam power throughout the long period [1780–1850] of transition from the domestic to the factory system in cotton textiles where sites near large towns could be developed.'[70]

He need not have been so cautious about the thoroughness of his analysis, because the conclusion he reached is virtually vouchsafed. So long as positive rents were earned on the scarce resource—water-power—it was inevitable that he would find instances of its profitability for entrepreneurs.

All that is needed for this result is the condition that competitive markets operate, for which there must be a strong presupposition in the present case. It is unlikely that one would always and everywhere find exact correspondence between costs of steam- and water-power. One important reason is that water rights were often rented on very long leases, sometimes a hundred years,[71] and rents could thus be 'sticky'—quite probable in the more rural areas where the supply of potential entrepreneurs could not be imagined to be infinitely elastic.[72]

---

[70] Op. cit., p. 19.
[71] e.g. P.P. 1833:XX. Chapman quotes sixty-five years at Bakewell. Even 999 years was recorded (Unwin, *Samuel Oldknow and the Arkwrights*, p. 121, for Stockport).
[72] Note how Robert Owen was able to fool some maiden sisters at New Lanark (Butt, *Robert Owen*, pp. 193, 217). Meagre evidence on mill rents shows that they were far from constant from year to year, e.g. the series for Ossett quoted by J. D. Goodchild, 'The Ossett Mill Company', *Textile Hist.* i, 1968.

## The Social Savings

The data provided by Chapman and others are unfortunately inadequate for examining this particular problem.

Instead I hope to show that in the early nineteenth century mill-owners were imputing to their water-power a price consistent with rational economic determination. The Ashworths in the early 1850s have the following entry in their Quarterly Stock Account:

Value of our Water power: During this quarter [ending 28 March 1851] the roof of the water tunnel fell in; this occasioned the stoppage of the water wheel, and during this stoppage which was $7\frac{1}{4}$ weeks, the works were driven by the steam engines only. At the end of this period it was found that we had consumed $51\frac{3}{4}$ tons of coals per week and that in the period of $7\frac{1}{4}$ weeks preceding, we had consumed only $36\frac{3}{4}$ tons per week; consequently our consumption had been increased at the rate of 15 tons per week. Or, in other words, we found out that 15 tons of coals @ $5/4d$. per ton or £4 per week of expenditure in fuel, was equivalent to the power derived from the water wheel, and that too, in the wet months of winter when we had command of as much water as the wheel could carry, and were using according to estimate from 15 to 18 cu. ft. per second.[73]

At Quarry Bank the Gregs carried out an even more elaborate computation:

... If the dam were double its present extent, we could work twice as much time without the engine on those days on which the supply did not exceed 15 000 gallons [per minute] and we could also work without it on a greater number of days in the year. I think we might safely take the number of days in the year that we should be benefited by a double extent of reservoir surface at 150 and to the average amount of $3\frac{1}{4}$ hours each day. This gives $487\frac{1}{2}$ hours or $47\frac{3}{4}$ days in the year. Our consumption of coal for working the engine after deducting for steaming the mill is about $1\frac{1}{2}$ tons per day at $9/5$ per ton. $47\frac{3}{4}$ days $\times$ $1\frac{1}{2}$ tons = $71\frac{5}{8}$ tons at $9/5$ = £33.14.6 per year the saving in coal ... [The wheel is then calculated at 172 horse-power]. For 172 horse-power at 5 lbs per horse-power per hour we should require 23 tons of coal per week or 1196 tons per year at

---

[73] Ashworth MS. 1201, p. 155. Cf. also Boyson, op. cit., p. 61, 'Henry Ashworth estimated that in 1851 the water-wheel at New Eagley saved an average of £4 of coal each week, but since the rent of the land, the waterfall, and the reservoir for this water-wheel cost £167.8.6d. a year, any savings in expenses was marginal'—Boyson misses the point.

9/5 per ton is equal to £563.2.4 per year in coal—Our present consumption at Quarry Bank is about 600 tons a year at 9/5 per ton is equal to £282.10.0 so that our waterpower is worth about £280 a year.[74]

In my view these demonstrate that some mill-owners understood the notion of opportunity costs in this context. Water-power was to be appraised in these terms rather than by whatever formal rent they may have been paying for or imputing to their water rights. The variable costs of steam are in effect being equated to the working plus rental costs of the resource in inelastic supply: water-power. Note especially that the Gregs were setting coal consumption at only 5 lb per horsepower per hour, as they were able to achieve once they had McNaughted their engine in the early 1850s. Before compounding the engine would have attained a consumption of about 15 rather than 5 lb per horse per hour, and as such the water-power would have been much more valuable. I shall return to this point in a moment.

There is some evidence that a market in water rights had developed in the Midlands by the last quarter of the eighteenth century. John Fothergill wrote to his partner, Matthew Boulton, in 1780: 'I could wish we could have the power of the different mills and watercourses in the neighbourhood [of Soho], if they might be obtained on reasonable terms by reason that if ever peace is established with America every power where mechanism can be established will be greatly wanted and might be let out to great advantage.'[75] Theory and the examples quoted appear to confirm the profitability of water-power up to at least the mid-nineteenth century, in line with the results got from case studies by others. Boulton & Watt's Soho works themselves continued to use water-power alongside their engines at least until the 1830s.

The question of the viability of water-power is less straightforward. In the terminology of orthodox microeconomic theory surveyed above, economic viability would be represented by equilibrium of the industry (or such disequilibrium as led to an expansion in aggregate size). Thus intramarginal firms

[74] Quarry Bank MS. C5/3/2, 4 Aug. 1856.
[75] Quoted from Boulton & Watt Assay Office MSS. by Court, *Rise of the Midland Industries*, pp. 250–1.

could readily remain profitable utilizing water-power even if steam-power were to become cheaper simply by paying reduced rents. However, the firms at the external margin would drop out of production, as stated before. In this case the use of water-power can hardly be regarded as viable for these particular firms in the long term, despite the possibility that firms at the internal margin might survive profitably for many years. A test for viability therefore comes from rents.

Regrettably there are no consistent data through time on trends in rents for water resources. The few observations we have do not suggest any sharp decline in the proportion of water costs represented by the capitalized rent,[76] but the heterogeneity of topography involved in comparing them does not make the exercise especially valid. For obvious reasons one really needs a range of locations of water-power observed at fairly regular intervals through time, but we do not as yet have even one. There is, however, little evidence that water rentals had collapsed with the early advent of steam-power. As late as 1846 R. H. Greg of the cotton family, in comparing Britain with the U.S.A., complained that in the former '. . . all water power is small and charged extremely high; that is to say, they [the manufacturers] either pay a high rent for it, or they have a very great outlay in improving it, so that it is very expensive.'[77]

Several indirect alternatives to rent information remain. For instance, it may be possible to obtain some impression of changes through time in the utilization of water-power, in order to assess whether equilibrium of the industry pertains. Again this exercise would have to be conducted at a low level of precision, because as stated earlier there is still very little known about total power supplied by water at any time in the eighteenth or early nineteenth centuries. Several authors seem to consider that until the last decade of the eighteenth century there was little pressure at all on water resources in those areas that were to achieve industrial predominance in the years to

---

[76] Chapman's eighteenth-century figures do not include rents. At Bakewell in 1827 the rental was under £40 p.a. for 60 HP; when again offered for sale in 1840 the rent was £44; the rent at Manchester £55 for 100 HP in 1840; op.cit., pp. 17–18.

[77] P.P 1846:VI, pt. I, q. 7588.

come.[78] Thereafter the take-up was rapid. Unwin described a race for water-power at Stockport in 1790–1.[79] Lancastrian spinners and manufacturers moved to Flintshire or Furness in search of water.[80] Samuel Crompton mentioned as many as 44 mills on the upper reaches of the River Aire at the time of his census in 1811. In Gloucestershire, 24 mills were concentrated on less than 5 miles of the River Frome.[81] Baines wrote of the congestion of water-wheels on the River Irwell by the 1830s—800 out of 900 feet of fall being fully occupied.[82]

The situation so far as textiles were concerned a little later in the century can be more firmly established, as shown in Table 6.3. Thus there was little decline before 1850. Even where there was some reduction for a particular industry, this has to be placed in the context not just of other textiles but also of all other industries, since the explanation may be that in certain areas pursuits other than textiles are becoming even more profitable for employing water-power.

A third method of examining viability is to use the previous work on rent theory and profitability. In no case studied (at least in the years 1800–50) have I yet found that costs of water-power were widely different from those of steam-engines. Assume for the moment that steam is the only competitor with water on cost grounds, say in the major industrial regions. Once the Watt patents expired in 1800, scarcity rentals on steam-engines can be regarded as unimportant. Then, as shown above, the total cost of water-power, including its scarcity rental, should equal the cost of steam-power. Thus

---

[78] e.g. Hills, *Power in the Industrial Revolution*, p. 102; Chapman, p. 2 ('The rent of water power sites might be thought to be a critical variant, but there is evidence to suggest there were plenty of cheap sites on the market in the later decades of eighteenth century.')
[79] Unwin, op. cit., pp. 123, 125, 128.
[80] Summarized by Chapman, pp. 8, 10–11, 15–16; see also Tupling, op. cit., pp. 204–5.
[81] Chapman, p. 19.
[82] E. Baines, jr., *History of the Cotton Manufacture in Great Britain*, p. 86 n. For indications of the kind of congestion that had arisen, note the list of mills occupied on the R. Wandle between 1800 and 1850, at Carshalton (Surrey), given by R. T. Hopkins, *Old Watermills and Windmills*, London, 1930, pp. 120–1 (10 mills, mostly 10–16 horsepower). Even more vivid is the 'Schedule of the number of horse-power used in the different mills &c. which are supplied with water from the Bradford Soke Mill Goit; Sepr. 1841' (Bowling Iron Works MSS., Bradford Reference Library, Box 7, Bundle 10). This lists 28 occupiers plus 3 mills unoccupied or in ruins, totalling 248 horsepower.

TABLE 6.3

*Horsepower from water in the textile industries, 1838–1856*[83]

|  |  | 1838 | 1850 | 1856 |
|---|---|---|---|---|
| England & Wales: | Cotton | 9677½ | 8182 | 6551 |
|  | Woollen | 7371½ | 6887 | 6261 |
|  | Worsted | 1313 | 1501 | 1301 |
|  | Flax | 1130½ | 871 | 1005 |
|  | Silk | 928 | 853 | 816 |
|  | Total | 20 420½ | 18 294 | 15 934 |
| Scotland: | Cotton | 2728 | 2842 | 2330 |
|  | Woollen | 1198 | 1853 | 1746 |
|  | Worsted | — | 88 | 34 |
|  | Flax | 1495½ | 1421 | 817 |
|  | Silk | — | — | — |
|  | Total | 5421½ | 6004 | 4927 |
| Ireland: | Cotton | 572 | 526 | 250 |
|  | Woollen | 523 | 229 | 404 |
|  | Worsted | — | 36 | 96 |
|  | Flax | 1052 | 1095 | 2113 |
|  | Silk | — | — | — |
|  | Total | 2147 | 1886 | 2863 |

water rents can be determined by subtraction: so long as the costs of steam do not change, or change at the same rate as water costs other than the rent component, then the latter should stay constant, etc.

Changes in the cost of steam-power through time have been considered in Chapter 4. In Table 4.11 it was shown that the annual cost of a 30-horsepower engine in Manchester should have been about £610 to £620 in 1800 and the same in the mid-1830s, so that there is no change in money terms. In real terms, steam-power costs would have risen quite sharply: the Rousseaux price index for principal industrial products (and also that for all goods) falls by one-third between 1800 and

[83] From Factory Returns, P.P. 1839: XLII; P.P. 1850: XLII; P.P. 1857 (Sess. I): XIV.

1834–8.[84] An index constructed in terms of the price of value-added in textiles would also fall over those years. The price of engines fell very little until the late 1830s or early 1840s, while variable costs fell more in the later 1840s. It is only then that the viability of water-power could have been much endangered on these grounds.

As explained, viability also depends on what is happening at the same time to costs of water-wheels, other than the rents. In spite of a technological revolution in the use of steam-power in the late 1840s and 1850s, the decline in water-horsepower recorded in the Factory Returns between 1850 and 1856 was not overwhelming—about 14 per cent in Great Britain and 9 per cent in the British Isles textile industry (Table 6.3). One reason may have been that as speeds increased in spinning and weaving by steam (see Chapter 7) the speed of water-wheels was also increased. Overshot wheels were generally recommended for working at 3 to 4 feet per second in the eighteenth century—Smeaton in fact found that efficiency rose when he slowed wheels down to this speed. But partly through the improvements in transmission and parts discussed in App. 6.2 the velocity of the circumference approximately doubled up to the 1850s.[85]

The variety of advances in the harnessing of water-power that occurred between the last third of the eighteenth century and the middle of the nineteenth, set out in detail in the Appendix, could well have acted to lower the real costs of supplying water-power. Even if some of the changes (e.g. the

---

[84] Mitchell and Deane, op. cit., p. 471.
[85] Speeds taken from Smeaton, Rees, Fairbairn, Morshead, and Rankine, opp. citt.; also R. Buchanan, *Practical Essays on Millwork and Other Machinery*, 2nd. edn., ed. T. Tredgold, London, 1823; Riche de Prony, *Nouvelle architecture hydraulique*, vol. i, Paris, 1790; Marshalls MS. 57. A. Stowers stated that 'Glynn adopted 6 feet per second in constructing several overshot wheels of iron, 30 feet in diameter or larger. By thus raising the velocity, the revolutions of a 30 foot wheel were increased from 2¼ to almost 4 per minute. The greater speed meant that less gearing was needed for driving machinery at the required rate, and reduced the load on the wheel and axle nearly in the inverse ratio of the speeds; furthermore there was a gain in the regularity of motion derived from the momentum of the quicker wheel. On the other hand, as Glynn stated, there was a limit to the diameter, for a very large overshot wheel was costly, cumbersome, and slow.' ('Water Mills, ca. 1500–1850', in Singer *et al.*, op.cit., p. 205). Fairbairn believed that variable supply of water was more important than economy in this speed-up (op.cit., p. 128).

substitution of iron for wooden components) simply held down any tendency for costs to rise, they as a group gave water-power plenty of leeway in competition with steam.

Strictly speaking, I should not be considering any improvements in water-power that were inspired by the technical development of the steam-engine. In other words, I ought to eliminate, as a first step, any 'secondary irreversible innovations' in water-power that derived from steam-power, in executing the social savings estimate. My belief is that irreversibilities of this particular kind were negligible before 1800,[86] though it is of course impossible to prove that. Thereafter, developments in water-power technology were undoubtedly promulgated by market competition with steam-engines, but direct flows of technical knowhow were still not very important.

### THE SOCIAL SAVINGS OF JAMES WATT

It is obviously illegitimate to proceed by making a simple calculation of the average difference in costs between steam- and water-power. In Chapter 4 it was discovered that the annual costs of steam-power varied substantially from one part of the U.K. to another. In theory, I should make a separate calculation for each locality in which an engine was erected. In practice, I do not have sufficient information on the price of coal in 1800 at every site on which a steam-engine was operating. What I have therefore done is to aggregate both horsepower and coal prices to the county level. Taking average coal prices over a whole county presumably imparts an upward bias, since for reasons that by now ought to be apparent, steam-engines flocked towards cheaper coal.

To establish the social savings, it will be recalled, involves taking the same pattern of output, according to both region and commodity, for the counterfactual world as for the real world. But if this is to mean replacing all the steam-engines with water-wheels, then I am likely to find that in many localities where the engines stood there was simply no satisfactory stream at hand; or even if there was, its power might be quite insufficient. It follows that assessing the least-cost alternative to the Watt engine may imply a complex and even insoluble

[86] Though note the possibility of a reverse flow of information, from the water-pressure engine to the steam-engine, as stressed in App. 6.2.

exercise in economic geography. As a first approximation, therefore, I have to be content with assessing what must have been a more expensive technology than the cheapest combination of the various alternatives discussed previously. I suppose that all Watt engines are replaced by atmospheric engines.

There are two practical problems associated with this hypothetical substitution.

(i) In several important industrial processes the atmospheric engine, driving a crank, proved unable to deliver the power with the requisite degree of regularity. In textile-spinning, for instance, any jerkiness would tend to be transmitted through the shafting and might break the thread. To cope with such situations manufacturers resorted to what they called 'returning engines'.[87] In this arrangement the engine was used simply to recirculate water back to the top of a water-wheel. As Smeaton wrote of this system in 1781

> ... I apprehend that no motion communicated from the reciprocating beam of a fire engine can ever act perfectly steady and equal in producing a circular motion, like the regular efflux of water turning in a water-wheel, and much of the good effect of a water-mill is well known to depend upon the motion communicated to the mill-stones being perfectly equal and smooth, as the least tremor or agitation takes off from the complete performance.
>
> Secondly, all the fire engines that I have seen are liable to stoppages, and that so suddenly, that in making a single stroke the machine is capable of passing from almost full power and motion to a total cessation ...
>
> By the intervention of water, these uncertainties and difficulties are avoided, for the work, in fact, is a water-mill ... .[88]

Many contemporaries thought the drive obtained from a water-wheel was smoother and better than from even a Watt engine, especially as the latter got older.

The irregularity of the action of the steam in ordinary low-pressure engines is very nearly counteracted by the use of a fly-wheel; nevertheless, in some of the cotton factories, (for instance, that of Messrs. Lane, of Stockport) two engines are employed to work the same machinery, the cranks being fixed at right angles to each other, as in marine engines. This arrangement equalizes the action of the steam still more, yet the motion is not so regular as that of an

[87] Farey, i. 275.   [88] Quoted by Hills, *Power in the Industrial Revolution*, p. 136.

## The Social Savings

overshot water wheel, where the supply of water is uniform, as it would be in this case . . .

So wrote Thomas Wicksteed in 1840.[89]

The first use of an atmospheric engine to drive a water-wheel was apparently at a brass battery works in Bristol in 1752.[90] At Coalbrookdale in the late 1750s the engine at the Horsehay works was originally used to pump water back to a reservoir from which it could be taken to the wheel to blow a furnace; Smeaton and others converted it in 1762 to work directly on the wheel.[91] Following on the work of Musson and Robinson and others, Dr. Tann has listed many Newcomen and Savery engines in Lancashire and Yorkshire that were being used in this way between 1780 and 1800.[92] Other examples are easy to come by in the literature.[93] The practice was known and advertised in London. P. Keir of Camden Town used a Savery-type engine to drive his overshot wheel: 'Mr. Keir considers it to have been a profitable engine to himself, and has no doubt that it will prove beneficial in many situations where coal can be obtained at a low price.'[94] Some early Boulton & Watt engines were in fact used in the same way, e.g. the earliest erected in a textile-mill—that at Papplewick in 1785.

(ii) The Newcomen engine, needing a much larger cylinder to supply the same power as a Watt, could not always attain enough power without an impracticably large cylinder. Thus Watt engines were being erected even at collieries in the late eighteenth century, simply to pump deeply enough.[95] The

---

[89] Writing in the *Civil Engineer and Architect's Journal*, iii, 1840. cf. also Marshalls MS. 57: 'J. W. [Joshua Wrigley] says the Boulton & Watt's Crank engines are the only ones that will produce a motion sufficiently regular for spinning.'
[90] Farey, i. 212. '. . . 25 years afterwards it became a common practice'.
[91] Raistrick, op. cit., pp. 116, 139 n., 144; Farey, i. 275.
[92] Tann (1970), p. 91. Also Marshall MS. 26.
[93] French notes at least three erected for this purpose in Bolton in 1792; see G. J. French, *The Life and Times of Samuel Crompton*, London, 1859.
[94] Birkbeck and Adcock, *The Steam Engine*.
[95] Mott (1962–3): 'In 1778 the first Boulton and Watt engine was erected on Tyneside at Byker, the site of the second Newcomen engine in this area. The prevailing problem on Tyneside at this period was to be able to sink to 600 feet and to deal with large quantities of water. The weight of the pump rods at this depth limited the quantity of water which could be lifted and efficiency became important not primarily to save coal, which was cheap, *but to limit the number of engines required*. The Boulton and Watt engine itself soon became inadequate for the tasks set and higher steam pressures became desirable.' (My italics.)

alternative was to bury another Newcomen engine at some depth, and pump up in stages. This entailed some loss in efficiency, through additional leakages, etc., and also of course the extra purchase and erection costs of the second atmospheric engine. Nevertheless, this was what Savery had originally intended for his engine,[96] and was what Cornish engineers did for Watt and later engines in the late eighteenth and nineteenth centuries. The supplementary engine was then said to be 'shammals' to the other, and the coupling was described as 'shammalling'.[97]

The cost of destroying all extant Watt and Watt-type engines of 1800 can be split into fixed and variable aspects. By fixed costs are meant purely cost of engine and boiler; i.e. costs of framework, erection, etc. are assumed to be the same for each type of engine for the same power, as contemporaries including Watt asserted.[98]

In addition there will be the capital costs of purchasing and erecting the water-wheel required for a 'returning engine' whenever a simple crank mechanism proved unsatisfactory. In many circumstances the crank would undoubtedly have been technically adequate. However, it is not possible to be sure. Therefore, and at the risk of greatly overstating the needs for smooth transmission across a range of industries, I shall suppose as a first approximation that *all* Boulton & Watt's rotative engines are hypothetically replaced by 'returning engines'. In such cases one must allow extra for the engine as well: because of the loss of one-third in effect through the overshot wheel, the entrepreneur had to acquire an engine that was proportionately greater to supply the power for

---

[96] 'Be the mine never so deep, each engine working it 60, 70, or 80 foot high, by applying or setting the engines one over another . . .' (*Miner's Friend*). There are several early illustrations of how this might be done.
[97] e.g. Farey, i. 190–1 and ii. 100, 183; also sources such as the *Engine Reporter* quoted in Ch. 9. In Cornwall it was unusual to bury the shammals engine (but see Barton, *The Cornish Beam Engine*, p. 233). However, at Yatestoop, Derbyshire, a 64½-inch atmospheric engine was fixed 100 fathoms underground by Francis Thompson; the excavation to receive engine-house and boiler cost £300. Because of strata difficulties the Whitehaven Colliery also had an engine working 80 fathoms underground in 1776 (Farey, i. 237–8).
[98] Farey, i. 378. Scraps of information confirm this, e.g. J. S. Allen, 'John Fidoe's 1727 Newcomen Engine at Wednesbury, Staffs.', *Trans. Newcomen Soc.* xxxvi, 1963–4.

TABLE 6.4
*Prices of Newcomen and Watt engines, c. 1800*[99]

| HP | Newcomen or Savery (£) | Watt Reciprocating (£) | Watt Rotative (£) |
|---|---|---|---|
| 2  | 100*  | 223  | 244  |
| 4  | 150*  | 320  | 350  |
| 6  | 270   | 358  | 385  |
| 8  | 320   | 435  | 470  |
| 10 | 394   | 498  | 520  |
| 12 | 465   | 530  | 560  |
| 14 | 534   | 629  | 600  |
| 16 | 600   | 660  | 710  |
| 20 | 700   | 753  | 820  |
| 24 | 809   | 954  | 1020 |
| 28 | 907   | 1060 | 1133 |
| 30 | 952   | 1208 | 1160 |
| 36 | 1088  | 1280 | 1280 |
| 40 | 1184  | 1393 | 1393 |
| 45 | 1297  | 1526 | 1526 |
| 50 | 1403  | 1650 | 1650 |

\* = Savery engines.

driving the wheel as well as the machinery driven by the wheel.[100]

The problem of limited size of atmospheric engines will be met by supposing that two (or more) Newcomens would be needed whenever a Watt engine supplied greater power than

[99] Atmospheric-engine figures from Table 4.1 where available, otherwise interpolated from costs per horsepower. For 40- to 50-horsepower engines a constant 85 per cent of Watt reciprocating engines is assumed.

[100] Wicksteed (1840) thought it took 76 horsepower of a steam-engine to drive a 50-horsepower water-wheel. Farey made several more precise calculations. At Carron, although some of the effect of the returned water was lost through the head of the wheel being 4 feet below the pond from which it flowed, the output of the wheel was rated at 15·76 horsepower, compared to 20·2 horsepower for the fire engine (i. 277–8). At Long Benton, a portable winding engine supplied 9·02 horsepower, the wheel only 5·22 horsepower, but the loss attributable to the fall of water was about 2·25 horsepower (i. 304); a larger engine at Long Benton in 1785 also lost about a quarter (i. 305). An engine at the Walker Colliery, Newcastle, gave a useful effect of only 8·44 horsepower from an engine of 36″ (17·1 horsepower) (i. 306).

they would normally be capable of. Smeaton[101] obtained 76 horsepower from the engine he supervised at Chacewater Mine, Cornwall, in 1795. Atmospheric engines of 100 horses or more were erected after the advent of the Watt machine, but for that reason will be ignored here.[102] Again I shall overstate and assume the need for two atmospheric engines for any Watt engines greater than 50 horsepower. Overlooking the likelihood that in many situations it would be cheaper to erect two engines side by side as compared with different locations, I have added 10 per cent to construction costs for erecting the additional engine within a mine.

Watt engines were first classified according to whether they were reciprocating or rotative, then into horsepower ranges in line with Table 6.4. The difference between column (ii) or column (iii) and column (i) in Table 6.4 is then multiplied by the number of engines in each class.

For reciprocating engines, there is an over-all additional cost of £68 000 from replacing the Boulton & Watt engines (208 of which were still standing in the U.K. in 1800) with atmospheric ones. All of this is due to having to erect additional Newcomen engines for horsepower sizes over 50, since as the Table shows they are cheaper to purchase at each level of horsepower up to 50. For the 263 rotative engines there was a net cost on engines alone of £8400. In addition there will be outlays on the water-wheel; if I put these at £15 per horsepower including erection, the over-all cost for rotative engines rises to nearly £71 000. Thus net capital costs come to £140 000; or £8400 at 6 per cent interest (as seems roughly correct for the year 1800).[103]

Naturally more important were the savings accruing to Watt's inventions arising out of operation. Here too we have to take account of the additional expense of the 'returning engine', in allowing for the power required just to turn the wheel.

As with the computation of thresholds in Chapter 4, I have

[101] *Reports*, vol. ii, p. 350.
[102] See Table 3.2 above (the 1810 engine noted there used a separate condenser).
[103] Bank Rate was set at 5 per cent in 1797; it was generally 4 per cent in the early nineteenth century when the conventional market rate quoted for industry was 5 per cent. Yield on Consols averaged 5·3 per cent 1796–1800 (4·7 per cent in 1800), and 4·9 per cent 1801–5 (Mitchell and Deane, op. cit., p. 455).

## The Social Savings 147

made the comparison for engines in optimal order. This amounts to the consumption of an additional 9·6 lb of coal an hour for each horsepower when a reciprocating engine of the Watt type has to be replaced with a Newcomen, but only another 6·8 lb hourly for the rotative cases. As in Chapter 4, one would ideally like comparisons between average practice rather than best practice. The difficulty of getting any consensus on average practice for atmospheric engines presents something of an obstacle to doing so. Can it be supposed that the lag of the average behind the best was the same for the two engine types? It would imply an average consumption of between 19 and 20 lb of coal per horsepower per hour in Newcomen engines, and comparison with data such as that in Table 4.9 suggests that this was readily achievable with a moderate amount of care.

In Table 6.5 I assume 4000 hours of use a year for reciprocating engines and 3600 hours for rotatives. Thus annual variable savings arising out of the Boulton & Watt engine come to almost £185 000. Aggregate savings (fixed and variable) on rotative engines are about £68 000; on reciprocating engines about £125 000. To these sums should be added the gains from those who pirated Watt's inventions, especially the separate condenser. They seem to have been distributed around the country in fairly similar patterns to the Boulton & Watt engines. If this is so, then straightforward multiplication by the average price of coal across the country obtained for Watt engines (15s. 6d.) should be close to the mark. It would be generous to assume that the pirates were as efficient as Watt's engines in optimal order, especially if they did not bother with the steam-case, but I shall err upwards and accept this assumption. If all the pirates are treated as reciprocating engines, the aggregate savings owing to them come to between £33 000 and £40 000.

Now the national income in 1801, according to estimates summarized by Deane and Cole, was £232 million in current prices.[104] If real national income in 1800 were to have been the same as in 1801, then in money terms 1800 national income would have come to £217 million. Whether it is appropriate

---

[104] P. Deane and W. A. Cole, *British Economic Growth, 1688–1959*, 2nd edn., Cambridge, 1967, p. 161.

## TABLE 6.5
### Fuel savings of Watt engines over Newcomens, by county, 1800[105]

| County | Price of coal pence per ton | Horsepower in Watt Rotative engines | Horsepower in Watt Reciprocating engines | Gross savings on Rotative engines (£) | Gross savings on Reciprocating engines (£) |
|---|---|---|---|---|---|
| Cornwall | 266 | 61 | 3250 | 1309 | 61 816 |
| Devon | 261 | — | 8 | — | 149 |
| Wiltshire | 236 | — | 5 | — | 88 |
| Hampshire | 396 | — | 30 | — | 850 |
| Berkshire | 360 | 2 | 4 | 58 | 103 |
| Surrey | 560 | 214 | 62 | 9702 | 2487 |
| Middlesex* | 458 | 486 | 322 | 18 021 | 10 538 |
| Kent | 554 | 27 | 78 | 1211 | 3074 |
| Cambridge | 238 | — | 80 | — | 1361 |
| Northants | 264 | 12 | — | 256 | — |
| Oxfordshire | 314 | 24 | — | 611 | — |
| Leicester | 130 | 22 | — | 232 | — |
| Warwickshire | 136 | 119 | 207 | 1310 | 2008 |
| Worcester | 128 | 10 | — | 104 | — |
| Gloucester | 231 | 49 | 61 | 916 | 1011 |
| Monmouth | 133 | — | 153 | — | 1457 |
| Glamorgan | 122 | 13 | 307 | 128 | 2680 |
| Shropshire | 163 | 220 | 923 | 2903 | 10 760 |
| Staffordshire | 119 | 507 | 570 | 4885 | 4848 |
| Anglesey | 336 | 3 | — | 82 | — |
| Caernarvon | 336 | — | 16 | — | 372 |
| Denbighshire | 105 | — | 189 | — | 1416 |
| Cheshire | 131 | 181 | 34 | 1920 | 318 |
| Derbyshire | 111 | 26 | 43 | 234 | 342 |
| Nottingham | 122 | 218 | 6 | 2153 | 52 |
| Lancashire | 121 | 929 | 479 | 9101 | 4145 |
| Yorkshire | 106 | 311 | 278 | 2669 | 2109 |
| Durham | 88 | 120 | 498 | 855 | 3134 |
| Northumberland | 76 | 136 | 340 | 837 | 1847 |
| Cumberland | 119 | 39 | 58 | 376 | 494 |
| Ayrshire | 138 | 30 | 70 | 286 | 689 |
| Renfrewshire | 130 | 42 | 22 | 442 | 204 |
| Lanarkshire | 130 | 102 | 96 | 1074 | 891 |
| Stirling | 84 | 20 | — | 136 | — |
| Argyll | 250 | 8 | — | 162 | — |
| Clackmannan | 96 | 24 | — | 187 | — |
| Edinburgh | 160 | 62 | 84 | 803 | 962 |
| Fife | 120 | 12 | 41 | 117 | 354 |
| Forfar | 183 | 20 | — | 296 | — |
| Ireland | 211 | 17 | 44 | 290 | 658 |
| Total |  | 4066 | 8357 | 63 666 | 121 217 |

* includes London.

[105] For England and Wales, most coal price figures in the first column are worked up from averages for 1842 and 1843 for the 480 specified Poor Law Unions (P.P. 1843: XLV), extrapolated on 1800 figures quoted in *Thesis*, Table I.12 and elsewhere (e.g. P.P. 1800: X, for London). Hampshire is taken from P.P. 1800: X (Apps. 46, 48: prices at Lymington and Rochester), extrapolated on London data; cf. also P.P. 1840: XLIV and P.P. 1845: XXX for the South Coast ports. Caernarvon is from T. S. Ashton and J. Sykes, *The Coal Industry of the Eighteenth Century*, Manchester, 1929; the 3 horsepower for Anglesey are priced similarly. Prices for Lanarkshire and Renfrewshire are from B. F. Duckham, *A History of the Scottish Coal Industry*, vol. i, Newton Abbot, 1970, p. 278; for Stirlingshire and Edinburgh from J. A. Hassan, 'The Supply of Coal to Edinburgh, 1790–1850', *Transport Hist.* v, July 1972; for Forfar from P.P. 1871: XVIII, App. 54; other Scottish figures are extrapolations. Ireland from P.P. 1819: XVII, pp. 661–70, and cf. Wood, 'Cumberland Colliery'; Sir Robert Kane, *The Industrial Resources of Ireland*, Dublin, 1844, esp. p. 49.

*The Social Savings* 149

or not to assume constant real incomes is difficult to assess. Hoffmann's index of industrial production does not begin until 1801. The few available indices of domestic output[106] for the two years give contradictory results; the same confusion arises with the principal items of consumption. To give only one example, cattle sold at Smithfield market rose 8 per cent between 1800 and 1801, whilst sales of sheep at the same place fell by 9 per cent. Agricultural production probably did better than in the disastrous years 1799–1800, though corn imports rose.[107] Net imports and exports rose both in official values and in declared values, though at a time of rather disarranged markets this may not prove very much.

In general, it may be reasonable to put 1800 national income at approximately £210 million. In that case, the social savings accruing to James Watt come to 0·11 per cent of current national income.

### THE SOCIAL SAVINGS OF ALL STEAM-ENGINES

The calculation just undertaken for Watt's engine may be put to answer a number of questions: for example, the role of the single inventor in industrial transformation, or the rapidity of that transformation. At the same time, it seems equally worth while to estimate the social savings on all engines; in keeping, for instance, with wider notions of the 'energy-crisis' kind. Naturally, the hypothetical removal of all species of steam-engine prevents the relatively straightforward 'soft option' used above (viz. the hybrid atmospheric engine plus water-wheel) being considered here. As a result, the computations hereafter will involve more attenuated extrapolations and end up with less precise estimates.

In any case, the social savings just produced for James Watt were excessively high, because I allowed for only one altern-ative to his engine. The familiar water-wheel, amongst other possibilities, must have proved cheaper than the somewhat clumsy 'returning engine' in certain circumstances (where waterfalls permitted, etc.). From what has been said earlier

[106] Figures were compiled from Mitchell and Deane, *passim*. Relevant tables include Agriculture 9, 10; Coal 1, 2; Iron and Steel 8, 10; Tin, Copper and Lead, 1, 2, 5; Textiles 1, 5, 10, 13, 18, 19, 23, 25; Transport 1, 2; Miscellaneous Pro-duction 1, 2, 3, 5, 7, 8, 9, 10, 11, 12; Overseas Trade 1, 2; Wages 5, 6.
[107] E. L. Jones, *Seasons and Prices*, London, 1964, pp. 154–6.

about the costs of water-power, the water resources unused in 1800 could have been only marginally more expensive than the Watt engine, if they happened to be sited so that they fell just beyond the external margin.

One could thus conceive of a technological alternative— whether to Watt's engines alone or to all steam-power—couched in terms of the various water-wheels. However, I have shown that without an exhaustive practical study in economic geography one could not begin to locate and cost water sites. Expressed in economic terms, we have very little information indeed about the shape and vertical alignment of the marginal cost curve for water-power. If however one could make a reasonable guess at the marginal cost function, the geographic objection could be finessed, at least over a moderate range. Since I have no time series on water rents as yet, I have to use data for the mid-nineteenth century, for it is only then that the costs of steam-power change sufficiently to give any real guide to changes in the use of water-power.

In the 1830s the cost of steam per horsepower in a Manchester cotton-mill was about £20·4 annually (see Table 4.11 above); at that time the total amount of water-horsepower used in cotton-mills in England and Wales was 9677½ (Table 6.3). By the mid-1850s costs of steam had fallen to about £11 a year in Manchester, while usage of water had dropped to 6551 horsepower. Thus the arc elasticity of supply of water-power was 0·69. However, a straight line connecting these two points and then produced would cut the vertical axis at a negative cost, which is clearly out of the question. Therefore, if all else stays constant, the relationship between power costs and water usage must be curvilinear, at least over some range.[108] The first estimate I make ignores this problem, though it is reintroduced later by dint of some guesswork.

Even the use of this 0·69 as a supply elasticity raises several almost intractable difficulties, and the best I can hope to do is to make some rough-and-ready allowances. Firstly, I have to

---

[108] Always granted the basic assumption—not exactly validated by previous discussion—that the curve did not shift downwards between 1838 and 1856. If, however, it did move down towards the horizontal axis (through technological advance or whatever), then my elasticity estimate is too pessimistic, and thus my reckoning of hypothetical costs too high.

## The Social Savings

assume that what was happening in cotton textiles between 1838 and 1856 was happening in equal proportions in other industries. As it turns out, the other textile industries quoted in Table 6.3 do not experience as great a reduction of water-power as cotton over the years 1838–56, but by the same token they may not have undergone as large a reduction in steam costs. For other industries, knowledge is virtually non-existent.

Secondly, practically nothing is known about the actual extent of water-power in 1800. I have tentatively suggested two situations: one assuming 50 000 water-horsepower and the other 100 000. Only if the former is too high will the social savings be understated.

Thirdly, it is conceivable that the elasticity I have derived, applying as it does to a shift down the supply schedule, is therefore larger than the elasticity 'upwards' that is really required. Fortunately, once I come to incorporate other options to steam than just water-power, I can assume a perfectly inelastic supply—or for that matter no water-power at all in 1800—as extreme upper bounds, and the result will not be greatly affected.

To begin then, take the (unlikely) supposition of a linear relationship between costs and employment of water-power. Even if this first-stage calculation cannot thus be believed, it may help in assessing how important the elasticity chosen is for the subsequent results. Water-horsepower would have to be increased by 29 000 to compensate for the lack of any steam-engine, i.e. 58 per cent if actual water-horsepower were 50 000 or 29 per cent if 100 000 in 1800. With the elasticity of supply quoted (0·69) the cost of water-power at the external margin would have risen to £49·6 in the former case or £38·3 in the latter from the actual of £27·0 per horsepower estimated for Manchester.[109] Total costs in the hypothetical situation would come to close to £4 million if actual water-horsepower were 50 000, or nearly £5 million if 100 000. Since actual costs

[109] Costs of a 10-horsepower Watt engine in Manchester are estimated at £106 annual capital expenses (engine £520, boiler £104, erection, etc. £165; all at 6 per cent interest), coal £94, oil, etc. £20, labour £50. On present assumptions the Newcomen engine would be too expensive to use in Manchester. Manchester is used because of its price leadership in cotton textiles, to which the elasticity estimate relates. If instead I use the average coal price for Watt engines taken from Table 6.5, the upper bound on the social savings rises from £1·1 m. to £1·4 m.

(including steam-power from Watt and atmospheric engines) would have been £2·2 and £3·5 millions respectively, the net increase in power costs comes to between £1·4 and £1·8 millions.

These last sums represent the additional bill that manufacturers and other steam-users would have been called upon to foot. As such, the figure has relevance for discussing the impact of steam-power on industrialization, and will be roughly what I have in mind when I move on to the forward linkages in the next three chapters. But it is a much more extensive concept than the customary 'social savings' idea. It includes not only additional resource costs that the counterfactual economy might have to meet, but also additional rents that manufacturers might have to pay to owners of water rights for the use of 'scarcer' water supplies. These outlays may alter the functional distribution of income within the economy, but do not themselves detract from its aggregate size. Indeed, as many manufacturers owned their own water-power supply, the *size* distribution of income might not have been greatly distorted. As a result, the loss in real resources to the economy as a whole through being forced to accept a higher-cost alternative to steam for driving its machinery is equal to from £165 000 to £330 000.

Next, allow for some curvilinearity. Suppose that the arc elasticity of supply falls to 0·55 as a result of having to withdraw to more-and-more remote areas to find water resources. Then hypothetical water costs after substituting for steam-power would rise to £55·5 per horsepower, and total savings to consumers of power would be £2·2 millions. Again, this will include a certain amount of redistribution of incomes. To find out how much, it is necessary to see how much curvature will be called for to induce an elasticity of 0·55 over the required range. Now from the findings of earlier sections of this chapter, the costs of acquiring and erecting a small water-wheel can be estimated at £15 per horsepower. If I assume 6 per cent rates of depreciation and interest, and operating costs of £1·07 per horsepower,[110] then annual costs amount to £2·87 per horse-

[110] Operating costs are taken from the figure Henshall estimated for a 100-horsepower wheel at Quarry Bank in 1847 (Edwards, *Growth of the British Cotton Trade*, p. 210; quoted by Chapman). Lower figures are scarcely plausible. Higher costs for the first horsepower would induce more curvature (i.e. a larger value for the coefficient of horsepower squared) and so reduce the net social savings.

power. For the very first horsepower, supplementary costs, as well as rents, should be zero. Thus the point (2·87, 1) lies on the curve. Assuming further that a second-degree polynomial describes the relationship, the required cost curve would be:

$$\text{Water} = 2\cdot 87 + 0\cdot 114\,\text{HP} + 0\cdot 0074\,\text{HP}^2 \qquad (6.1)$$

Where: Water = Costs of water per horsepower in £s;

HP = Horsepower, in thousands.

Of the total savings, under £½ million represents increased costs of real resources (obtained by integrating the curve on the assumption that actual water-power = 50 000 horsepower); the rest is again increased rents on intramarginal water resources.

The figures used to obtain these estimates, and especially those that are rather plucked out of thin air, can readily be altered to see how sensitive the answers are to them. The indications are that plausible alterations are unlikely to revise the results very radically.

As stated, the hypothetical usage of water-power is permitted to rise to either 79 000 or 129 000 horsepower. In geographical terms, what would this have involved? Sir Robert Kane made some calculations for Ireland in 1844:

... We had for the total quantity of rain falling in a year 100,712,031,640 cubic yards [on the whole of Ireland]; of this one-third flows to the sea, that is 33,237,343,880 cubic yards, or for each day of 24 hours, 91,061,216 cubic yards, weighing 68,467,000 tons. This weight falls from [an average of] 150 yards, and as 884 tons falling 24 feet in 24 hours is a HP, the final result is that on average we possess distributed over the surface of Ireland a water power capable of acting night and day, without interruption, from the beginning to the end of the year, and estimated at the force of 3227 HP per foot of fall, or for the entire average fall of 450 feet, amounting to 1,452,150 HP.

But mechanical power is never this uninterruptingly driven, and if we reduce this force to the year's work of 300 working days of 12 hours each, we find it to represent 3,533,565 HP . . .[111]

---

[111] Kane, op.cit., pp. 70–1. An equivalent calculation for Lancashire gave 72 600 horsepower working continuously (ibid., p. 101).

Even if the computation is overstated by several hundred per cent, it follows that Ireland alone could have supplied all the necessary increment in resources conjectured, with lots to spare.

It may be objected that this is too superficial a way of looking at the potential difficulties that might have had to be surmounted if water-power had been more widely used (apart from the data shortcomings, which are acknowledged). However, my definition of the external margin ought to take into account extra costs such as transportation of raw materials and finished goods; of having to provide one's own accommodation for the operatives (a factory village), etc.[112]

Even more worrying, at first sight, are the problems that owners of water-mills suffered through being subject to a variable power supply; with natural causes such as droughts and floods proving especially bothersome. This is a sufficiently important matter to demand special study, and accordingly I have devoted Appendix 6.3 to it. I reach two conclusions: (i) that the scale of such interruptions was rarely great enough to invalidate the above results very seriously; (ii) if the interruptions had become more acute through continuing industrialization in a world deprived of the steam-engine, more could easily have been done by way of alleviation, along the lines of schemes that were mooted in the early nineteenth century.

The calculation of the social savings of all steam-engines thus suffers from a number of weaknesses. In the rest of this section two of the least plausible assumptions will be further examined. These are the necessity to extrapolate water-power costs well beyond the actual employment of water-power in 1800, and the neglect of other alternatives to steam-power. The former may be a less serious drawback than it seems, because water-horsepower probably went on rising for some time after 1800.

---

[112] This calculation does not allow for any social external diseconomies, e.g. Farey, sr. thought that watermills were a dreadful nuisance to the agriculturalist (*Derbyshire*, vol. iii, p. 492). But by the same token I have ignored the external diseconomies of the steam-engine, and most authors would implicitly regard these as much greater (see F. Engels, *The Condition of the Working Class in England*, Panther edn., ed. E. J. Hobsbawm, 1969, e.g. p. 83; A. Briggs, *Victorian Cities*, London, 1963; etc.). Ironically, a complete calculation might well show that James Watt's greatest contribution (as compared with the voracious fuel consumption of atmospheric engines) was in *reducing* pollution and thus these external diseconomies. I intend to follow up the issue of pollution and the steam-engine at an early date.

## The Social Savings

As I cannot be certain of this, I approach it from the latter direction.

In the early sections of this chapter I considered the use of animal and windmill in industry. In the early eighteenth century Desaguliers reckoned the fifty horses at the Griff colliery to cost £900 a year.[113] On this basis, the cost of powering the customary 10-horsepower Arkwright mill would be £400. In the 1770s, I showed previously, Smeaton's figures can be used to get an estimate of £650. However, with the costs given for the late years of the century, i.e. about £4·50 a year for capital and labour and £40 for feed per horse, powering the mill would cost about £980 annually. To replace the atmospheric and Watt engines with horses would thus have cost nearly £2 millions in 1800. On my calculations, it should have been preferable everywhere to choose water rather than animals for power, once feed costs had climbed to the levels attained in the period of the French Wars. The disappearance of animal-powered mills at this time of course agrees with historical reality.

Windmills provided another possible alternative in areas with sluggish streams—recall that several authors considered them cheaper than horses. Out of a total of 8760 hours in a year, suppose that windmills were able to work 5·2 per cent of the time at full power and 20 per cent of the time at a quarter power.[114] Then it would require 4 windmills rated at 10 horsepower at full capacity to supply as much power as a 10-horse steam-engine working 3600 hours. This is probably erring on the cautious side. Charles Partington, seeking to advocate the use of steam rather than wind for draining the Fens, and thus unlikely to overrate its relative cost, quoted a certain Mr. Savory of Downham—not far from Mildenhall—that 'one steam-engine, with equal powers, would do as much execution in the course of a season, as three windmills'.[115] As £10 per horsepower emerges as a general allowance for costs of such windmills from my earlier discussion, the equivalent amount of

---

[113] Desaguliers, *A Course of Experimental Philosophy*, p. 482. Landes misprints the 50 horses as '500' and thus comes to some surprising conclusions (*The Unbound Prometheus*, pp. 96, 100).
[114] From the Mildenhall data, n. 31 above.
[115] Communications to the Board of Agriculture, vol. iv, p. 52; quoted by Partington (1822), pp. 22–3.

work would be done on my more conservative assumptions for about £40 per horsepower.

For both animal- and wind-power, in contrast to water, the marginal and average costs are likely to have been about the same; that is, entrepreneurs so motivated could have gone on building horse-mills or windmills *ad infinitum* at much the costs I have estimated. So even if the supply curve of water-power had become even less elastic at the hypothetical margin than I gave credit for, windmills or (sometimes) horses could have been used at about this point to prevent any undue escalation of costs.

### SUMMARY

In measuring the direct impact of the steam-engine during the take-off period, I have studied two situations. In the first case, I have computed an upper bound on the social savings that comes to little more than one-thousandth of 1800 national income, by positing that the cheapest alternative technique to every Watt engine that existed in 1800 was an atmospheric engine—driving a water-wheel if required to produce a very even transmission. This estimate is of special interest because the Watt engine probably comes closest to meeting Rostow's specifications for technologies developed during the 'take-off'. Moreover, the limitation of attention to Watt's engines seems most appropriate for probing the 'energy crisis' of the late eighteenth century which has implicitly entered into many historians' views of the Industrial Revolution. From his own day up to the present Watt's work has been hailed as decisive in the evolution of industrializing economies in his era. The normally restrained and balanced engineer, John Farey, jr., wrote:

We may compare his [Watt's] rotative engine, in the state above described, with Newcomen's engine in its greatest perfection, as made by Mr. Smeaton, and considering the degree of invention requisite to produce all the new properties of the improved machine, from its original state, and also taking into account the great perfection with which it is adapted to its purpose, the rotative steam-engine will be found to be the greatest step of useful invention ever made by one individual, at any time in the history of the arts.[116]

[116] Farey, i. 473; Usher, *History of Mechanical Inventions*, p. 353.

This, which explicitly excludes the (larger!) savings from reciprocating engines, would contrive to make the contribution of individual men look puny indeed.

The problem is secondly specified as supposing that no steam-engines of any kind had existed. By assuming a competitive pricing system for determining power costs—which, on the basis of admittedly limited evidence, seems not unreasonable to describe water-power at the beginning of the nineteenth century—I make some crude guesses at what water costs might have been if engines had never been invented by reference to the rate at which water-power declined as all power cheapened in the nineteenth century. With a fairly generous allowance for increasing supply inelasticities as more and more distant or unsatisfactory water resources had to be hypothetically brought into production, the marginal cost of water for textile manufacturing rises from £27 per horsepower to £55·5. Even this would raise the cost of power to its immediate consumers by £2·2 millions only, of which over £1·7 millions would be a distributional change towards those who possessed water rights. Thus the aggregate social savings under these assumptions come to about 0·2 per cent of 1800 national income.

The existence of direct water-power of course also lowers the estimate that we have already made for Watt-only engines. By the same token, it is too large even for all atmospheric engines, because still other techniques for providing power could be adopted. In East Anglia, at least, windmills could have kept the marginal costs of power in the counterfactual case down to about £40 per horsepower, even allowing for deficiencies in wind to drive them. Animals, especially horses, were frequently used in eighteenth century textile-mills, etc. for certain functions. However, the available figures suggest that by the 1790s the cost of animal feed was so high that horses would not have been economically viable except perhaps in very small concerns, where steam-engines might be uneconomic; and this accords with actual observation.

Allowing windmills (or animals at pre-1790 prices) as alternatives lowers the social savings below £1·1 millions, and indeed since their average and marginal costs must have been approximately equal, means that my estimate is relatively immune to a range of (reasonable) choices of the supply elasticities.[117]

All estimates are still upper bounds since the counterfactual model adopted exaggerates the inelasticities of both supply and demand. Some locations that happened to have been chosen because the Watt engine rendered them economic propositions would probably never have been chosen with only the array of power possibilities that my least-cost alternative allows. With the Newcomen or Savery engine only, as profligate of fuel as they were, one might expect that few sites where coal was remote in access would have been considered. The pattern of production would have shifted nearer the coalfield, accepting the brunt of greater transportation charges on raw materials and finished goods for the cost benefits obtained from having much cheaper coal. Thus on the supply side the whole location of production might be predicted to shift in my hypothetical circumstances—definitionally shifting in such a way as to ease the burden of particularly crippling fuel bills at extreme points. A similar argument applies for the other alternatives, i.e. with entrepreneurs moving to sources of water-, wind-, or animal-power. Obviously, the degree of supply response varies between industries—in the case of naturally occurring commodities, such as lead- or copper-mines, it would be quite small.

An analogous problem arises on the side of demand. Final demand (for power) is taken as it actually existed in 1800. My calculation requires the production of the same goods for customers located in the same places with a high-power-cost alternative, be it atmospheric engines or a combination of

[117] A brief mention of the general-equilibrium consequences should perhaps be made at this point. I assume that the replacement of Watt engines by Newcomens, with or without water-wheels, would have minimal feedbacks on prices of inputs. The additional investment required comes to £140,000; the single firm of Walkers of Rotherham had a capital half as large again by 1796 (Ashton, *The Industrial Revolution*). The severe constraints I imposed on the sizes permitted for the hypothetical engines mean that there was no question of having to diffuse any more modern technology in producing them. The price of coal at the pithead in Co. Durham in 1800 was 5s. 8d. a ton; at Newcastle 8s. 4d. f.o.b.; and in the Pool in London 26s. in 1800 (P.P. 1st Ser. X, p. 549 and App. 9). Thus a rise of 10 per cent or so in the demand for coal would make very little proportional difference to the London (Cornwall, etc.) price even if the supply of coal were highly inelastic; assuming that harbour facilities, etc. allowed expansion of traffic at more or less constant costs.

## The Social Savings

water-, wind-, and animal-power. Intuitively a reasonable argument could be made that the demand for power is quite price-inelastic, so that on this account my upward bias may not be very large.

However, all this assumes general equilibrium to prevail, so that resources become reallocated to other uses. This may be too strong an assertion for the U.K. in 1800, e.g. with the need to retain export competitiveness.

Some further references should be made to the problem of irreversible external economies. By confining myself to a comparatively short period of time, I have probably eliminated the worst of such effects. Many associated developments that might be considered as falling in this area will be dealt with explicitly in the ensuing chapters on forward linkages. By way of brief summary, those results suggest little need to alter the conclusions hitherto attained, at least in the case of the textile industries (though industries such as brewing might be less easily swept aside). It was water-power that inaugurated large-scale factory organization, first with the work of Sorocold and others in the late seventeenth and early eighteenth centuries; followed by a sequence of factories in the 1760s such as the Carron Company (1760), Soho (1764), and Kilnhurst Forge (1765). In 1800 most of the industrial establishments that were really large by the standards of the time were still principally water-powered, and this continued for some time, as at New Lanark, Catrine, Quarry Bank, etc. Moreover, the supposed flexibility that steam-power gave to industrial location was still somewhat illusory: water was probably still far and away the leading determinant of location (partly because condensing steam-engines needed plenty of it),[118] even if the supply at particular points had been outstripped by industrial demand. As will be shown in several of the later chapters, the old 'tyranny' of

---

[118] Cf. Ch. 3. Farey, ii. 318, notes that deficiencies in the supply of cold water limited the application of Watt's engine in some manufacturing towns. Also W. C. Taylor, *An Illustrated Itinerary of the County of Lancaster*, London, 1842, p. 30: 'Though water may not be wanting to drive a wheel, the vicinity of a river or canal is almost essential to a mill, in order to facilitate the conveyance of fuel, to supply the boilers, and to afford good drainage. Hence, most of the mills in Manchester are close either to the Irwell or the Medlock; and the noble Mersey is studded with factories for miles upon miles of its course.' If cold water was lacking, one alternative was to install a high-pressure non-condensing engine, e.g. at Bolton (P.P. 1871: XVIII).

abundant water was replaced by the scarcely less binding 'tyranny' of abundant coal in fixing industrial sites.

With regard to individual innovations, one can note that virtually all the celebrated eighteenth-century inventions in textiles were created for either animals or simply man-power. Hargreaves and Crompton were avowedly improving the lot of female spinners in cottage industry. Paul and Wyatt, Arkwright, and Cartwright all began with animals. Even for the spinning-mule, water-power was applied in incorporating the invention into factory industry before the steam-engine.

# APPENDIX 6.1

*Comparative cost of steam- and water-power in the U.S.A., France, and England, in £ per horsepower*

|  | Water | | | | Steam | | | |
|---|---|---|---|---|---|---|---|---|
|  | £ | s. | d. | % | £ | s. | d. | % |
| **(A) U.S.A.** | | | | | | | | |
| Engine/wheel, etc. | 2 | 7 | 6 | 24 | 1 | 17 | 6 | 17 |
| Foundations | 1 | 2 | 8 | 11 |  | 5 | 0 | 2 |
| Heating |  | 16 | 8 | 8 |  |  |  |  |
| Total capital cost | 4 | 6 | 10 | 44 | 2 | 2 | 6 | 20 |
| Total rents | 2 | 10 | 2 | 25 |  |  |  |  |
| Transportation | 1 | 13 | 4 | 17 |  |  |  |  |
| Coal | 1 | 9 | 2 | 15 | 7 | 6 | 1 | 67 |
| Wages |  |  |  |  | 1 | 9 | 3 | 13 |
| Total variable costs | 3 | 2 | 6 | 31 | 8 | 15 | 4 | 81 |
| Total costs | 9 | 19 | 6 | 100 | 10 | 17 | 10 | 100 |
| **(B) France** | | | | | | | | |
| Total capital cost | 6 | 8 | 0 | 14 | 5 | 5 | 3 | 11 |
| Total rents | 18 | 0 | 0 | 39 | 8 | 1 | 9 | 17 |
| Transportation | 14 | 8 | 0 | 31 |  |  |  |  |
| Town expense | 7 | 4 | 0 | 16 |  |  |  |  |
| Maintenance and repairs |  |  |  |  | 5 | 4 | 2 | 11 |
| Enginemen |  |  |  |  | 5 | 11 | 1 | 12 |
| Coal |  |  |  |  | 17 | 5 | 0 | 37 |
| Total variable costs | 21 | 12 | 0 | 47 | 28 | 0 | 3 | 60 |
| Total costs | 46 | 0 | 0 | 100 | 47 | 1 | 11 | 100 |
| **(C) England** | | | | | | | | |
| Engine/wheel, etc. | 2 | 6 | 8 | 13 | 3 | 9 | 2 | 20 |
| Foundations |  | 15 | 4 | 4 |  | 15 | 0 | 4 |
| Tunnel, wheelrace, etc. | 2 | 13 | 4 | 15 |  |  |  |  |
| Heating |  | 5 | 6 | 2 |  |  |  |  |
| Total capital cost | 6 | 0 | 10 | 33 | 4 | 4 | 2 | 24 |
| Water power | 7 | 15 | 4 | 43 |  |  |  |  |
| Cottages | 2 | 2 | 6 | 12 |  |  |  |  |
| Total rents | 9 | 17 | 10 | 54 |  |  |  |  |
| Coal | 1 | 14 | 4 | 9 | 9 | 14 | 8 | 55 |
| Oil, tallow, gasking |  |  |  |  | 1 | 2 | 8 | 6 |
| Labour |  | 10 | 4 | 3 | 2 | 10 | 0 | 14 |
| Total variable costs | 2 | 4 | 8 | 12 | 13 | 7 | 4 | 76 |
| Total costs | 18 | 3 | 4 | 100 | 17 | 11 | 6 | 100 |

162    *The Evaluation*

*Notes:*

(A) U.S.A. Derived from P. Temin, 'Steam and Waterpower in the Early Nineteenth Century', *J. Econ. Hist.* xxvi, June 1966, p. 197. Temin bases his calculation on data from a tract by 'Justitia', i.e. Charles Tillinghast James, though the propagandist nature of this tract leaves much to be desired in relying on its information in this way. Dollars are converted to sterling at the rate of £1 = $4·79; this being obtained from L. E. Davis and J. R. T. Hughes, 'A Dollar-Sterling Exchange, 1803–1895', *Econ. Hist. Rev.* N.S. xiii, 1960–1—to the official rate of £1 = $4·444 was added the average premium on sterling 1830–3, measured quarterly. Water rights were deducted from capital costs and inserted under rents. Temin discounted engine, wheel, and foundations all at 6 per cent. His item 'heat for factory' was split up between capital and variable costs by supposing that coal consumption for heating was 20 per cent of that for steam-power. The low overhead costs of steam are probably accounted for by the adoption of a high-pressure non-condensing engine, but such engines—common enough in the U.S. by this date—were not economical of fuel. Since fuel was costly in Massachusetts (cf. Ch. 10) the expenditures on coal appear much too low (the steam-engine is supposed to be located in a cotton-mill in Newburyport, and the water-wheel in Lowell, Mass.). It seems possible that Massachusetts fell outside the feasible fuel-cost range for American steam-engines in this period, and that James was deluding either himself or his (subsequent?) audience (see pp. 268–9 for further comments).

(B) France. From P. Grouvelle and A. Jaunez, *Guide du chauffeur et du propriétaire de machines à vapeur*, 4ᵉ édn, Paris, 1858–9, pp. 66–8. The engine is of 2 cylinders and is rated at 12 horsepower. Interest and rent are not separated by the authors for this particular calculation. I took the average cost differential for 14-horsepower condensing engines in France compared with England as reported at the Great Exhibition of 1851 (*Exposition universelle de 1851: Travaux de la Commission française, IIᵉ Groupe*, Paris, 1857), viz. 1340 francs per horsepower for the French-built engine as compared with 1210 fr. per horsepower for the English-constructed one: and computed this addition to the cost of the English engine used in Section (C) of the above table. Discount was 10 per cent, from ibid. Simply taking 1340 fr. @ 10 per cent would give £5 11s. 6d. rather than £5. 5s. 3d. Rents for steam were then calculated by subtraction (this item of course does not appear for the other two countries). The transportation charge represents the cost and upkeep of a team of 4 horses, plus driver. Town expenses, consisting of an office, clerks, etc., also do not appear for other countries, and might well be cancelled out against steam rents if one wished to compare total power costs across countries. In a hypothetical calculation, the Gregs costed a town establishment, presumably in Manchester, at £500 discounted at 5 per cent, or £2. 10s. a year (Quarry Bank MS. C5/8/93/2, Dec. 1828). All water costs for France were multiplied upwards since the water-mill was estimated to work 295 days a year, compared to the 340 days (!) of the steam-mill. The steam-mill was hypothetically located in Metz, and the water-power 6 km outside.

(C) England. Steam costs are as for Manchester in the 1830s, summarized in Table 4.11. As with the other countries, the interest rate is not calculated. The water-power is intended to reflect costs per horsepower at Styal Mill (Quarry Bank), Wilmslow. The wheel and gearing were rated at £26. 16s. 8d. per horsepower (£2,400 for 90 horsepower); the same as at Bakewell, Notts, in 1830 (Quarry Bank MS. C5/1/1/2, Nov. 1831; Chapman states £1,600 for 60 horsepower at Bakewell). This was charged at 8¾ per cent (C5/8/16/2, etc.) and the foundations at 7 per cent (ibid.). Tunnel and wheel-race were valued at £5,000 in 1831, the remainder presumably being weirs, etc. Thomas Benton assessed the

## The Social Savings

value of the water-power at £14,000 (@ 5 per cent) in 1841, and though he makes an addition for the tunnel of £2,000 it is possible he includes weirs, etc. in this item, so it may be too large. Rent on the cottages was put at £168 in 1831, to which was added the apprentices' house (£500 @ 5 per cent)—the result is very close to the equivalent item at Bakewell (£202 p.a.). Variable costs are taken from Edwards, op.cit., p. 210, his figure being for Styal in 1847; I have deducted rents. Transportation has been omitted. A much more serious omission—which operates in the other direction—is that the Gregs confessed themselves to be the lowest-paying mills in the country (P.P. 1833: XX, evidence of R. H. Greg in App.D2; F. Collier, *The Family Economy of the Working Classes in the Cotton Industry, 1784–1833*, ed. R. S. Fitton, Manchester, 1965).

## APPENDIX 6.2

*Technical improvement in the harnessing of water, 1760–1850*

I NOTED on page 140 that speeds at which water-wheels turned were increased to the point that by 1850 best practice was considered to be about double the velocity recommended for 1760. This of course could not apply to undershot wheels, whose speed was regulated by the current. For breast and overshot wheels, speeds could be regulated by a governor, which, though popularly associated with the steam-engine and James Watt, had in fact earlier in the eighteenth century found use on water-wheels and in windmills. At Papplewick the centrifugal governor was being used on the large wheels erected there in the late 1770s and early 1780s, before it became the first locality to apply the steam-engine to any aspect of cloth production.[1] Also employed there was some iron gearing taking the drive off the wheel, in lieu of the traditional wood. In 1769 Smeaton had experimented with the use of iron for axles, though not entirely successfully.[2]

In the early nineteenth century the use of iron became more 'structural' than simply axles and gearing. The arms came to be built of cast iron, and the shroudings (the circular framework boxing in the buckets).[3] (The 1800 wheel at Quarry Bank has been claimed as the first wheel in the country to be built mainly of iron.[4]) T. C. Hewes replaced the heavy cast iron arms with light cross-rods of wrought iron, by reorganizing the drive.[5] By 1850 it was noted that wood was fast disappearing in the construction of buckets.[6] Axles for small wheels were of wrought iron, and for large of hollow cast iron.[7] Fairbairn's 36-foot wheels were 67 per cent cast iron, 32 per cent wrought, and the rest ($2\frac{1}{2}$ cwt) brass; the 16-foot ones were 73 per cent, 26 per cent, and 1 per cent respectively.[8]

---

[1] Smeaton, *Experimental Enquiry* . . ., pp. 31 ff.
[2] Hills, *Power in the Industrial Revolution*, p. 99; Stowers, in Singer *et al.*, *History of Technology*.
[3] Morshead (1856); Fairbairn *Mills and Millwork*.
[4] Lazenby, *Social and Economic History of Styal*. Note also the wheel mentioned by Svedenstierna at Cyfarthfa works (*Tour of Great Britain 1802–3*, new edn., Newton Abbot, 1973).
[5] Tann (1970); Hills, *Power in the Industrial Revolution;* Fairbairn, op. cit. Note also the suspension wheel invented by Hewes, e.g. Fairbairn, pp. 117, 150.
[6] *Engineer and Machinist's Assistant*, p. 213.
[7] Morshead (1856).
[8] *Engineer and Machinist's Assistant*, p. 244.

It is conceivable that these developments prevented marginal costs from rising much rather than leading to any substantial reduction in costs. However, other advances could well have lowered costs by raising technical efficiency. The design of the buckets (or floats on undershot wheels) was shown to be crucial by French engineers. These ideas were taken up in Britain, e.g. by Robison.[9] The escape of air from the buckets as the water rushed in was shown to be affected by the design. An early remedy was simply to bore holes in the starts of the buckets, though the loss of water obviously meant this was inefficient. From the 1820s William Fairbairn devoted much of his energy to resolving this difficulty. His first proposal was the use of 'inner buckets' to carry the air away. Later he employed independent soles to prevent any serious loss in strength arising out of the advance. By the sum of these accomplishments he reckoned to gain one-quarter in power.[10]

Out of the combination of (often wrong-headed) science and engineering peculiar to the French in this period evolved much improved shapes of floats, minimizing turbulence especially for undershot wheels. The name most frequently associated with these developments was Poncelet.[11] He was concerned about the loss of head of water suffered from the centrifugal force generated by the rotation of the wheel itself. (Smeaton's decision to slow down the velocity of the circumference arose out of this inadequacy of filling the buckets with water, as they slipped too quickly past the head.) Poncelet therefore designed curved floats to accept a smooth and rapid inflow of water without setting up counteracting forces.

Arthur Morin, a leading French civil engineer, wrote, "Les roues à aubes courbes, établies d'après les règles posées par M. Poncelet, utilisent 0.60 du travail moteur lorsque la chute totale est de 1m.50 et au dessous, et 0.50 à 0.55 pour les chutes plus grandes. Elles peuvent marcher à une vitesse considérable, ce qui permet de faire faire à la roue un plus grand nombre de tours par minute que dans les autres systèmes."[12]

These efficiency ratings seem valid—Rankine stated that the available efficiency of the Poncelet wheel had not been found to exceed 60 per cent, though theoretically they could be as efficient as overshot wheels.[13] As Morin states, they were particularly effective for low falls and hence for undershot wheels, the inefficient

[9] In *Encyclopaedia Britannica*, 1815 edn.; also Hills, *Power in the Industrial Revolution*, pp. 100, 112.
[10] *Engineer and Machinist's Assistant*, p. 217.
[11] e.g. J. V. Poncelet, *Introduction à la mécanique industrielle*, 3ᵉ édn, Paris, 1870.
[12] A. J. Morin, *Aide-mémoire de mécanique pratique*, 3ᵉ édn, Paris, 1843, pp. 225 ff.
[13] Rankine, *Manual of the Steam Engine*, p. 177; Fairbairn, p. 153.

type necessitated by the sluggishness of many British streams—here they nearly doubled efficiency. It is hardly surprising to find Fairbairn stating that 'A curvilinear form for the buckets has been adopted [since Smeaton's time], the sheet iron of which they are composed affording great facility for being moulded into the required shape.'[14] This sheet iron vane in place of the old wooden bucket seems to have been a principal reason permitting the rise in speeds of rotation mentioned earlier.

Poncelet's wheel, combining impulse with reaction, verged on yet another sphere of developments, namely horizontal wheels working by reaction. The turbine with its curved vanes was the logical successor, in effect applying similar hydrodynamical principles to the horizontal water-wheel (so long out of favour in England). The first working turbine was one erected by M. Fourneyron in Alsace in 1827; by 1834 he had a turbine rated at 7–8 horsepower driving an ironworks, though this was only a prelude to the great strides taken thereafter.[15] At St. Blasien in the Black Forest a 26-inch wheel was rated at 40 to 65 horsepower, and drove 8000 throstle spindles.[16] Arago found that a turbine designed to work with a head of 6 ft 6 in was still working at 58 per cent efficiency with as little as 10 in of head[17]—indeed this ability to cope with a highly variable water supply was one of the main attractions of the turbine. As early as 1838 French experts claimed that turbines were at least as efficient as ordinary water-wheels.[18] The Report of the Juries at the 1851 Exhibition stated the advantages of a turbine entered by M. Fontaine Baron to be:

1st. It occupies a small space; 2nd. Turning very rapidly, it may, when used for grinding flour, be made to communicate the motion directly to the millstones; 3rd. It works equally well under great and small falls of water; 4th. It yields, when properly constructed, and with the supply of water for which it was constructed, a useful effect of 68 to 70%, being an efficiency as high as any other hydraulic machine; 5th. The same wheel may be made to work at very different velocities, without materially altering its useful effect.

This is quoted with apparent approval by Fairbairn.[19]

Unlike the overshot water-wheel, the turbine also worked satisfactorily when 'drowned'. In spite of its great versatility the turbine was still relatively most effective for high falls of water without any

[14] Fairbairn, p. 112.   [15] Glynn (1853).
[16] Fairbairn, p. 159; Usher, *History of Mechanical Inventions*.
[17] Glynn (1853).
[18] A. J. Morin, *Expériences sur les roues hydrauliques à axe vertical appelées turbines*, Paris, 1838; and *Aide-mémoire*.
[19] Fairbairn, pp. 156–7.

## The Social Savings

great volume of water, the kind of supply encountered at St. Blasien. Because of this the so-called high-pressure turbine found comparatively little use in the manufacturing districts of England at this stage, as water resources were not of this kind. Fairbairn was cautious of approving of them except for high falls: 'Certain advantages, it must be admitted, are obtained by the turbine in certain localities under certain conditions; but it is very doubtful whether they are equal, either on the score of expense or ultimate efficiency, to well-constructed water-wheels.'[20]

The Gregs made notes on the turbine, probably in the 1840s:

Water wheel on Fournereau's [sic] plan of 'Turbix'. Netts 50% of fall on all heights of fall. Common wheels 30% on falls 1 to 6'. Cost of both kinds same up to 6'. Common kinds above 6' nett 70–80%. Above 60' fall, common wheels become impracticable. Fournereau's plan does for 60 or 600 equally well. At St. Blaize in the Black forest, 65 horse-power, 8,000 throstles, 60' fall. Wheel 2' wide × 2" deep, proved by long and repeated trials—65 horse-power cost £100-0-0 including masonry.[21]

At this price, turbines ought to have been far more cost-effective on high falls even if their efficiency was a little lower than ordinary breast or overshot wheels. While topography dictated otherwise in most parts of England, some use was found for such wheels in Scotland, where the firm of Messrs. Whitelaw & Stirrat built their reaction wheels (sometimes known as the Scotch turbine). One such reaction wheel was erected at Greenock for £300 when it was estimated that an overshot wheel would have cost £1,500—with an efficiency reputedly as high at 90 per cent.[22] A 5-horsepower reaction wheel on this plan would have cost some £70 for a 30-foot fall in the mid-1850s (£60 if the fall was 100 foot), while supplementary costs were lower than for wheels since the water was *piped* into the centre of the turbine (to be discharged outwards)—so that capital costs were indeed similar to small wheels, *cet. par.*

An alternative to the turbine was the vortex wheel, which reversed the action by taking water from the outside and sucking it into the centre. Such wheels were first marketed in the U.K. by James Thomson. Williamson & Bros. of Kendal were said to be the only manufacturers of vortex machines in Britain by 1861.[23] By 1881 they had built 439 such machines plus 63 waterwheels.[24] Thomson's wheel at Ballysillan was designed to work at $292\frac{1}{2}$ rev/min, while Fourneyron's turbine at St. Blasien operated at

---

[20] Ibid., p. 174.   [21] Quarry Bank MS. C5/8/24.
[22] Morshead (1856).   [23] Fairbairn, p. 163.
[24] J. D. Marshall and M. Davies-Shiel, *The Industrial Archaeology of the Lake Counties*, Newton Abbot, 1969, p. 58.

168                    *The Evaluation*

2200 to 2300 rev/min; Joseph Glynn thought the turbine superior to the vortex.[25]

One further development deserves some notice, and I single it out here because it preceded Watt's patents. This was the water-pressure engine.

> Engines of this description were first employed in Cornwall about 80 years ago, but were not much used, owing to the difficulty which the incompressible nature of the water occasioned in turning the centres, the flow of water to and from the cylinder being then for an instant stopped. In the hydraulic engines at present constructed this is remedied by an air-vessel placed at the foot of the column of water, while two small plungers, each working in a stuffing-box at either end of the cylinder, and pressed on by a spiral steel spring, equalise the pressure on the piston, and enable the engine to pass its centres as smoothly as any steam engine. Engines on this principle are frequently employed in mines and factories (at Fowey Consols, for example, there are three of them).[26]

William Westgarth of Northumberland is usually credited with erecting the first in England, in 1765 for draining a mine.[27] However, the basic principle of using the weight of water for reciprocating motion (rather than rotative, as on the wheel) is age-old.[28] At its simplest the water-pressure engine was not very efficient—its motion was extremely slow and friction high. However, the water could be utilized under some pressure, and this engine was using a forerunner of the expansive principle before James Watt patented it for steam.[29]

Like the turbine, it had special advantages with high falls of water; it was reckoned to be cheaper than a series of overshot wheels. It did not have to be placed at the foot of a fall, since a clear fall of water *below* was just as effective in drawing its piston. In the 1850s a 4-horsepower water-pressure engine was estimated to cost £85 and a 6-horsepower £105 for a fall of 30 ft to 50 ft; £80 and £100 respectively if the fall were over 50 ft (since the cylinder diameter could be reduced with a higher fall). Erection costs were very light—about £10 in these sizes.[30] An engine of 168 horsepower erected in 1842 at the Alport Mines, near Bakewell, Derbyshire, was claimed

---

[25] Fairbairn, pp. 161, 281; Glynn (1853). Fairbairn thought that a turbine working at 4 to 500 rev/min was going too fast to work pumps efficiently (pp. 178–9).
[26] Morshead (1856). See Farey, *Treatise* ii. 169–73 for further descriptions.
[27] e.g. by Smeaton. Similar engines were earlier in use on the Continent, e.g. at Schemnitz (Cardwell, *From Watt to Clausius*).
[28] See G. Downs-Rose and W. S. Harvey, 'Water-Bucket Pumps and the Wanlockhead Engine', *Industrial Archaeology*, x, May 1973, for a clear discussion of the evolution.
[29] Cardwell, *From Watt to Clausius*.    [30] Morshead (1856).

## The Social Savings 169

by its engineer never to have cost more than £12 a year to run.[31] In 1804 Richard Trevithick had installed a double-acting water-pressure engine working at 75 lb p.s.i. (having built his first such in 1798). With a cylinder of 25" × 10', it was calculated that the engine exerted a total force of 16 tons, compared to only 14 tons by Watt's famous 63-inch steam-engine at Dolcoath Mine, Cornwall.[32]

Water-pressure engines were best for small amounts of power used somewhat irregularly; in this their operating cost was estimated in the early 1850s at 6*d.* per horsepower per hour.[33] However, the drive was often rather erratic. In some cases a water-wheel had to be interposed between the engine and the machinery it drove, and this jeopardized its economy.[34] It is no surprise that it gave way later to the turbine.

[31] Glynn (1853).
[32] Trevithick, *Life of Richard Trevithick*, vol. i, pp. 84–6. He may be exaggerating.
[33] Glynn (1853). A list of several large water pressure engines is given in Phillips and Darlington, *Records* (Darlington was probably the engineer referred to in n. 31).
[34] Fairbairn, p. 178.

## APPENDIX 6.3

*Water-power and the elements*

MUCH HAS been heard, then and since, of how water-power was incapacitated in competing with steam through being subject to the whims of natural forces, such as droughts, floods, and freezing. Yet the screeds of complaints heard by one Parliamentary Commission after another do not necessarily imply that water-power was for ever doomed.

For one thing, if the engine had not been invented, then all firms using water-power would have been affected (though naturally in varying proportions and times); hence it is conceivable that the penalties of slightly erratic working would not have been so great. More speculatively, one might suppose that mill-owners might have been able to negotiate insurance schemes to cover themselves in the event of exceptional losses; these costs could then be appended to the social savings. As it was, individual mill-owners of course complained bitterly whenever they struck a poor year—but as in agriculture this was a fact of existence.

Even in the situation where the (atmospheric) engine had been invented, it is important to distinguish between the individual firm and the industry. If adverse weather were exactly consistent from one year to the next, then by the theory already expounded one would suppose that rentals on water supply would adjust themselves accordingly. Naturally English weather is not possible to forecast as precisely as this, but one can invoke the central limit theorem and suppose a normal distribution, even to describe the English winter.

Ice was particularly a problem in north-east Scotland. A 'gorge' of ice in the river at Aberdeen sometimes caused stoppages for several days at a time. Gordon, Barron & Co. of Old Machar, Aberdeenshire, lost 8·6 per cent of their 3900 annual hours in the particularly cold year of 1826, though they worked 95 per cent of the time on average from 1828 to 1832[1] (these figures include heavy losses from flooding, etc., and 3900 hours is a much longer year than further south). Losses from freezing are unlikely to have been anywhere near as great elsewhere in the U.K.

Floods had the effect of 'drowning' the wheel, and the overshot wheel especially became awkward to manage (this was one of Fairbairn's main reasons for preferring the high breast wheel).

[1] P.P. 1834: XX, pp. 18–19.

## The Social Savings

Samuel Greg at Styal noted that he occasionally lost a day or a day and a half through floods.[2] On the other side of the Pennines, Benjamin Gott's mill at Armley (Leeds) claimed that there were records of having lost as much as a week, though it was unusual to lose more than a day or two.[3] Thomas Cook at Dewsbury thought he might lose 15 to 20 days a year,[4] though this seems excessive unless he did have an overshot wheel.

More serious in frequency and result was the loss incurred by low water and droughts. In the thirteen years up to 1833 the Gregs lost a total of 728 hours from droughts at Quarry Bank, excluding time lost that was made up in the same week—say 5 per cent loss a year if the latter were included, though they thought this was quite a good stream.[5] The problem was often exacerbated by the existence of several wheels on the same stream. Unless one was 'head of the river' one was at the mercy of those higher up for releasing the limited supplies of water. In practice, it became increasingly common on less dependable streams to supplement the water-power with an auxiliary steam-engine. The widespread nature of this solution in the 1830s suggests that it must still have been economic to incur the extra capital and rental costs on water for the sake of the variable-cost savings on engines. Of those textile firms responding to a Parliamentary Commission in 1833, thirty stated they were using both steam and water, whereas only twenty-six were using water alone.[6] Most firms outside the immediate Pennine valleys of Derbyshire, Nottinghamshire, Lancashire, Yorkshire, and Cheshire adopted auxiliary steam-power. Near Bingley a steam-engine was installed because of the apparently catastrophic loss of 32 days of work in one particular year.[7] But even this is not much over 10 per cent of total time. Given that capital costs of steam-power are about one-third of total power costs in an average-sized mill, and power in turn is under 15 per cent of total value-added,[8] it would presumably pay to install an auxiliary steam-engine whenever the mean shortfall of water involved a loss in cotton yarn production of 5 per cent or more. The threshold point would in fact differ among industries, not only because power formed a differing proportion of total value-added, but also because some industries could react more elastically to water short-comings. In woollen-mills,

---

[2] Ibid., p. 780.
[3] Ibid., p. 808. All these figures are probably upward-biased to impress the Commissioners.
[4] Ibid., p. 418.
[5] P.P. 1833: XX. In 1856 they computed that a doubling of the size of the dam would add 487½ hours of power p.a. (Quarry Bank MS. C5/3/2).
[6] P.P. 1834: XX.   [7] P.P. 1833: XX.   [8] See Ch. 7.

for instance, it was normal to keep the spinning going when power dropped to a quarter or less of the usual level, whilst activities such as fulling were relegated to better conditions.[9]

If steam-engines had not in fact been invented, it requires little imagination to suppose that more orderly efforts to regulate water supply would have been undertaken. Baines describes a project for the Irwell basin, costing £59 000 and designed to permit 6600 horsepower to be controlled.[10] A decade later Fairbairn's associate J. F. Bateman was collecting rainfall and evaporation data with an eye to ambitious hydraulic schemes for Manchester.[11] Bateman also undertook two major reservoirs on the River Bann catchment in Ulster. Perhaps the most effective for industry of those schemes actually implemented was Shaw's for Greenock in 1824. With a total head of water of 512 feet it was designed to cope with all the water-power needs of Glasgow; it was supposed to have sufficient storage for six months during the dry season.[12] Other schemes are known to have been planned in the mid-nineteenth century.[13]

The complaints of water-mill owners that were being vented especially loudly in the 1830s reflect not so much a grave level of disability in their competition with steam-mills—eventually rents on water should have taken care of that—as a squeeze on costs they feared would come as the institutional framework changed. Many were in the habit of night-work and work at odd hours to compensate for the irregularity of their water supply—especially those low in the pecking order on well-exploited streams.[14] The proposed Factory Acts would prevent them working so irregularly. It was this sudden change in the legal environment they all feared. By theory the rents would in due course readjust to suit these circumstances; but caught as many were on leases of 20 to 100 years this was rather cold comfort.

Memorandum as to the effect of the Proposed Prohibition of Night Work upon the Business of William Pirie & Co., Woollen Manufacturers, Gordon's Mills, near Aberdeen. The gentlemen concerned in this company are the lessees of a valuable waterfall and other property upon the banks of the

[9] P.P. 1834: XX. Cf. also Raistrick, *Dynasty of Ironfounders*, pp. 107–10, showing how production at Coalbrookdale was geared to the available supply of water-power.
[10] Baines, *History of the Cotton Manufacture*, p. 86 n.
[11] Fairbairn, *Mills and Millwork*, pp. 70 ff.
[12] Kane reckoned the supply at Greenock at 2000 NHP (*Industrial Resources of Ireland*, pp. 94–7). Also Glynn (1853), Wilson (1954).
[13] e.g. that noted by Wilson (1954) for Kendal in the 1840s. It was held over then scrapped c. 1850, presumably because of the cheaper cost of steam-power at the later date (see Ch. 4).
[14] P.P. 1834: XX, *passim*; Hills, *Power in the Industrial Revolution*, Ch. VI.

## The Social Savings

river Don, for the long period of 100 years at a very heavy rent. When the company took upon themselves the burden of this rent, when they erected expensive buildings, and filled them with machinery, excavated their reservoirs and water courses, and embarked their whole capital in the undertaking, they fully calculated upon being obliged at certain seasons, when the water should be low, to have recourse to night-work; and it is needless to say that no restrictive law was in operation at the period, and that none such could, by any foresight, have been contemplated. The experience of several years has shown that the average extent of night-work may be about 3 months in the year, of about one-third part of the whole machinery which is attached to, and driven by, one particular wheel; and should the prohibition alluded to be carried into effect, this wheel, and the machinery attached thereto, would be thrown utterly useless upon their hands.[15]

Thus this particular firm predicted a rise in overhead costs (treating rents as a medium-term overhead) of about one-twelfth; though, as previously stated, this was in an area with an extremely harsh climate. It would be appropriate to include this in an estimate of the social savings which gave full weight to *social* factors—here, the extent to which the steam-engine was responsible for curtailing hours of work. In this chapter, however, I have limited myself to more narrowly *economic* factors. Related aspects of the wider social issues are taken up in Chapters 7 and 11.

[15] P.P. 1834: XX, p. 24.

# PART III. STUDIES IN THE APPLICATION OF STEAM-POWER

## 7. Steam-power and the Cotton Industry

ON ANY reckoning, the cotton industry has to be accounted as one of the more spectacular phenomena of the Industrial Revolution in Britain. Schumpeter no doubt went to extremes in asserting that 'English industrial history [from 1787 to 1842] can . . . be almost resolved into the history of a single industry'.[1] 'Whoever says Industrial Revolution says cotton', says Hobsbawm.[2] Rostow in the first edition of his *Stages of Growth* described cotton as the original leading sector of the first Industrial Revolution.[3] Even if these views are wildly overstated, there is a strong case for placing primary stress on the forward linkages from steam to cotton.

I intend to establish how much power was applied to this sector and therefore what scope there was for exerting any impact. Even this, as it turns out, has to be taken further than a superficial comparison with the growth of cotton production; for changes in the quality of that production emerge as quite fundamental to understanding the diffusion of power and machinery. And, as is obvious, it is only by detailed study of the machines in depth that the paths along which the linkages ran can be understood.

[1] J. A. Schumpeter, *Business Cycles*, vol. i, New York, 1939, pp. 270–1.
[2] E. J. Hobsbawm, *Industry and Empire* (Pelican Economic History of Britain, vol. 3), London, 1969, p. 56.
[3] Rostow, *The Stages of Economic Growth*, pp. 53–4.

## STEAM-POWER AND COTTON TO 1800

In Chapter 6 it was stated that the early developments in the field of spinning by means of roller-drawing—developments associated with the names first of Paul and Wyatt, and later of Arkwright—were hitched to animals to provide the power. Conversely, Hargreaves's jenny and Crompton's mule were both planned to be manually operated; their inventors wanted to bolster labour productivity without bringing about technological unemployment. In the form in which it spread widely in the 1780s (possibly improved by Highs), the jenny normally came equipped with 60 to 80 spindles, and this size could still be driven by a woman. Some larger jennies of around 120 spindles were actually built in the 1790s, but they required a man to power them; and since men usually cost more than half as much again as women to employ, this solution did not enjoy great popularity. Thus so long as low-wage female labour could easily be found there was not much temptation to search for other sources of power.[4]

The mule, however, soon benefited from having artificial power applied. By the late 1780s mules were commonly being built of up to 144 spindles each, and were manually operated one at a time. Minor technical advances, e.g. in the rollers, permitted one man to drive more spindles in the mid-1790s, but still on one mule. With the application of water-power—conventionally dated to its use by William Kelly of Lanark in 1790—the number of spindles could rise still further. But much more important was that water-power could allow installation of the mules in pairs; one in front of the spinner and one behind.

---

[4] The traditional arguments about 'challenge and response' may best be seen in this light. Because weaving (say) speeded up in the wake of Kay's flying shuttle and so required greater quantities of yarn, there is no overpowering reason why spinning should have had to be improved, since more yarn could have been spun simply by recruiting more spinners. True, transport costs would have risen, but these were never a large component of the costs of cloth. The problem seems to have been that adult males wove and females spun; so when the former speeded up, there were not enough women around to keep them all supplied—men could have converted from weaving to spinning, but only at the cost of a substantial rise in spinning costs. Thus the challenge lay as much in the structure of the family as in the structure of production. See N. J. Smelser, *Social Change in the Industrial Revolution*, London, 1959; also M. M. Edwards and R. Lloyd-Jones, 'N. J. Smelser and the Cotton Factory Family: A Reassessment', in Harte and Ponting, *Textile History and Economic History*.

Power from the water-wheel was then used to push the carriage out, while simultaneously the spinner, facing the opposite mule, backed it off, returned its carriage, and wound the yarn on. When that was finished, he turned around and repeated these processes on the first mule, now out at full stretch. So from the first this doubling-up of the mules was adopted to compensate for the larger costs of power. Steam-power was applied in the same way later in the 1790s.

Richard Arkwright himself quickly turned from horses to water in order to power his water-frame—from which, of course, the name is derived. According to widely held convention, water-frames required 1 horsepower to drive about 100 spindles (or 80 when Arkwright began).[5] By contrast, the mule was said to need only one-fifth of 1 horsepower to propel as many (mule) spindles on coarse yarns.[6] This meant that the additional power costs for mules were much lower than for frames of similar size, provided that both were working at the same speed.

Unlike the mule, most frames continued to be driven by water well into the nineteenth century. For this one may conjecture several reasons.

(i) The usual argument found in recent studies is that water was cheaper than steam.[7] This is difficult to accept in its entirety on a number of grounds. I assumed in the last chapter that in 1800 the average costs of steam- and water-power for cotton production were identical. Even in 1800 the farther-flung regions could probably still offer the more adventurous entrepreneur the opportunity for cheap pickings in the way of water resources.[8] Market imperfections (i.e. restrictions on the free flow of capital, skilled labour, information, or whatever)

---

[5] Hills, *Power in the Industrial Revolution*, pp. 197, 202; R. Hills and A. J. Pacey, 'The Measurement of Power in Early Steam-Driven Textile Mills', *Technology and Culture*, XIII, no. 1, 1972.

[6] i.e. 500 mule spindles per horsepower on coarse yarns and 1000 for fine yarns, e.g. Brunton, *Compendium of Mechanics*.

[7] 'To retain two separate power sources must have been very expensive and again seems to indicate that steam engines were more expensive than water-wheels . . .' (Hills, *Power in the Industrial Revolution*, pp. 102–3). Also Tupling, *Economic History of Rossendale*, p. 205, and Rees (*Cyclopaedia*, 'Manufacture of Cotton') : 'The trade of Manchester is chiefly mule spinning, whilst the water twist is mostly spun in the country by water mills because the great power it requires is too expensive for steam engines' (N.S. iii, 1970).

[8] Unwin, *Samuel Oldknow and the Arkwrights*, pp. 123–8. Cf. p. 138.

could have arrested the workings of the Ricardian model of rents, somewhere short of the external margin. Yet it is unlikely that this can explain the prevalence of water-power in spinning with the frame as late as the 1840s and 1850s. By then much of the spinning on the throstle (successor to Arkwright's frame) was conducted, using water-power, in areas where water resources were overtaxed.

(ii) Even if power costs bulked less with mules than frames, it would still be worth *something* for a mule-spinner to go over to cheaper power—if water was in fact cheaper. As he did not usually do so, it may be that the *composition* of costs was more influential than the relative totals. As is well known, water-power had high fixed and low variable costs compared to steam (see App. 6.1). It was thus more likely to attract the prospective entrepreneur who envisaged a relatively low return from any further demands placed on the power source. In the 1820s and 1830s it was a common practice to purchase mills with what Matthews described as 'semi-reserve capacity';[9] i.e. only partially filled with machinery and possibly power supply at the beginning, but extending these rather than the buildings as a whole in the fullness of time. If power requirements had to be augmented in this way, or alternatively by running the machinery faster—the significance of which will become apparent below—then the more power-intensive technique would increasingly favour the use of water-power, with its low marginal costs.

(iii) There may have been externalities arising out of the employment of water-power for frames, or steam for mules. The latter seems very plausible. The spinning of fine cotton yarns on mules required a highly heated mill—several Parliamentary Commissioners in the 1830s found the overheating of fine spinning mills unbearable for the physical well-being of the employees. Dr. Tann[10] has investigated a debate over the economy of steam-heating at the beginning of the century, and has defined a size of firm above which steam-heating became economical in terms very similar to the threshold analysis adopted in Chapter 4 of this book. Her discussion was not

---

[9] R. C. O. Matthews, *A Study in Trade-Cycle History*, Cambridge, 1954, p. 133.
[10] Jennifer Tann, 'Fuel Saving in the Process Industries during the Industrial Revolution: A Study in Technological Diffusion', *Business Hist.* xv, 1973.

specifically directed at textile industries, but it can easily be reworded to show up this situation. The fact of already having a steam-engine adds to the economies of scale of the kind with which Tann is most concerned. The cost disadvantages to a water-mill of having to install steam-boilers for heating purposes alone are well illustrated by the figures quoted for costs of steam- and water-power in the U.S.A., in Appendix 6.1.

From the data collected by Lord, Musson and Robinson, and others, it seems likely that by 1800 somewhere between 1500 and 2000 horsepower in the form of steam-engines was used to turn textile machinery.[11] Under some reasonably charitable assumptions steam-power could have been involved in the spinning of up to one-quarter of the cotton processed in that year.[12]

Hargreaves's original jenny was not only unsuited to the spinning of warps, but in addition suffered a sharp decline in cost-efficiency when used for anything above quite coarse grades of weft.[13] The water-frame of Arkwright solved the problem of a strong warp, through the principle of roller-drawing, but proved equally expensive for producing finer qualities. At the same time there is some evidence that demand was shifting towards higher grades of cotton cloth; away from the old coarse fustians and calicoes into finer calicoes and muslins.[14] At first these demands could be met only by importing fine cloths from India, or by the laborious hand-spinning of

[11] Lord quotes 84 engines totalling 1382 horsepower supplied by Boulton & Watt to cotton-mills from 1775 to 1800 (Lord, *Capital and Steam Power*, pp. 167–71); Musson and Robinson did not challenge his subtotal of 55 engines in Lancashire, and in any case the company made a separate list of its sun-and-planet engines, which has survived as a check. The pirate Bateman & Sherratt engines include 2 of 93 horsepower altogether plus 1 of unknown size for cotton-spinning; the same firm also built ordinary atmospheric engines for cotton firms. Other builders cannot be ignored, e.g. the engines of Francis Thompson in Manchester in 1794 (Musson and Robinson, *Science and Technology*, pp. 415–16; Farey, *Treatise* i. 662; W. H. Chaloner, 'The Cheshire Activities of Matthew Boulton and James Watt . . .', *Trans. Lancs & Cheshire Antiq. Soc.*, 1949; Chapman (1967)).
[12] The assumptions are: (i) that two-thirds of the steam-power was used for mule spinning, (ii) that on average each horsepower in 1800 drove 750 mule spindles, (iii) that average output per spindle in 1800 was $1\frac{3}{4}$ hanks of 40s per day. Total annual output from steam sources would then be about 13 m. lb of yarn. Total raw cotton imports in 1800 were 56 m. lb, of which about 11 per cent would have been lost in spinning.
[13] C. Aspin with S. D. Chapman, *James Hargreaves and the Spinning Jenny*, Preston, 1964, pp. 42 ff.
[14] Edwards (1967), Ch. 3.

finer yarns. It was Crompton's mule that broke through the technico-economic barrier to permit the economical spinning of fine yarns by machine methods.

The effect in economic terms may be traced in the prices of various qualities of weft and warp over the last quarter of the eighteenth century and beyond, as shown in Table 7.1.[15]

The impact of the mule is reflected both in the appearance of high-quality yarns on the market at all, and in the rapid decline in their price. According to Kennedy there were no more than 1000 mule spindles in existence before 1785;[16] thereafter the commercial crisis of 1788 is the proximate cause of prices of fine yarns tumbling, but these low prices then get 'locked in', despite generally inflationary tendencies in the economy elsewhere. Unfortunately, it is not possible to tell from these price trends alone whether the mechanical improvements described were most important in explaining the fall in prices, or whether cheaper sources of best-quality raw cotton should be given primacy.

There is some reason to believe that the influx of top-quality cotton wool cannot account for the price trends shown in Table 7.1.[17] If this is so, then any evaluation of the historical role of the steam-engine ought to give some weight to changes in the quality as well as just the quantity of output.

[15] Sources: 1778–88: from an unpublished MS. by Patrick Colquhoun, discovered by the author in the Baker Library, Cambridge, Mass. (Baker MS. E-46); hereafter *Diary*. 1795, 1799: H. Ashworth, *Cotton: Its Cultivation, Manufacture, and Uses*, Manchester, 1858, quoting John Simpson & Co.'s prices for machine twist at six months' credit (5 per cent discount). 1798: James Dymock, *The Manufacturer's Assistant*, Glasgow, 1798. 1810: R. S. Fitton and A. P. Wadsworth, *The Strutts and the Arkwrights, 1758–1830*, Manchester, 1958, plate opp. p. 323. 1816: S. C. Pigott, *Hollins: A Study of Industry 1784–1949*, Nottingham 1949, p. 42. Baines, *History of the Cotton Manufacture*, p. 357, has a slightly different series for prices of 100s wefts from 1786 to 1807. Cf. also G. Lee, 'To the Committee at Manchester, meeting to oppose the exclusive Trade of the East India Company' (B.M. Add. MS. 12131): '. . . In the year 1782, Cotton Twist by Sir Richard Arkwright's invention, which was the precursor and parent of the subsequent Improvements, exceeded the cost of the raw material, 20s. per pound for No. 60. It now exceeds it, by the Mule, only 1s. 6d. per pound . . .'; also Unwin, op. cit., Ch. 5.

[16] J. Kennedy, 'A Brief Memoir of Samuel Crompton', *Mem. Lit. & Phil. Soc. of Manchester*, 2nd Series, vol. V, 1831.

[17] The raw cotton for fine yarns was imported first in limited quantities from Brazil, the East Indies, and sometimes Réunion('Bourbon'); later the famous Sea Island long-staple type dominated, but it was not until the surge of the latter in 1798 or thereabouts that supplies really became generous. See Edwards (1967), Ch. 5, and App. C.

Steam-power and the Cotton Industry

TABLE 7.1
Price per lb of cotton wefts and warps, according to count

(A) Wefts

| | Count | | | | | | | |
|---|---|---|---|---|---|---|---|---|
| | 18 | 24 | 40 | 60 | 80 | 100 | 150 | 200 |
| 1778 | 2s. 10d. | 3s. 8d. | 6s. 2d. | 10s. 7d. | 18s. 0d. | — | — | — |
| 1779 | — | — | 6s. 2d. | 10s. 7d. | 18s. 0d. | — | — | — |
| 1780 | 2s. 11d. | 3s. 8d. | 6s. 2d. | 10s. 7d. | 18s. 0d. | — | — | — |
| 1781 | — | — | 7s. 3d. | 12s. 6d. | 26s. 8d. | 37s. 6d. | — | — |
| 1782 | — | — | 7s. 3d. | 12s. 6d. | 26s. 8d. | 37s. 6d. | — | — |
| 1783 | — | — | 6s. 5d. | 12s. 3d. | 23s. 5d. | 36s. 9d. | — | — |
| 1784 | 2s. 8d. | 3s. 2d. | 6s. 3d. | 10s. 0d. | 21s. 8d. | 36s. 9d. | — | — |
| 1785 | 2s. 10d. | 3s. 6d. | 5s. 10d. | 10s. 3d. | 22s. 1d. | 37s. 0d. | — | — |
| 1786 | 3s. 0d. | 3s. 6d. | 5s. 2d. | 10s. 0d. | 21s. 6d. | 37s. 4d. | — | — |
| 1787 | 3s. 4d. | 4s. 0d. | 5s. 11d. | 10s. 0d. | 16s. 0d. | 26s. 6d. | — | — |
| 1788 | 2s. 10d. | 3s. 4d. | 4s. 9d. | 6s. 10d. | 12s. 10d. | 22s. 0d. | 58s. 8d. | — |
| 1795 | — | — | — | — | — | — | — | 51s. 0d. |
| 1798[b] | — | — | 5s. 10d. | 7s. 6d. | 9s. 4d. | 12s. 1d. | — | — |
| 1798[i] | — | — | — | — | — | 16s. 6d. | 33s. 2d. | 65s. 8d. |

(B) Warps

| | Count | | | | | | | |
|---|---|---|---|---|---|---|---|---|
| | 15 | 25 | 40 | 60 | 80 | 100 | 150 | 200 |
| 1782 | 4s. 11d. | 6s. 1d. | 10s. 8d. | — | — | — | — | — |
| 1783 | 4s. 1d. | 5s. 4d. | 9s. 11d. | — | — | — | — | — |
| 1784 | 3s. 6d. | 4s. 7d. | 9s. 11d. | 21s. 5d. | 43s. 6d. | — | — | — |
| 1785 | 3s. 6d. | 4s. 6d. | 9s. 9d. | 23s. 7d. | 42s. 6d. | 63s. 0d. | — | — |
| 1786 | — | — | — | 21s. 1d. | 37s. 0d. | 57s. 6d. | — | — |
| 1787 | — | 5s. 3d. | 9s. 3d. | 18s. 4d. | 35s. 1d. | 52s. 0d. | — | — |
| 1788 | 3s. 4d. | 4s. 1½d. | 6s. 10½d. | 14s. 0d. | 20s. 11d. | 28s. 6d. | — | — |
| 1798[b] | — | — | 6s. 6d. | 8s. 4d. | 10s. 7d. | 14s. 7d. | — | — |
| 1798[i] | — | — | — | — | 14s. 0d. | 19s. 0d. | 36s. 6d. | — |
| 1798[w] | — | 4s. 7½d. | 6s. 6d. | 8s. 7d. | — | — | — | — |
| 1799[m] | 4s. 7d. | 5s. 1d. | 7s. 5d. | — | — | — | — | — |
| 1810 | 4s. 1d. | 4s. 8d. | 6s. 7d. | 9s. 7d. | — | — | — | — |
| 1816 | 3s. 11d. | 4s. 7d. | 6s. 1d. | 8s. 1d. | — | — | — | — |

Key:
b = Brazil cotton (mule-spun) ⎫
i = India cotton (mule-spun)   ⎬ in Glasgow
w = water twist                ⎭
m = mule twist, in Manchester

Later in the chapter I shall demonstrate more exactly how improvements in steam-power influenced the mechanization of the cotton industry in the mid-nineteenth century. At this juncture I shall merely indicate what degree of underestimation might result from failing to take account of rising quality of the goods produced. Table 7.2 summarizes some of the best-known estimates of the number of spindles and compares them with cotton consumption.

TABLE 7.2

*Spindleage and cotton consumption, 1788–1850*[18]

| Date | Spindles, million | Raw cotton consumed, m. lb | lb of cotton per spindle per year |
|---|---|---|---|
| 1788 | 1·94 | 18·67 | 9·6 |
| 1811 | 5·07 | 89 | 17·6 |
| 1817 | 6·65 | 110 | 16·5 |
| 1834 | 12 | 321·6 | 26·8 |
| 1850 | 20·47 | 635·7 | 31·0 |

The average count of yarn in 1788, according to 'a Committee of intelligent Spinners at Manchester',[19] was 27, so that the average rate of output would amount to 5 hanks per spindle per week, accepting the figure of 1·94 million spindles in all. By 1812 5 hanks of 27s would have been spun in about two days, according to another Committee of Manchester.[20] Thus for the same quality of yarn, spindle productivity was approximately trebled; whereas Table 7.2, where no account is taken of improving quality, it seems to have scarcely doubled.

In summary, the steam-engine had a belated effect on the spectacular expansion of the cotton industry in the last quarter

[18] For general nineteenth-century figures see H. D. Fong, *Triumph of Factory System in England*, Tientsin, 1930, p. 35. Specific sources are: (i) 1788. Colquhoun made similar estimates in several places, including his *An Important Crisis in the Calico and Muslin Manufactory in Great Britain Explained*, London, 1788 (cf. P.R.O. B.T. 6/140). The present numbers are taken from his *Diary*; they are probably overstated as a guide to machine spinning since, for example, water-mills are averaged at 2000 spindles apiece, whereas most were of 1000 spindles on a standard Arkwright plan (cf. Ch. 6). On the other hand, hand-wheels are excluded. The consumption of 18 670 000 lb of cotton is taken from ibid.; total imports of raw cotton in 1788 were 20·4 m. lb (Edwards (1967), p. 250). (ii) 1811: A partial revision of the census by Crompton given in B. M. Egerton MS. 2409; see G. W. Daniels, 'Samuel Crompton's Census of the Cotton Industry in 1811', *Econ. Hist.* ii, 1930. (iii) 1817. Kennedy's estimate; see Baines, *History*, p. 369. (iv) 1835. Ure's estimate, quoted by Blaug (1960–1), p. 371; who also quotes 10·8 m. on p. 379. (v) 1850. From P.P. 1850: XLIII, excluding spindleage on smallwares, etc.
[19] This is Colquhoun's way of putting it. I have computed the result as a weighted average of different grades: 5·33 m. lb spun into nos. 8–16 (average no. 12), 3·7 m. into nos. 17–20 (av. 18), 5·05 m. into nos. 21–8 (av. 24), 1·5 m. into nos. 29–40 (av. 35), 2 m. into nos. 41–70 (av. 56), and 1·09 m. into nos. 71–150 (av. 80). *Diary*.
[20] G. Lee (B.M. Add. MS. 12131).

of the eighteenth century, all the major technical breakthroughs in cotton-spinning having been originally developed for other forms of power. The coupling of steam-power to the major processes of cotton manufacture on a sizeable scale can be dated to the last few years of the eighteenth century. Its subsequent career can be charted not just in the number of spindles—largely mule spindles—that it drove, but also in the rising average quality of British cotton cloth that the mule facilitated.

### COTTON QUALITY IN THE EARLY NINETEENTH CENTURY

By 1820 cotton-spinning was fully fledged as a factory industry, though weaving by power was still little more than experimental. The spinning branch had come to be recognized as consisting

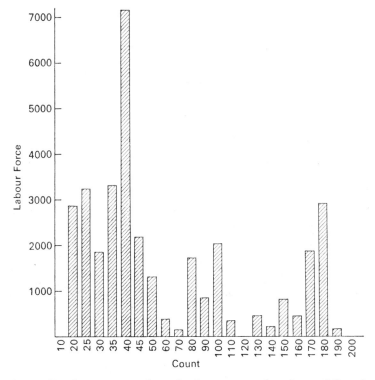

Fig. 7.1. Employment on each grade of cotton yarn, based on 178 firms in the SELNEC region, 1833.

of three main subgroups, ranked according to the quality of their yarn as 'fine spinning', 'coarse spinning', and an intermediate subgroup that had no special name.[21] Figure 7.1 depicts a cross-section of part of the industry in 1833.[22] It sums together those firms with about the same average count of yarn, according to the number of employees they had.

On inspection, the three important production bands emerge from the sample shown in this figure. At the top end are the prestigious 'fine spinning mills', with counts generally averaging from 140s up to 180s, and sometimes beyond. These are concentrated in Manchester. Beneath them are the medium counts, in the main from 70s to 100s; these were pre-eminent in the Bolton area. The largest band, however, comprises the coarser counts from 16s to 50s. The marked peak at 40s helps to confirm the opinion of contemporaries that the average count spun in Britain in the 1830s and 1840s was 38s to 40s. Much of this coarse branch was located in the districts to the south and east of Manchester, running from Stockport around to and past Oldham.[23]

At the very beginning of the century the fine-spinning branch had represented the white heat of technology, as then understood. Headed as it was by several large firms with comparatively familiar names, like McConnel & Kennedy, and somewhat oligopolistic in structure,[24] it was frequently in the forefront of applying artificial power to the new textile machinery. In the 1820s its influence waned and the branch stagnated.

---

[21] Angus Bethune Reach, *Manchester and the Textile Districts in 1849*, ed. C. Aspin, Helmshore, 1972, p. 12.

[22] Samuel Stanway issued a circular to firms in the SELNEC (South-East Lancs., North-East Cheshire) region, and sifted the responses from 350 or so firms; the 178 whose average quality of yarn are included in the figure are taken from those whose reply Stanway considered reliable. See P.P. 1834: XIX, and Baines, *History*.

[23] Other specializations arose in other cotton districts, e.g. in Scotland Renfrewshire and areas to the west spun higher counts, Lanarkshire the coarser (A. J. Taylor, 'Concentration and Localisation of the British Cotton Industry, 1825–1850' Univ. of Manchester Ph.D. thesis 1947).

[24] McConnel & Kennedy wrote to their agents in Glasgow in 1805 '. . . we do not wish to reduce the price except we are obliged to do it from being undersold by Messrs. Murrays or the like' (McConnel & Kennedy MSS.). In 1816 there were said to be 1666 employed in the spinning of very fine counts (150s and up); P.P. 1816: III. In 1834 the state of knowledge was said to be advanced enough to carry some of the improvements over from fine to coarse spinning—see P.P. 1834: XIX.

A letter to Inspector Horner in 1841, penned by an 'extensive' fine-cotton-spinner in Manchester, complained that fine spinning had been depressed 'with little intermission' since 1827: 'In 1829, the fine spinning trade in Manchester numbered 24 firms: they are now reduced to 16. The number of spindles then employed was near 1 000 000: it is now not more than 800 000. The number of operative spinners then employed was about 1028. It is supposed their number does not now exceed 500, or at most 600. The number of other persons employed was then estimated at 10 000; it is probably now about 9000.'[25] The reasons for this decay are to be sought in both demand and supply factors. On the supply side, fine spinning found it difficult to integrate forward into weaving, for reasons that will soon be evident. A further supply-side factor, and one more important for my particular purposes, has to do with the fact that the finer the quality of yarn spun, the less the amount of power required. An indirect way of showing this is by looking at the 1841 Factory Returns, as set out in Table 7.3.[26]

TABLE 7.3
*Hands per horsepower in the Lancashire cotton industry, 1841*

| Branch | Horse power | Number of employees | Hands per horsepower |
|---|---|---|---|
| Fine spinning (60s and over) | 2417 | 14 000 | 5·79 |
| Coarse spinning | 10 807 | 37 988 | 3·52 |
| Combined fine and coarse spinning | 416 | 2262 | 5·44 |
| Power weaving | 1211 | 7843 | 6·48 |
| Integrated spinning and weaving | 19 862 | 97 453 | 4·91 |
| Miscellaneous* | 689 | 4108 | 5·96 |
| Total | 35 402 | 163 654 | 4·62 |

\* = doubling, smallware weaving, waste manufacture, etc.

[25] P.P. 1842: XXII. Kay-Shuttleworth also noted 19 mills, employing 837 mule spinners, who in turn employed another 3233, in Manchester fine spinning.
[26] Aggregated from P.P. 1842: XXI. See also H. B. Rodgers, 'The Lancashire Cotton Industry in 1840', *Trans. & Papers Inst. British Geographers*, xxviii, 1960, for work on this rather neglected survey.

Each labourer in the coarse sections of the industry therefore worked with more horsepower, on average, than his counterpart in the finer grades; though this was less marked when power-weaving was included in the mill's activities. Two technical factors behind this greater 'power-intensity' of coarse spinning are of some relevance. In the first place, the ratio of spindles to horsepower was greater in fine than coarse spinning. Brunton's *Compendium of Mechanics* in 1825 used the proportion—by then conventional—of twice as many fine as coarse spindles per horsepower, while a witness before the Select Committee on the Exportation of Machinery in 1841 put it at two and a half times as many.[27] Since the slower rate of production of fine yarns was more than offset by their higher value per hank, the significance of power costs to total value-added was less than half as great for the very fine grade of 200s as that for the standard grade of 40s. Secondly, these last comparisons are for yarns spun on the hand-mule. For this type of mule it was customary to allow 500 spindles per horsepower on 40s in the 1820s and 1830s. Arkwright's old waterframe, as noted above, could support only 100 spindles from the same amount of power, and even its successor, the throstle, only 180.[28] The self-acting mule, when it was introduced, also needed 60 per cent more power than the hand-mule. Now both throstle and self-actor were to be found spinning coarse yarns only at this date, so that their presence reinforced the 'power-intensity' of coarse yarns.

The purpose of this excursion into technicalities is to underpin the conclusion reached at the end of the last section that the diffusion of machinery and power in the cotton industry cannot be understood if factors of yarn and cloth quality are not taken into account. But whereas in the first few years of the nineteenth century it was the finer qualities that bore the brunt of the impact of steam-power, twenty or thirty years later it was the coarse grades. Since coarse spinning was so 'power-intensive', any improvement in the application of power should have had a disproportionately large effect on costs of production.

[27] Brunton, *Compendium of Mechanics* (comparing 48s with 110s); P.P. 1841: VII.
[28] These are Ure's figures; Hoole quoted 600 mule spindles and 200 throstle spindles per horsepower in 1833 (P.P. 1833: XX).

POWER COSTS AND MECHANIZATION IN THE 1830S

The cost of steam (or any other form of power, for that matter) might be expected to have profound effects on the quantity of machinery it drives. In the first place, it is evident that the cost of imparting motion to the machine may be influential in any decision by an entrepreneur as to whether to adopt that particular machine.

Shortage of information makes it very difficult to establish with any degree of certainty those cases in which the cost of power was critical in the eighteenth or nineteenth centuries. On the other hand, one can easily point to several types of machine for which the question of power cost could have contributed significantly to the decision of adoption or rejection, even though it may not have been decisive.

I have tried to be more precise about two of the most notable developments of the early part of the nineteenth century: the self-acting mule and the power-loom.

(a) *The self-acting mule.* I have already said that on the hand-mule artificial power was used only for pushing out the mule carriage, on which the spindles were mounted. Attempts were soon under way to render automatic the remaining operations ('backing-off', returning the carriage, winding-on').

Dr. H. J. Catling has recently clarified the process of technical development and shown why it took so long.[29] The various movements had to be geared to one another in such a way that any fluctuations or irregularities in speed at any stage would be corrected for in the other stages. Roberts's improvements to the winding nut in 1830 finally rendered the machine cybernetic.

Even so the diffusion of the self-acting mule after 1830 can scarcely be called rapid or ubiquitous. There were said to be some 300 000 or 400 000 self-acting spindles in existence by December 1834[30]—this would be about 3 per cent of the total spindleage in Britain at that date. There are no accurate counts of self-acting spindles thereafter until late in the century. However, all authors seem to agree that the self-actor made little headway in superseding hand methods in the spinning of

[29] H. J. Catling, *The Spinning Mule*, Newton Abbot, 1970.
[30] Ure, *Cotton Manufacture of Great Britain*, vol. ii, p. 155.

finer counts of yarn (no. 60 or more) until about the 1860s, while some have maintained that even in the coarse counts, better adapted to automated techniques, the supremacy of the self-actor was not universal until the 1850s and 1860s.[31] Moreover, this occurred in spite of the fact that self-actors were from the beginning producing a fifth to a quarter more yarn (of the same quality) in a given time as the hand-mule.[32]

Was it ignorance on the part of the mill-owners that these new technical opportunities were disregarded? Contemporary writers—who undoubtedly had certain vested interests—seem to have thought that the introduction of the self-actor was justifiably cautious. More recently, authors such as S. J. Chapman have largely emphasized some technical drawbacks of the early self-acting mules. My intention is not to deny these but to reinterpret them in economic terms. In my opinion, few of the drawbacks were technically absolute—by which I mean that nobody could, for instance, have spun a higher count of yarn on extant self-actors. Only yarns up to a certain count were normally spun on self-actors in the 1830, but for higher counts I shall try to show that it was not impossible, but rather just uneconomical, to spin them automatically.

This links with the orthodox view, in that the high expense of spinning fine yarns on self-actors of course had its root in certain technical details. But where I part company is that with the orthodox view it is necessary to invent a way around the then short-comings; with my reinterpretation that does not *necessarily* follow. It may be enough for other economic circumstances to change in the right sort of way to make self-actors economic on finer yarns.

For my economic slant I make use of capital theory to collate the miscellany of data at hand. What I am trying to capture is the economic rationality of a 'typical' entrepreneur, in the middle of the 1830s, deciding whether or not to invest in self-

---

[31] S. J. Chapman, *The Lancashire Cotton Industry*, Manchester, 1904; S. Andrew, 'Fifty Years in the Cotton Trade' (paper delivered to the British Association for the Advancement of Science, Oldham, 1887); J. A. Mann, *The Cotton Trade of Great Britain*, London, 1860; etc. Hand-mules were still being erected for the medium-fine counts of Bolton as late as 1868. In the finer grades the more common form of technical advance in the 1830s (e.g. in Glasgow) was the 'coupling' or 'double-decking' of hand-mules; see P.P. 1837–8: VIII.
[32] *Thesis*, Table IV.8.

acting rather than hand-mules.³³ I assume perfect competition —surely reasonable for coarse spinning—so that his individual decision will not affect the prices paid by him or his rivals for either inputs or outputs. I further assume that his decision is founded on the belief that the relevant economic conditions will stay the same for as long as his new machine is expected to last. That is, in reckoning the profit he is going to make he might as well take 1835 prices as given for the lifetime of the mule: he has no strong expectations of bigger (or smaller) payoffs in later years. I am thus not computing his actual profit (or loss) from going ahead with the decision to install self-acting mules, so much as the profit he may *reasonably* have envisaged in the mid-1830s; and by 'reasonably' I mean on the supposition that on average the 1835 prices were going to prevail before his machine became obsolete.³⁴

Several conceivable situations deserve consideration.

(i) The master spinner may be viewed as intending to begin or expand activities, and thus having a free choice over investing in hand-mules or self-actors. The source materials permit an assessment of the differences in cost (according to the type of cost) between the two techniques. Hence it is possible to calculate the internal rate of return which would equate the over-all cost differences to zero over the average life-expectancy of the mule. The rate of depreciation conventionally applied to cotton machinery in this period was $7\frac{1}{2}$ per cent, so I suppose an average life of $13\frac{1}{3}$ years.³⁵

---

[33] Similar to most of the calculations of the profitability of negro slavery in the *ante-bellum* American South, see R. W. Fogel and S. L. Engerman (eds.), *The Reinterpretation of American Economic History*, New York, 1971.

[34] Naturally, in real life entrepreneurs could be more inclined either to caution or to gambling than I have allowed for. There is some evidence that on average they were 'risk-averse' in their behaviour, and so would hesitate to act if the anticipated gains were small. Charles Babbage, famed as the inventor of the calculating machine, wrote in 1832: 'But machinery for producing any commodity in great demand, seldom actually wears out; new improvements by which the same operations can be executed either more quickly or better, generally superseding it long before that period arrives; indeed, to make such an improved machine profitable, it is usually reckoned that in 5 years it ought to have paid itself and in 10 to be superseded by a better.' (*On the Economy of Machinery and Manufactures*, London. 1832, p. 231); contrasting with a life of $13\frac{1}{3}$ years estimated below.

[35] This has been rounded to 13 in most calculations, to ease the computational burden. The last 4 months are generally subject to quite a high discount. The calculation is undertaken for a self-actor of 600 spindles, the largest size readily available at this date. Line (i): net additional cost of self-actor from *Thesis*,

TABLE 7.4
*Costs and benefits of self-acting versus hand-mules,
c. 1835 (all figures in £s)*

| Year: | 1 | 2 | 3 | ... | 13 |
|---|---|---|---|---|---|
| Capital costs: | | | | | |
| (i) Mule | −105 | | | | |
| (ii) Power | − 59·01 | −10·69 | −10·69 | ... | + 0·55 |
| (iii) Materials | − 1·5 | − 1·5 | − 1·5 | ... | − 1·5 |
| Labour costs: | | | | | |
| (iv) Overlooking | − 6·25 | − 6·25 | − 6·25 | ... | − 6·25 |
| (v) Piecing | − 17·76 | −17·76 | −17·76 | ... | −17·76 |
| (vi) Spinning | + 32·07 | +32·07 | +32·07 | ... | +32·07 |
| (vii) Extra output | + 22·45 | +22·45 | +22·45 | ... | +22·45 |
| (viii) Net gain of self-actor | −135·00 | +18·32 | +18·32 | ... | +29·56 |

The amount of additional piecing (i.e. joining of broken threads) required in operating with a self-actor as opposed to a hand-mule (Table 7.4, line 5) is based on the experience of Birley & Kirk, one of the largest spinners in Manchester, and their average product was yarn no. 25.[36] For this quality of yarn the entrepreneur who based his decision on the costs and prices prevailing at the time would have expected to earn a rate of return that was almost 9 per cent higher from self-actors than the hand-mule would have offered.

But this was quite a low grade of yarn. I was able to give some confirmation to the contemporary opinion that the

[36] P.P. 1833: XXI.

Table IV.5—other situations are considered later. (ii) Suppose 500 hand-mule and 300 self-acting spindles per horsepower, e.g. Ure, vol. i, pp. 310–11; power costs then taken from Table 4.11 above. (iii): from J. Montgomery, *A Practical Detail of the Cotton Manufacture of the U.S.A.*, Glasgow, 1840, pp. 76–7, by deducting 'power' from 'extra charges for power, oil, and banding'. (iv): Montgomery (includes extra mechanics). (v): from attainments at Birley & Kirk's; see next n. (vi): Stanway's return of average wage of a mule spinner (Supplement E, P.P. 1834: XIX). (vii): Increased productivity from *Thesis*, Table IV.8. Prices of 25s obtained by interpolation of the series in Ure, vol. ii, p. 337 (prices of best 2nds mule 20s and 40s). Cost component of this price arising from the mules obtained from Montgomery, for 36s wefts.

average count spun in England was 38s or 40s in Figure 7.1. On wefts of 36s the information presented by a one-time Glaswegian spinner, James Montgomery, suggests that additional piecing involved on self-acting mules over that for hand-mules came to £63 annually.[37] At this rate the hand-mule should have proved superior to the self-actor not only in first cost but also for running costs in each subsequent year. This allows me to reinterpret the practical limitations of the self-acting mule as the cost of correcting for its shortcomings; in other words piecing together the breakages that occurred at finer yarn counts.

(ii) An alternative possibility was the situation of the entrepreneur who might have been confronted with the decision whether to scrap the hand-mules he had in order to acquire self-actors. In this case, 'bygones are bygones', i.e. the scrapping would be remunerative only if the total costs of producing by means of self-acting mules were greater than the variable costs of using hand-mules.[38] If the hand-mules had had to be dismantled, the net benefits from utilizing self-actors in year 1 would have come to $-£315$, and the summed net benefits over costs until the self-actor was itself scrapped would be negative. Thus there was no economic incentive for taking down hand-mules while still in good condition for the sake of erecting self-actors.

(iii) However, it was not necessary to dismantle the hand-mule completely in order to install the self-acting principle. It was possible in fact to convert the hand-mule by replacing the headstock, which contained the cybernetic mechanisms. Ure estimated that this would cost only a fifth of the full expense required to erect self-actors *de novo*.[39] More precisely, Montgomery quotes the cost of conversion to Roberts's self-actors at 4s. 1d. per spindle.[40] In this case net costs of conversion in the first year would be £152. 50, *cet. par.*, and conversion would earn a return of almost 7 per cent p.a. on 25s yarn.

[37] Op.cit., p. 78. Montgomery expresses piecing wages on this quality as 1s. 9d. per 1000 hanks, with an output of 21 hanks per spindle per week. Note that Montgomery was trying to advocate the self-actor on grounds of its economy.
[38] The theory is spelled out by W. E. G. Salter, *Productivity and Technical Change*, 2nd edn., ed. W. B. Reddaway, Cambridge, 1966.
[39] Ure (1861), op. cit., vol. ii.    [40] Op. cit., p. 76.

(iv) In practice, self-actors were often installed in mills that had hitherto spun their yarn on 'throstles' in the 1830s. Many spinners thought throstles were superior to hand-mules in spinning coarse counts of yarn, and certainly it was in these grades that the throstle thrived—the same low counts that the self-actor was being applied to. A rough calculation shows the self-actor to have been considerably cheaper for its power and labour costs, but that this gain was almost exactly offset by the higher price fetched by the kind of product ('water twist') spun on the throstle, compared to mule warps.[41] Thus self-actors were really a complement to throstles rather than a substitute, with market forces allowing them to survive conjointly at roughly the same level of profitability.

(v) I can now trace the consequences of the given level of power costs. It was customary for manufacturers in the 1830s to suppose a rate of interest of 5 per cent (imputed if necessary) on their capital, both fixed and floating.[42] I shall do likewise. On this basis it follows from the calculations (whose precision, it must be admitted, is more than a little spurious) that the self-actor would have earned at least a competitive rate of return: (i) as an investment alternative to hand-mules for 25s yarn, from the beginning; (ii) as a technical transformation of the hand-mule, on the same grade; (iii) as an adjunct to throstles. Coal at Manchester costing 9s. a ton was assumed in assessing the costs of power. However, suppose coal were purchased elsewhere, at other prices.

By an iterative calculating procedure I have established that a rise in annual costs of £1. 4s. per mule would be just sufficient to reduce the return on hand-mules that had been transformed into self-actors below the competitive rate. This could, for example, have been brought about by a rise in coal costs from 9s. a ton to about 10s. 4d. Now in 1842–3 the price of household coal in England and Wales varied from about 4s. 3d. a ton if sold commercially in the North-Eastern mining districts to 42s. a ton in Witney, Oxon.,[43] i.e. a range of about ten times the price of the cheapest domestic coal. The range easily

---

[41] Ure, op.cit., vol. ii, p. 104.
[42] In the only published series of profits for a nineteenth-century cotton firm, Boyson uses 5 per cent (*The Ashworth Cotton Enterprise*, Ch. 3).
[43] P.P. 1843: XLV. Cf. Ch. 4.

encompasses the hypothetical increase being contemplated to make the decision to convert to self-actors unprofitable. In fact, most steam-mills in the cotton industry were located in areas of moderately cheap coal,[44] but this in itself is no doubt partly testimony to the consequences of power costs.

Another example can be found in the reduction in power costs around 1850, of which more will be said later in this chapter. In Manchester in the early 1850s the annual costs of power were much less than in the 1830s.[45] This fall in costs would have rendered economic a considerably higher quality of yarn on the self-acting mule, even if other cost and technical items had not changed.

This conclusion is possibly a little odder than it looks, because it partly contradicts the last section. There I was saying that the effect of improvements in power supply would be more pronounced for coarse than fine cottons, because of the greater share in costs occupied by power. Here, however, I am taking it as read that the self-actor already dominated at low counts (at least for new buildings or extensions). As power costs fell further, fine spinning came into the ambit of the self-acting mule not so much because power costs had been especially inhibiting as because breakage rates on the automatic machines remained inhibiting.

---

[44] See Table 4.8.
[45] 'George Branson's Estimate of the cost of Power &c for weaving, Manchester, August 23, 1850. "I consider the annual charge for supplying power for 288 looms of 45″ wide running 140 picks per minute and supplying steam for dressing (The Weaving department finding its own building and shafting) should be £252 p.a., as follows:

The Looms with Dressing Machines I call 18 HP (say 16 Looms per HP);
Coals p.a. .. .. .. .. .. .. .. .. £7 per HP
Steam engines, Grate Bars, Foundations, Boiler Seatings, Boiler House and Engine House, including Wear & Tear and repairs .. .. .. .. .. .. .. .. 5 ,, ,,
Engine Tenter, Oil, Tallow &c .. .. .. .. .. 2 ,, ,,
£14 per HP

× 18 HP = £252 rental."

Note: by this calculation the annl. charge for room and turning will be 39/6 per Loom, for our Looms. The annl. charge made for rent at Stockport is 45/0 do. The rent charge at Stockport for room and Turning is 1/0 to 1/2 per spdle. for self actors, 7½′ to 10′ do. for Hand Mules.' (Ashworth MSS., Quarterly Stock Accounts, 28 Sept. 1850, p. 151).

Both contemporaries and modern authors have argued that factors other than costs contributed to the diffusion of the self-actor. Richard Roberts originally invented the machine to forestall any labour problems in the event of a strike, which appeared not unlikely in the boom of 1825.[46] However, there were few strikes after 1825 initiated for the purpose of raising wages, until the great Preston strike in the early 1850s. Such activity as there was tended to be rearguard operations attempting to prevent any diminution of wages in particular areas, rather than conspiring to raise them. For fairly apparent reasons these were generally short-lived and failed to have the effect of inducing automation in quite the same way as the other situation. Some spinners, such as William Graham of Glasgow, did install the self-acting mule for just this reason,[47] but there is no evidence that this was widespread.

I conclude that a narrow cost-benefit assessment of the kind I gave numerical shape to in Table 7.4 cannot explain everything about the diffusion of the self-actor, nor was it ever intended to do so. It does show that what we know about the adoption of the self-actor, little as it may be, does not clash with my evaluation of economically rational decision-making. Hence it acts as a backdrop to establishing the importance of power—and power costs—for mechanization in this guise.

The self-actor did have a number of qualitative advantages that were not recognized in Table 7.4. It produced a cop that was more evenly wound on, with favourable effects on weaving.[48] (i) This meant less wastage of yarn in the weaving, because of the uniformity of the yarn and the cop. (ii) It was particularly important for weaving with the power-loom, which had its own problems in mechanizing the handling of fibres, and so especially benefited from a more regular input. Thus the

---

[46] Catling, op.cit., Ch. 4; N. Rosenberg, 'The Direction of Technical Change: Inducement Mechanisms and Focussing Devices', *Econ. Development & Cult. Change*, Oct. 1969.

[47] P.P. 1833: XXI; A. J. Taylor, op.cit. See also H. A. Turner, *Trade Union Growth, Structure and Policy*, London, 1962, for suggestions of self-actors replacing hand-mules in strikes at Preston in 1836 and 1853.

[48] See esp. A. Ure, *A Dictionary of Arts, Manufactures and Mines*, 4th edn., London, 1853, p. 533.

diffusion of the self-actor was partially interconnected with that of the power-loom.

(b) *The power-loom*. Practical efforts to mechanize the operations of the hand-loom date back at least to the efforts of Edmund Cartwright in 1787. It is usually believed that Horrocks's power-loom, patented in 1803 and improved in 1813, was the first truly workable machine. According to standard sources there were 2400 power-looms in Great Britain in 1813; 14 140 in 1820; about 40 000 in 1825 and double that number by 1830; 108 210 by the time of the first Factory Returns in 1835; and 249 627 in the next full Return in 1849.[49] It took some fifty years from the time of Horrocks's first patent before the hand-loom weaver in the cotton trade was finally confronted by the threat of extinction, and considerably longer in other textile industries such as linen and woollens.

As with the self-acting mule, it has been common to blame this prolonged diffusion process largely on technical factors. Again I feel these should be integrated with economic factors. James Kennedy, partner in one of the largest and most famous of Manchester cotton-spinning firms, stated in 1819 '. . . one person cannot attend upon more than two power-looms, and it is still problematical whether this saving of labour counterbalances the expense of power and machinery, and the disadvantage of being obliged to keep an establishment of power-looms constantly at work.'[50]

Although the demise of the hand-loom weaver took a painfully long time, it is clear that the balance between hand-loom and power-loom weaving was in disequilibrium for most of the period from the early 1820s onward. The exact chronology of the decline in numbers of hand-loom weavers is shrouded in ideological debate.[51] Until the late 1820s it would appear that the decline was limited, principally by two factors: (i) the technical-economic limitations of the existing power-loom, as hinted at in Kennedy's statement, and (ii) the willingness of

---

[49] E. J. Donnell, *Chronological and Statistical History of Cotton*, New York, 1872; Baines, *History*; P.P. 1836: XLV; P.P. 1850: XLII.
[50] Quoted by H. J. Habakkuk, *American and British Technology in the Nineteenth Century*, Cambridge, 1962, p. 29.
[51] See D. Bythell, *The Handloom Weavers*, Cambridge, 1969, and refs.

the hand-loom weaver to accept wages that were steadily being pushed down to subsistence or starvation levels in order to maintain his independence. Under some very strong assumptions, especially on the demand side, one can very crudely approximate the technical progress achieved by the power-loom from the fall in earnings attained by hand-loom weavers on products being superseded by the powered technique. The assumptions involved would not stand up to much scrutiny, yet it seems reasonable to assert that the process of grimly accepting reduced incomes for the sake of tenaciously holding on to the hand-loomer's independence had by and large come to an end by the early 1830s, at least for these kinds of cloth. Incomes would hardly fall any further without making existence at subsistence levels virtually inconceivable. Hence numbers declined—if not for the first time, probably at a faster rate than previously.

By the middle 1830s the power-loom engaged in weaving plain calicoes in Stockport or Oldham, or 50-reed cambrics in Bolton, would turn off nearly three times as much in given time as a hand-loom weaver. It is difficult to be more precise, because the hand-loomer vacated the weaving of these coarser grades of fabric. Thus power-looms began with the coarsest cloths, such as plain calicoes and fustians, in the early decades of the century. By 1834 it was stated that they were capable of outcompeting hand labour in the weaving of cloth as fine as 100s, measured in the Manchester or Stockport count, though not economic for 120s.[52] Another witness before the same Parliamentary Committee, however, claimed that the large Bolton manufacturers, Ormrod & Hardcastle, had not found the weaving of 60-reed cloth (equal to 100 on the Manchester count, or 1800 on the Scottish count) profitable.[53]

On a somewhat coarser product, 5/4 50-reed Cambric, I have estimated the profitability of scrapping hand-looms in favour of the power-loom by a Bolton manufacturer in the

[52] P.P. 1834: X, q. 6748. 'The 40's warps are woven commonly through a reed containing from 72 to 80 threads per inch, constituting in the language of Stockport, a 72 to 80 Reed' (Ure, *Cotton Manufacture*, vol. ii, pp. 253–4). 100 on the Stockport count equalled 60 on the Bolton count.
[53] P.P. 1834: X, q. 5728. Ormrod's had 600 power-looms at this time. Another witness stated that 56-reed plain cambrics were then the limit in Bolton (ibid., q. 4593).

Steam-power and the Cotton Industry 197

middle 1830s, as set out in Table 7.5.⁵⁴ For this quality there was thus very little return on power-looms over and above that earned on hand-looms, and this is consistent with the qualitative evidence already presented. Within the margin of error of the data, which is large, the decision could have gone either way.⁵⁵ The calculation is unfortunately particularly sensitive to the assumption about the precise productivity of power-looms on this grade of cloth. Another half a piece a week, bringing

---

⁵⁴ 5/4 means a cloth 1¼ yd wide. Line (i): From *Thesis*, Table IV.5. (ii): Assumes one dressing-machine costing £35 per 21 looms; ratio from Montgomery, op.cit., p. 117. (iii): Total cost of plant and equipment for most of the century was stated to be £24 per loom by T. Ellison, *The Cotton Trade of Great Britain*, London, 1886. Depreciation put at 4 per cent p.a. (iv): Assumes coal costing 6s. a ton in Bolton, and each loom required 0·1 horsepower (Ure (1861), op. cit., vol. i, pp. 310–11). Other assumptions as for self-actor. (v): R. H. Greg stated that the total consumption of flour in the 80 000 power-looms he believed (reasonably correctly) then existed in Britain would be 44 572 loads of 280 lb each, costing £92,601 annually (P.P. 1833: XX). Other estimates are slightly higher. (vi): Obtained from average expenditure on fuel (oil), soap, and candles by Bolton hand-loom weavers (P.P. 1840: XXIV, p. 601), revised downwards to allow for economies of scale in lighting per loom. Hand-loom weavers of course paid for these items out of their net wages. Cf. also G. von Schulze-Gaevernitz, *The Cotton Trade in England and on the Continent*, London and Manchester, 1895, for comments on the delays to mechanization offered by sizing, dressing, etc. (vii): from Montgomery, p. 120. Slightly higher dressing costs were given by Philadelphian spinner James Kempson, discussing the English industry (P.P. 1833: VI), but his data were criticized by Montgomery on other grounds. (viii): Assumes one overlooker per twenty-five weavers—ratio and earnings from Stanway's Returns (P.P. 1834: XIX, p. 427, Supplement E (General)). (ix): Wages for hand- and power-looms on 50-reed and 60-reed from P.P. 1835: XIII, p. 232 (evidence of Jonathan Hitchin, qs. 3059–63). (x): Assumes 5 pieces of 50-reed cloth turned off power-looms weekly; from P.P. 1835: XIII, qs. 2110–11. Value taken from wages by assuming labour costs were 60 per cent of the value-added. Value of hand-loom cloth taken as gross wages paid to Hitchin's employees in Bolton (P.P. 1840: XXIV, p. 601)— in the mid-1830s he had been singled out as typical of the industry. Labour costs come to about 75 per cent of the value-added in hand-weaving (cf. also P.P. 1834: X, q. 675 for Glasgow).

⁵⁵ At the margin, it might still have been tempting to invest in power-looms for reasons that had a more subtle effect on profits than those explicitly considered. Richard Muggeridge summarized the position in 1840: 'One manufacturer, who had recently adopted power-looms, gave as the influencing causes of the change, that "He could turn over his capital in much quicker time. He could command a much better price for his cloth, by its being more regular in the weaving, and more honestly manufactured. That he was neither robbed of warp or weft; and that now he could take an order, and know *when* he could execute it, which, while he employed hand-loom weavers, he could never do." ' ('Report on the Condition of Hand-loom weavers of Lancashire and the North', P.P. 1840: XXIV, p. 627). For general confirmation of these remarks, see P.P. 1834: X, q. 5204, etc.: also S. Pollard, *The Genesis of Modern Management*, London, 1965, pp. 33–4.

TABLE 7.5

*Costs and benefits of power-looms versus hand-looms for 50-reed cambrics in Bolton, mid-1830s*

(All figures in £s)

| Year: | 1 | 2 | 2 | | 13 |
|---|---|---|---|---|---|
| Capital costs: | | | | | |
| (i) Power-loom | − 9 | | | | |
| (ii) Dressing machine | − 1·67 | | | | |
| (iii) Buildings | − 7·27 | | | | + 3·39 |
| (iv) Power | − 7·05 | − 1·01 | − 1·01 | ... | + 0·79 |
| Materials costs: | | | | | |
| (v) Flour | − 1·16 | − 1·16 | − 1·16 | ... | − 1·16 |
| (vi) Oil, lighting | − 1·00 | − 1·00 | − 1·00 | ... | − 1·00 |
| Labour costs: | | | | | |
| (vii) Dressing | − 2·58 | − 2·58 | − 2·58 | ... | − 2·58 |
| (viii) Overlooking | − 1·31 | − 1·31 | − 1·31 | ... | − 1·31 |
| (ix) Weaving | − 5·44 | − 5·44 | −5·44 | ... | − 5·44 |
| (x) Extra output: | +14·57 | +14·57 | +14·57 | ... | +14·57 |
| (xi) Net benefits: | −21·91 | + 2·07 | + 2·07 | ... | + 5·68 |

loom output up to five and a half pieces, would convert this $3\frac{1}{2}$ per cent rate of return into one of about 14 per cent p.a. On the other hand, the output of power-looms on 60-reed cambrics was four pieces a week.[56] At this rate the power-loom would have earned a net loss over its lifetime, despite a small positive return in each year after the first; and this is consistent with my earlier qualitative findings.

As already stated, the situation persisted for an unduly long time partly because the hand-loomer was prepared to reconcile himself to lower and lower wages rather than abandon his skilled trade. Had the wage rate earned on hand-looms in the early 1830s been the same as in 1816, the labour costs of a power-loom would have been lower than on hand-looms, instead of £5 a year greater (in the improbable circumstances of all other things being equal).

The limitation on the power-loom, therefore, was its inability to compete economically with either fine or figured work

[56] P.P. 1835: XIII, q. 3061. A dandy loom would produce almost two pieces: ibid., 1966, 1777.

produced by hand-looms. Many witnesses to the 1834 Parliamentary Inquiry believed that, technically speaking, the power-loom was or would be capable of weaving nearly any fabric then woven on hand-looms.[57] However, to do so would have required one attendant per steam-loom (the so-called 'one-loom system'), and in addition the rate of production could be little greater than on hand-looms. For Lancashire at this date the two-loom system was probably the norm. If this is so, then wage data indicate that 50-reed fabrics required 4 weavers for every 5 power-looms;[58] and on the results of Table 7.5 this must have been as 'labour-intensive' as power-weaving could afford to be. The pure one-loom system simply could not save a sufficient quantity of labour. Conversely, going over from the one-loom to the two-loom system meant annual savings in labour costs of about £13. 10s. a loom, and the significance of this gain should be evident from the Table.

In sum, the powered technique could be applied to finer or fancier cloths, but could not yet be profitable on them. Mechanical knowledge was, of course, steadily improving. The dropbox principle, developed by Robert Kay for the hand-loom in the mid-eighteenth century, was extended to allow the power-loom to use several coloured threads simultaneously in the shuttle. There were a succession of improvements to the common power-loom in the 1840s, notably (i) devices to stop the machine automatically when the weft broke or the shuttle needed refilling; (ii) devices to set the 'temples' in such a way as to counteract the tendency of the cloth to contract in width as weaving progressed; and (iii) adjustments to the take-up motion so that the speed of weaving was kept constant. Alderman Baynes believed that these nearly halved the labour costs of weaving.[59] Evidently these improvements would tilt the balance towards the adoption of the power-loom for finer products, since even on these items the wages of hand-loomers had little scope for further reduction in the 'hungry forties'.

[57] P.P. 1834: X, e.g. q. 5488.
[58] Average power-weaving wages from Stanway's returns, loc.cit., adjusted to Bolton levels from Supplement F. The average for Lancashire was about 2:3. These adjustments have not been made to Table 7.5.
[59] R. M. Hartwell, 'The Yorkshire Woollen and Worsted Industry 1800–1850', Univ. of Oxford D.Phil. thesis, 1955; Ellison, op.cit., pp. 36–7.

Once again it is straightforward to assess the importance of the costs of power in dictating the adoption of the power-loom. Merely on achieved fuel savings, Bolton cotton manufacturers would have been paying no more than £14 per horsepower per year for their steam in the early 1850s, excluding other types of advances in the engine.[60] This alone would increase the rate of return from $3\frac{1}{2}$ per cent to about 9 per cent on the above-mentioned cloth.[61] On the basis of the above arguments, this reduction in power costs also gave considerable leeway for extending the economic realm of the power-loom to the finer and fancier qualities.

The economic and technical factors hindering the spread of the power-loom were not limited to those discussed above. A further element was the introduction of certain kinds of 'intermediate technology'.

(i) The Jacquard loom was imported from France in about 1821.[62] This loom was too expensive to be purchased outright by a hand-loom weaver; expensively mounted Jacquards were said to sell for as much as £200, though less complicated versions could be bought for not much more than the price of a power-loom. The usual procedure was for an entrepreneur to acquire them and rent them out to the weavers for use in a weaving shed or their own homes (in silk-weaving, one-quarter of the value of output was often deducted for loom rent of Jacquards). The hand-driven Jacquard provided at least a temporary refuge for some hand-loomers squeezed out of plain fabrics. The depressing of the price of fancy goods presumably would have made the cost advantage of power-looms in the coarse, plain grades still weaker. Later it became more common to apply

[60] Based on cost of coals consumed by Ashworths in 1851 (44 tons per week for 186 IHP, @ 5s. 4d. per ton); their former partners Thomas Thomasson were reported as paying only £2. 3s. for coals per HP p.a., or £13 for power. These costs exclude gains from cheaper engines, such as were appearing on the market around the time of the Great Exhibition. Eckersley's of Wigan paid as little as 16s. 1d. per HP per year in 1853, using steam at 54 lb p.s.i. and presumably burning slack. All these figures (and others confirming) from John Rylands Library MS. 1201, p. 165. Using Branson's figure for Manchester in 1850 (see n. 45) would reduce the sum to about £12 p.a.
[61] This calculation does not include the gains of extra output from running the machinery at high speeds, cf. below.
[62] P.P. 1835: XIII, q. 2467. The Jacquard was introduced into Coventry in 1823 (P.P. 1840: XXIV) and Yorkshire in 1827 (W. B. Crump (ed.), *The Leeds Woollen Industry, 1780–1820*, Leeds, 1931).

artificial sources of power to the Jacquard loom. It has been argued that the ease of adapting it to mechanical power was a reason encouraging the adoption of this type of loom in Britain,[63] though adoption was never very rapid. Barlow's patenting of the 'double lift' principle in 1849 however kept it competitive at a time when conventional power-looms were becoming less expensive to operate.[64] For simpler patterns the so-called 'tappet' power-loom was developed, using two tappets or eccentrics to generate a treadle motion.[65] Intermediate between these was the dobbie loom.

(ii) The so-called 'dandy loom' was designed by William Radcliffe in 1802, and better machines flourished especially in Blackburn from about 1820 onwards.[66] In this hybrid the shuttle was still thrown by hand but the loom incorporated machinery to wind the yarn on, so sparing the weaver from doing this himself and having to stop his actual weaving as on the ordinary hand-loom. Some data for 6/4 60-reed cambrics (the quality still marginally unprofitable on power-looms in 1834) indicate that they may have increased weaver productivity by up to 50 per cent on that product.[67] However, their major use was on the heavy fabric known as 'Blackburn Greys'. On such heavy goods the costs of power per steam-loom were greater than on lighter fabrics such as muslins and calicoes.

---

[63] Sigsworth, *Black Dyke Mills*, p. 37.
[64] A. Barlow, *The History and Principles of Weaving by Hand and by Power*, London, 1878; J. G. Jenkins, *The Wool Textile Industry*, p. 154.
[65] J. H. Clapham, *The Woollen and Worsted Industries*, London, 1907.
[66] W. Radcliffe, *Origins of the New System of Manufacturing, commonly called 'Power Loom Weaving'*, Stockport, 1828; W. A. Abram, *A History of Blackburn, Town and Parish*, Blackburn, 1877; P.P. 1834: X, q. 5734. 'A hand-loom, on a new construction has been recently introduced, which has received the appellation of the Dandy Loom. Its principal advantage over the common hand-loom consists in its being much smaller, and in the application of a crank by which, as in steam looms, the number of picks of weft in an inch is regulated, and the cloth consequently made more even. We understand, also, that the new hand-loom weaves the yarn without dressing, which is an expensive process, whilst, by the use of a cap shuttle, the necessity of winding the weft is superseded. The loom measures only about 30″ in depth from the cloth to the yarn beam; and its cost, in wood, is not more than 35*s*. or 36*s*.; or, in iron, than 52*s*. 6*d*. A fair weaver, with tolerable exertion, will weave a piece of 25 yards in 8 or 9 hours.' (*Mechanic's Magazine*, i, 1823.)
[67] Radcliffe himself obtained three pieces of printing cambric weekly as early as 1812, before the further improvements (Radcliffe, p. 184). The quotation in n. 66 makes the dandy loom as fast as a steam-loom. Also P.P. 1835: XIII, esp. q. 1777; and P.P. 1834: X.

Hence in these items the cost of power supplied by steam was great enough to tilt the balance towards manual operation. However, the weaver had to be exceptionally powerful to throw this shuttle at the requisite pace all day long.[68] Because of the capital and power costs dandy looms, like Jacquards, were generally owned by capitalists and installed in weaving sheds.

Neither Jacquard nor dandy loom may ever have dominated huge sections of the industry but within their own spheres they cramped the style of the power-loom still further, and gave some hand-loom weavers a short breathing space.

To combine the results for each sector, I conclude that, until the 1840s, the spinning of average yarns on self-actors and the weaving of above-average cloths on power-looms do not appear to be lucrative uses of capital resources. Lower costs of power in the 1850s permitted the economical spinning of yarns and weaving of cloths of substantially higher qualities. This occurred even though in both cases the limitation in the 1830s was in a way due to labour costs rather than power costs: spinning higher-grade yarns on self-actors would have accelerated breakage rates, and weaving finer cloths on power-looms implied a power-weaver for every loom. This complication is still not enough, for I shall demonstrate in the next section that cheaper power led to a faster rate of output from the machines, and this consolidated their advantage.

## THE PRODUCTIVITY OF MACHINERY

It has already been shown how sensitive the diffusion of new machinery was to how fast it happened to work; exemplified by the much higher rate of return for the power-weaving of 50-reed cambrics (as compared with hand-weaving) when the output of the power-loom rose from five to five and a half pieces a week. One can generalize from this result: the costs of power *directly* influenced the pay-off to mechanizing particular grades through altering the most profitable speed for operating the machine. This proposition will be illustrated here for mule-spinning, but generally holds good for other forms of spinning and for weaving.

[68] P.P. 1835: XIII, q. 13.

For hand-mules working on the standard grade of 40s yarns, conventional sources provide the productivity figures set out in Table 7.6. For other machinery the data are more scrappy

TABLE 7.6

*Productivity in hand-mule spinning, 40 hanks per lb*[69]

| Date | Hanks/spindle/week | Source |
| --- | --- | --- |
| 1790s | over 6 | Crompton |
| 1800s | 12 | Ashton |
| 1810s | 12 | Crompton |
| 1823 | 14 | Babbage |
| 1829 | 16·5 | Kennedy |
| 1830s | 18 | Hoole, Ashton |

but one can find commensurate increases in capital productivity on roving frames, etc.[70] These would have resulted partly from improvements in the construction of roving frames (introduction of the stretching mule; replacement of stretcher and can frame by the bobbin and fly fame; introduction in 1825 of Dyer's tube frame) and partly from faster speeds of operation of frames of the same type. Similarly, the power-loom not only became cheaper to purchase in the 1830s[71] but also quicker in weaving the cloth. In 1823 a weaver about 15 years old was said to be able to weave 7 pieces of 9/8 shirting, 24 yards long, weekly, working on 2 looms; ten years later a weaver supplemented by a 12-year-old girl attending 4 looms could weave 18 and even 20 pieces of the same fabric in the same time.[72] The 4-loom system was still uncommon at this stage, but even on 2 looms there is evidently a productivity rise of 30 to 40 per cent on the assumption that the other facts are correct. Certainly older vintages of power-looms found themselves becoming rapidly obsolete in the later 1820s.

There is little economic theory which can help to explain changes in capital productivity. By and large such matters are

[69] Sources: Crompton, B. M. Egerton MS.; P.P. 1833: XX, App. D.2 (Ashton); Kennedy (1831); Babbage, op. cit., 3rd edn.
[70] e.g. Babbage; P.P. 1833: VI; P.P. 1833: XX.   [71] *Thesis*, Table IV.5.
[72] Donnell, op. cit.; R. Guest, *A Compendious History of the Cotton Manufacture*, Manchester, 1823 (B.M. copy with MS. additions).

relegated to the penumbra of exogenous technical change; partly, one would imagine, through lack of acquaintance with the basic technical problems. In the present case it is not especially clear what the technical difficulties were. It is apparent that at a very early stage of mechanization there were problems with bearings, rollers. etc. that might have hindered any attempt to operate the machinery more intensively by speeding it up. Yet such problems, at least for the hand-mule, would seem to have been largely resolved by 1800. It is not easy to evade the conclusion that if faster machinery speeds had been desired after 1800, the mechanical obstacles to achieving them would have been readily overcome over quite a wide range.[73] Moreover, at first sight it would appear that such an increase in machinery speeds and thus the rate of output would be highly desirable, since by that means it would be possible to spread the capital costs of mill and equipment over a larger total output and thus reduce their incidence per unit of output. As contemporaries frequently remarked upon the high capital intensity of cotton-mill in the 1830s,[74] this ought to have been a desideratum.

Why was it not done more intensively? Look first at what data there are on the number of machines per horsepower. The rule of thumb used at the end of the eighteenth century and beginning of the nineteenth was between 500 and 1000 mule spindles to the horsepower—Catling indeed averages the two.[75] By 1835 the mean number of spindles per horsepower in the

[73] At a time when the advised speed for throstles was 4500 rev/min the Orrell's new factory in Stockport regularly worked them at 5000 rev/min, and it was said to be possible to raise their speed to 7000 rev/min in spinning 24s (Ure, op.cit., pp. 104–5). For power-looms, Thomas Ashton of Hyde was asked in 1833, 'Then would not the decreased profits caused by the Ten Hour Bill stimulate the invention of new machinery, or increase the speed of the old, so as to place you in the same position as before?' To which he replied, 'We have it not in our power to increase the speed of the machinery sufficient to make up the difference; the speed could not be much increased with advantage in either spinning or weaving . . ..
In some mills, where they had increased their speed in weaving to 150 picks a minute, it has been found necessary to diminish the speed to 130 upon the same machine, which may consequently be considered as the maximum of speed . . .'—this at a time when the customary speed for power-looms was said to be between 90 and 112 picks per minute (P.P. 1833: XX).
[74] Pollard (1964); Matthews, op.cit., p. 130.
[75] Op.cit., pp. 53–4. In 1864, during the 'Cotton Famine', James Hyde quoted a complete range, from 380 spindles plus preparation per HP on 10s to 672 on 300s (*The Science of Cotton Spinning*, Manchester, n.d. [1867]).

U.K. was probably just under 400; agreeing with the convention of 350-plus found in the literature.[76] The Factory Returns of 1850 yield a figure of 358 spindles per horsepower in October 1849—certainly no more than in 1835 and likely to be 5 to 10 per cent less.[77]

Is the inference that the efficiency of power was declining at a fairly substantial rate through the first half of the nineteenth century? Some authorities clearly believed that workmanship on steam-engines was not what it had been in James Watt's time, but even so this is not a necessary conclusion.

A more profound explanation lies embedded in an identity. Since we are looking at a vertically aligned process rather than at the 'horizontal' association of inputs which is implicitly established in the conventional production function, there is little to learn of real explanatory power from direct comparisons of the various quantums of inputs. Instead the inputs can be compared individually with final output. Or, if one prefers it this way, the output that is relevant for gauging the capabilities of a power source is not its immediate 'output' (the number of spindles it drove) but ultimate output—the quantity of yarn, cloth, etc. produced (always assuming no change in *machine technology*).

Power productivity is then identically equal to output per spindle times spindles per horsepower. Thus it is possible to draw a kind of 'isoquant' between these ratios; an isoquant

---

[76] Extrapolated horsepower for the industry in 1835 is 42 600 (Table 7.15 below). From this I deduct the power required for 109 000 power-looms with dressing, to obtain horsepower in spinning as 31 600. Extrapolating from output and productivity figures, I conclude in App. 7.1 that there were the equivalent of 12·7 million mule spindles in the U.K. in 1835, leaving an average of 402 mule spindles per HP. However, not all spindles were on hand-mules. The convention of 350 appears in the McConnel & Kennedy MSS., e.g. Edwards (1967), p. 188.

[77] The 1850 spindleage is obtained from a multiple regression using the Factory Returns:

$$HP = 131\cdot5 + 2\cdot53\, F + 0\cdot00259\, S + 0\cdot0596\, PL \qquad R^2 = 0\cdot994$$
$$(1\cdot39) \quad (0\cdot00011) \quad (0\cdot0037)$$

Where $F$ = no. of firms, and is included as an extra 'constant term', because of the nature of the aggregation, across counties;

$S$ = no. of spindles;
$PL$ = no. of power-looms.

Total spindleage of all kinds was 20 977 015, and there were 1596 firms either spinning or combining spinning with weaving, so total horsepower in spinning was assessed at 58 474.

taking the form of a rectangular hyperbola. It will be unique for any given state of the arts. At any one point of time, corresponding to the given level of knowhow, the appropriate level of output per horsepower can be maintained at any point along the hyperbola, by increasing output per spindle while reducing the number of spindles per horsepower, or conversely. This is essentially tautologous.

Practical significance is given to the tautology by my belief that these trade-offs were physically possible over a sufficiently wide range. Since records relating to production costs before and after increases in speeds have not survived, I have conducted a simulation study for an indirect test. The mill chosen is a hypothetical one, devised by James Montgomery as characteristic of Glasgow in the early 1830s.[78] I take it to be driven by a 30-horsepower steam-engine, of which 12 horsepower went to drive 2160 throstle spindles and 6 horsepower on 2400 mule spindles (including preparation); the rest being for weaving.

I then assume, quite arbitrarily for purposes of exposition, that the speed of the spindles is increased from the 4200 rev/min that Montgomery notes for the mule spindles (4400 rev/min for throstles) to 5000 rev/min, or 5238 for throstles. Loom speeds are increased proportionately. Requirements of materials per unit of output are initially assumed to be unchanged. The results of the simulation are given in Table 7.7.

On the figures in the Table fixed capital costs for items other than power decline by 0·13$d$. per lb, but power costs rise as much as 0·206$d$. The explanation is that at the spindle speeds being attained in Lancashire in the 1820s and 1830s, the amount of power required increased approximately as the cube of the speed (in rev/min), largely through aerodynamic factors.[79] This cube-law has been adopted to derive the hypothetical power costs at 5000+ rev/min.

If the number of employees does not change, however, labour

---

[78] Montgomery, pp. 144 ff. Montgomery rated his mill at 25 horsepower in this work, but admitted in later correspondence with C. T. James that 30 horsepower was more appropriate. Other figures in Table 7.7 have been checked; for example, I have utilized the conventional ratios of £1 of plant and equipment per mule spindle and £2 per throstle spindle, etc. in the column entitled 'revised costs'.
[79] Catling, p. 54. Power costs at 4200 rev/min are therefore taken from Table 4.11; at 5000 rev/min = £617·9 (1·19)$^3$.

costs per lb of output will also fall; on the basis of Montgomery's data, from 0·653$d$. to 0·549$d$. But the assumption of no increase in employment is a strong one; it was seen earlier how significant for wage costs was any increase in quality, and it has been traditionally assumed that the same would have been true of a rise in speeds, with the quality kept constant. Additional piecing

TABLE 7.7
*Effects on costs of hypothetical increase in machinery speeds*

|  | Montgomery | Revised | Annual costs |
|---|---|---|---|
| Factory buildings | £1530 | £1679 | £151·1 |
| Machinery and gearing | 6863 | 6863 | 857·9 |
| Total | 8393 | 8542 | 1009·0 |
| Total including engine | 9053 | 9792 | 1165·3 |
|  | At 4200 rev/min | | At 5000 rev/min |
| Total output, lb | 300 275 | | 357 470 |
| Fixed capital costs, per lb | 0·807$d$. | | 0·677$d$. |
| Power costs, p.a. | £617·9 | | £1042·6 |
| Power costs, per lb | 0·494$d$. | | 0·700$d$. |

would have been required in aggregate, and quite possibly per unit of output. Hence it is unlikely that the decline in labour costs from the hypothesized increase in spindle and loom speeds would have entailed a reduction in labour costs of as much as 0·104$d$. In any case, there is a small offsetting factor in the rise in inventory requirements to match the higher speeds. If one-eighth of the yarn input is desired as stock, then the additional inventory costs will amount to 0·01$d$. per lb.[80] Moreover, no rise in the quantity of machinery or gearing has been allowed for, yet in the short run (at least) it seems probable that such will be wanted in certain departments. In general it seems plausible to represent the cost structure as being portrayed schematically as in Figure 7.2.

[80] Fragmentary data for the period suggest about 15 per cent of the output was held as inventory at any one time, e.g. FR (Factory Returns) 1842.

The optimal speed is dictated by the location of the point of minimum total cost along the horizontal axis.[81] It is a corollary of the above argument that if power costs fall (or other plant costs rise) the optimal speed for running the machinery will rise, and, reverting to the isoproductivity schedules, that the number of spindles per horsepower will fall, unless the whole isoquant shifts outwards.

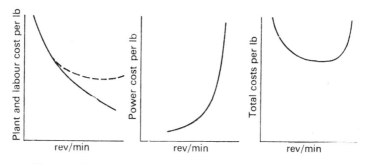

Fig. 7.2. Cost of yarn per lb spun according to spindle speeds

Of the many assumptions needed to produce these results, two in particular deserve emphasis. In the first place, I have assumed so far that the quality of output did not change. In the next chapter we shall find that it was possible for quality to change so much (in spinning flax yarns, as it happens) that the predictions are turned inside out. Fortunately, the quality of cotton yarns and cloth changed very little over the period I am principally concerned with (see Table 7.12 in App. 7.1). In the second place, I supposed that machine technology remained fixed as well. This would rule out just the kind of changes examined in the previous section: the evolution of the mule

[81] This is not to suggest that at every point of historical time within the period the actual optimum was attained. There is extensive evidence that in depressions of prices and profits, speeds were increased and led to some diminution in costs, though it is not entirely certain that the schedules graphed in Fig. 7.2 did not shift commensurately as well. Cases of speed-up through pressure on profits are noted, e.g. in P.P. 1816: III; P.P. 1833: VI; and P.P. 1833: XX. However, a rapid preceding fall in the price of coal (altering Fig. 7.2 (c)) is noted in P.P. 1833: VI; while in P.P. 1834: XX it is mentioned explicitly that 'Since the above period [1821–2] the improvements in machinery, fall in wages (about 20 per cent), and the reduction in the price and freight of coals (about 25 per cent) have reduced the cost of production fully as much as the above reduction of working hours (81 to 74½) raised it . . .' (p. 99)—in Dundee.

from being partially hand-powered to being fully automatic, and the variety of improvements in the power-loom. The evidence[82] shows that self-acting mules regularly spun about one-fifth more yarn in a given period of time than hand-mules, once quality is allowed for. If my model is appropriate, therefore, the higher capital costs associated with the self-actor (amounting to some 20 to 25 per cent per spindle)[83] must have swamped both the lower labour costs and the larger power costs in dictating optimal operating speeds, since both of the latter should have reduced speeds in isolation. The same goes for the throstle; the conventions applied in both the 1790s and the 1830s were £1 of plant and equipment for each mule spindle, but £2 for each throstle spindle.[84]

## POWER PRODUCTIVITY IN THE MID-NINETEENTH CENTURY

The model is designed to show how, even in the absence of technological progress of the kind noted in the last paragraph, power costs can strategically influence rates of production. It is time to confront it with the evidence. In Appendix 7.1 I have generated annual series on horsepower, employment, output, and costs—amongst other things—for the cotton industry from the middle or late 1830s to the mid-1850s. Here I shall concentrate on three key dates: 1835, 1850, 1856. If necessary, interpolations on the annual series can establish that these are good enough indications of the trends I shall be discussing.

The productivities of power and employment are shown separately for spinning and weaving in Table 7.8. Along with figures on spindleage, such as those given in Table 7.2, these results can provide us with expansion paths for the industry for these years. One such path, that between horsepower and spindles in cotton-spinning, is charted in Figure 7.3, p. 212.

The model enunciated in the last section can be used to clarify why the expansion path took on this 'flat' shape: it would be particularly interesting to find that reductions in

---

[82] *Thesis*, Table IV.8.
[83] This is simply the extra cost of the spindles: hand-mule spindles cost about 4s. 6d. each and self-acting ones up to 9s. 6d., while the mill itself cost £1 per spindle in the former case (*Thesis*, Table IV.5).
[84] *Thesis*, Table IV.6.

## Table 7.8
### Power and labour productivity in cotton-spinning and cotton-weaving, 1835–1856[85]

(A) Cotton-spinning

| Year | (i) Quantity of yarn spun m. lb | (ii) Total power 000s of horsepower | (iii) Product per horsepower 000 lb. p.a. | (iv) Employment 000s | (v) Product per man 000 lb p.a. |
|---|---|---|---|---|---|
| 1835 | 277 | 31·7 | 8·7 | 144 | 1·92 |
| 1850 | 571 | 58·3 | 9·8 | 182 | 3·14 |
| 1856 | 753 | 68·9 | 10·9 | 201 | 3·75 |

(B) Cotton-weaving

| Year | Quantity of cloth woven m. lb | Total power 000s of horsepower | Product per horsepower 000 lb p.a. | Factory employment 000s | Factory product per man 000s lb p.a. |
|---|---|---|---|---|---|
| 1835 | 200 | 10·9 | 14·7 | 72 | 2·24 |
| 1850 | 421 | 21·2 | 19·4 | 135 | 3·05 |
| 1856 | 588 | 27·2 | 21·4 | 173 | 3·37 |

[85] Col. (i): Yarn spun taken from consumption figures (App. 7.1), deducting wastage. Cloth woven deducts yarn exports, etc.

Col. (ii): For 1835, requirements of power-looms deducted from total U.K. horsepower (Table 7.15) essentially on the basis of 10 looms per horsepower including dressing. This figure of 10 appears in Ure, op. cit., vol. i, pp. 309 ff.; House of Lords Record Office MSS.; etc. Montgomery quotes 10 at Orrell's mill but only 6 at Peile & Williams's in Manchester. The nature of the product greatly affected the proportion (cf. P.P. 1841: VII—more power required for fustian waving than calicoes; also cf. linen in Ch. 8). A large number of regressions were carried out on the data in P.P. 1836: XLV and elsewhere, but to little avail. 1850 and 1856 figures split HP in combined mills according to the ratio of spindles to looms. See also App. 7.1 on 1856 spindleage.

Col. (iii): For spinning, Table A = col. (i) divided by col. (ii). For weaving, cloth woven on hand-looms was first deducted. This was established from data on *relative* productivities of hand-loom and power-loom assessed in Table 7.5; see also Habakkuk, op.cit., pp. 147–9. Number

## Steam-power and the Cotton Industry

power costs induced the movements down the isoquants. It will, however, also be recalled that the model assumed an unchanged level of machine technology, whereas the major development that actually took place at this time—the adoption of the self-actor—would have helped increase output per spindle and reduce spindleage per horsepower by itself. Nevertheless, even the diffusion of the self-acting mule was shown in the preceding section to be partly responsive to falling costs of power.

It is illuminating to study the background to these trends in the context of changes in the cost of production of cotton yarn and cloth. The top panel of Figure 7.4 graphs the value-added in spinning (given the way the numbers are calculated I could have taken weaving or both together and got the same pattern). The figures are taken from the relevant column of Table 7.13 but deflated using the implicit GDP deflator from Phyllis Deane's recent studies of nineteenth-century national

---

of hand-loomers taken from G. H. Wood, *The History of Wages in the Cotton Trade during the Past Hundred Years*, London and Manchester, 1910, but their low productivity means that a sizeable error will not greatly affect factory productivity (series also in Mitchell and Deane, *Abstract*, p. 187). Hand-loom product comes to 200 lb per man p.a. in 1834, 212 lb in 1849, and 222 lb in 1855. Since they were resorting to increasing fine cloths as a refuge from the competition of the power-loom, this slow growth in productivity measured in volume term seems not inappropriate.

Col. (iv): For 1834, Stanway extrapolated factory employment of weavers as 75 055, from the experience of 3 firms. Baines criticized this ratio as being too high, and with an arbitrary alteration of the proportion gave the number as 56 840 (P.P. 1834: XIX; Baines, *History*, p. 382), but then decided that his computation was perhaps too low. My assessment simply assumes 1½ looms per employee, which accords with miscellaneous data. For 1850 and 1856, I have taken numbers of spinners in spinning-only firms and weavers in weaving-only firms as given in FR, then extrapolated on the proportions of spindles to looms in combined firms. The result gives a constant ratio of weavers per horsepower over the period (6·6). Oddly, the number of power-looms per weaver falls a little in the 1850s (this occurs in weaving-only firms as well as in the extrapolated total); the reason is probably increasing loom speeds demanding greater application of labour—compare the American situation noted in Ch. 10. Spinning employment obtained by subtraction.

Col. (v): Col. (i) divided by col. (iv); with weaving output adjusted as for col. (iii).

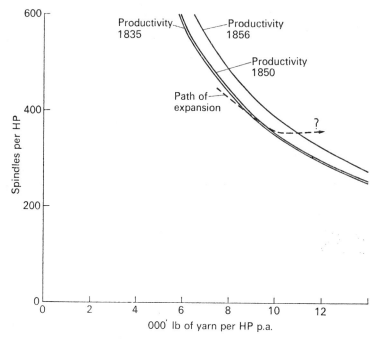

Fig. 7.3. Isoproductivity schedules and a possible expansion path for U.K. cotton-spinning, 1835–1856.

income.[86] The first decline in real value-added per lb in the late 1830s and early 1840s may have been achieved through harder driving of the engines. To quote the Factory Reports for 1842:

... the fall in prices stimulated mill-owners to lessen the cost of production by making their machinery, by various improvements and increased speed, turn off more work in a given time. To what extent it has been accomplished I have no means of forming an estimate, except from a few individual instances; and I have learned that, in one mill, the same number of spindles which produced 12 100 lbs. in 1834 [20 000 spindles?], produced 13 300 lbs. in 1841, and that one drawing frame, attended by one person, produced, in 1841, 2 700 lbs. weekly, while two drawing frames, attended by 6 persons, in 1831 produced only 900 lbs. weekly.[87]

[86] P. M. Deane, 'New Estimates of Gross National Product for the United Kingdom, 1830–1914', *Rev. Income and Wealth*, June 1968.
[87] P.P. 1842: XXII.

## Steam-power and the Cotton Industry

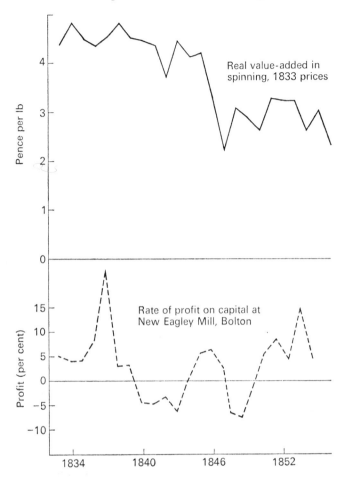

Fig. 7.4. Average value-added in cotton yarns, 1833–1856.

It is likely that power costs fell somewhat at this time, but unlikely that this was the main cause bringing about the rise in spindle speeds. Thomas Ashton, a large cotton-spinner and manufacturer from Hyde, was asked in 1842:[88]

What, in your opinion is the proportion of fixed capital to each horse-power, which in the case of new establishments being erected, may be considered as the usual outlay?

[88] P.P. 1842: XXXV.

Ashton: 'It will be more or less, according to the intended application of the power; but on the average I should say that at this time it would be £400 per horse power; in 1835 it was about £500.'

Q.: 'What has led to the difference?'

Ashton: 'The altered costs of the mills and machinery, and the application of more power to the same number of spindles or looms.'

The cost of hand-mule spindles fell from the 4s. 6d. or 5s. that had prevailed in the 1830s to 3s. 6d. or less in the 1850s.[89] Nevertheless, the fact alone is a little peculiar, since, as Figure 7.2 shows, reduced plant costs should have *lowered* working speeds (i.e. weakening the pressure to spread overheads). It seems highly likely that there were short-run advantages to raising rates of output even if the long-run optimum fell; e.g. ignoring the chance that harder working would cause the machines to wear out rapidly, for the sake of any cost-cutting to meet the rigours of depression in the early 1840s.

Yet there is no indication that the position was reversed during the partial recovery of the mid-1840s. On the contrary, the anonymous correspondent of *The Economist*, in his 'Review of the Cotton Trade' for January 1845, claimed that '. . . with the present demand for yarn and goods, and the low prices of cotton wool, we have no other limit to the consumption of the latter but the quantity of machinery. And as the speed has now been increased *to the utmost extent, and rather beyond what can profitably be used*, there is now no possibility of increasing the consumption of cotton wool except by new mills and additional machinery.'[90] In arguing that speeds of operating the machinery had overreached themselves in 1845 the correspondent was to be confounded by the speed-up that took place over the next few years. Figure 7.4 shows that much the most dramatic decline in real value-added per lb came about in the late 1840s, when it fell by about one-third. This is to deal in quite a short period, historically speaking, but the fall that occurred in that time was far and away the greatest since the initial mechanization of spinning and weaving, and would have few subsequent parallels (except perhaps in the 1930s slump).

[89] *Thesis*, Table IV.5.   [90] *The Economist*, iii, 25 Jan. 1845, p. 71. My italics.

Mill-owners were compelled to come to terms with both shifts in the demand curve for cotton textiles and movements along the demand curve imposed by shifts in the supply schedule. In the former respect one may note the readjustment of international demand following the repeal of the Corn Laws in 1846, the Irish famine, a panic in North America, competition in the Far East, and so forth. Foreign demand typically took around 60 per cent of total British cotton output by value and even more by volume in the mid-1840s. However, most of the reduction according to my figures arose in sales to the home market. Export values fell by approximately £2·7 m. in 1846–7 and by another £0·7 m. in 1847–8; home sales dropped about £7·75 m. in 1847 alone, and did not recover to the levels of 1844–6 until 1852.

Estimates of changes in production costs (with profits or losses deducted from the value-added series) will help to sort causes on the supply side from those of demand. The Factory Returns suggest positive (and perhaps supernormal) profits in 1845 and again by about 1852. These qualitative remarks can be given greater precision by using estimates published in *The Economist*. The figures in Table 7.9 of costs per lb are carefully considered annual averages.[91]

Another way of eking out the qualitative evidence is by examining the profits of one firm, the Ashworths of Bolton,[92] for whom we do have fairly continuous records of profits. In no sense were the Ashworths a typical firm: they produced goods of too fine a quality to represent the average of the industry outside Bolton, and they were principally water-powered. The data are provided in the bottom panel of Figure 7.4 for better or worse. A rate of profit of 6½ per cent such as they were earning positively in 1845 and negatively (as losses) in 1847 constituted on average about 24 per cent of the value they

[91] Ibid., 1 Dec. 1849, pp. 1328–9; 15 Dec. 1849, p. 1348.
[92] From Boyson, op.cit. This suggests that water-powered mills must have recovered alongside steam in the early 1850s. The Ashworths explicitly revalued their water-power in the light of revised steam costs (see p. 135). In addition, there were changes in the ratio of the price of water twist to that of mule-spun yarn: between 1846 and 1852 the strong shift in the ratio towards favouring the mule product that had occurred in the late 1820s was largely reversed. According to Mann's series, 30s water twist was actually more expensive than 40s mule twist from 1850; though other sources, such as T. Ellison (*A Handbook of the Cotton Trade*, London, 1858), still show a slight advantage for the finer mule-spun yarn.

TABLE 7.9
*Cost of yarn and coarse cloth, 1845–1849*

|  | Cost of cotton | Workmanship and waste | Total cost | Selling price | Profit |
|---|---|---|---|---|---|
| (i) 20s water twist | | | | | |
| 1845 | 4·1*d*. | 3·0*d*. | 7·1*d*. | 8·47*d*. | +1·37*d*. |
| 1846 | 4·8 | 3·1 | 7·9 | 8 | +0·1 |
| 1847 | 6 | 3·25 | 9·25 | 8·4 | −0·85 |
| 1848 | 4 | 3 | 7 | 6·65 | −0·35 |
| 1849 | 4·25 | 3 | 7·25 | 7 | −0·25 |
| (ii) 40s mule twist | | | | | |
| 1845 | 4·25 | 4 | 8·25 | 10·6 | +2·35 |
| 1846 | 5 | 4·1 | 9·1 | 9·55 | +0·45 |
| 1847 | 6·25 | 4·25 | 10·5 | 9·7 | −0·80 |
| 1848 | 4·1 | 4 | 8·1 | 7·5 | −0·60 |
| 1849 | 4·3 | 4 | 8·3 | 8·4 | +0·10 |
| (iii) Medium domestic cloths | | | | | |
| 1845 | 4·1 | 5·9 | 10 | 11·75 | +1·75 |
| 1846 | 4·8 | 5·95 | 10·75 | 11·25 | +0·5 |
| 1847 | 6 | 6 | 12 | 11·25 | −0·75 |
| 1848 | 4 | 5·75 | 9·75 | 10·25 | +0·5 |
| 1849 | 4·25 | 5·75 | 10 | 10 | 0 |

added to the raw materials. Thus the decline in value-added in 1845–7 was taken out mainly in a rapid fall in profits in the short run.

The recovery into the black in the early 1850s was, however, not achieved by any return of the value-added to its old mid-1840s levels. It is the primary task of the remainder of this chapter to explain how it was that the Ashworths and firms like them managed to restructure their costs in those years in order to earn normal profits on much reduced gross margins.

So far as the supply side was concerned, a number of factors out of the control of the individual millowner had acted autonomously on him in 1846–7 to force a response. Karl Marx, in Volume I of *Capital*, paid special attention to the limitation of factory hours to ten a day by the Factory Act of 1847—the first really effective and comprehensive factory act in Britain—and to the repercussions of that Act on the cotton-spinners in particular.[93] There were other exogenous influences on the supply curve; for instance, a sharp rise in the supply price of raw cotton from the U.S.A. and elsewhere late in 1846 which undoubtedly triggered off the slump in the first place, and a great coal strike beginning in 1846. All these were such as to shift the supply schedule upwards and so raise costs or reduce demand, *cet. par.* To disentangle all these factors would require the construction of an elaborate model, including the simultaneous determination of overseas supply and demand. This is far beyond my immediate intentions.[94] These shifts are here regarded as given, so that I can examine the medium-term response of British industry to them.

As Marx immediately recognized, the consequence of these exogenous shifts was rapid mechanization in the textile industries, and especially labour-saving mechanization. Marx may or may not be correct that the shortening of factory hours was primarily responsible—personally I believe that it was not so exogenous a factor as Marx implicitly did[95]—but his remarks on the techniques adopted seem to be remarkably accurate.

So soon as that shortening [of hours] becomes compulsory, machinery becomes in the hands of capital the objective means, systematically employed for squeezing out more labour in a given time. This is effected in two ways: by increasing the speed of the machinery, and by giving the workman more machinery to tent.

[93] Ch. 15, sect. 3c. M. W. Thomas (*The Early Factory Legislation*, Leigh-on-Sea, 1948) sees the passing of the Act essentially as a victory for reasoned liberalism as purveyed to the House of Commons by Lord Ashley (Shaftesbury) and later John Fielden. Without gainsaying the power of their oratory, a more cynical view would have it that the Act went through as a result of massive support from the aristocracy, in direct revenge against the manufacturers for the Repeal of the Corn Laws in 1846—as a casual glance at the voting pattern suggests.
[94] See G. Wright, 'An Econometric Study of Cotton Production and Trade, 1830–1860', *Rev. Econ. Stats.* lii, no. 2, May 1971, for one such attempt, and cf. P. Temin, 'The Causes of Cotton-Price Fluctuations in the 1830s', *Rev. Econ. Stats.* xlix, no. 4, Nov. 1967.
[95] There is some indirect evidence that certain manufacturers may have been aware of the possibilities of speedier operation to compensate for shorter hours, and if so may have weakened their resistance to the Act fairly deliberately, e.g. Hansard, 3rd Ser. xv, cols. 785–7, quoting James Marshall of Leeds. Cf. also the experiments carried on by the influential firm of Horrockses, Miller & Co. of Preston, reported in the *Manchester Guardian* and in *The Economist*, iv. I, 11 Apr. 1846.

Improved construction of the machinery is necessary, partly because without it greater pressure cannot be put on the workman, and partly because the shortened hours of labour force the capitalist to exercise the strictest watch over the cost of production. The improvements in the steam-engine have increased the piston speed, and at the same time have made it possible, by means of a greater economy of power, to drive with the same or even a smaller consumption of coal more machinery with the same engine. The improvements in the transmitting mechanism have lessened friction, and, what so strikingly distinguishes modern from the older machinery, have reduced the diameter and weight of the shafting to a constantly decreasing minimum. Finally, the improvements in the operative machines have, while reducing their size, increased their speed and efficiency, as in the modern power-loom; or, while increasing the size of their framework, have also increased the extent and number of their working parts, as in spinning-mules, or have added to the speed of those working parts by imperceptible alterations of detail, such as those which ten years ago [i.e. 1855] increased the speed of the spindles in self-acting mules by one-fifth.[96]

The correspondence between depression and technical change is underlined by Sir William Fairbairn, in a lecture delivered in 1849:[97]

To the skill, ingenuity, and careful attention of the Cornish engineers, we are indebted for many of the improvements connected with the use and application of expansive steam. . . . Although the Cornish miners set an excellent example and exhibited a saving of more than one-half the fuel, there were nevertheless few, if any attempts, made to reduce what is now considered an extravagant expenditure in most if not the whole of our manufactories. But in fact the subject was never fairly brought home to the millowners and steam navigation companies, until an equalization or reduction of profits directed attention to the saving attainable by a different system of operation.

Ten years ago, the average or mean expenditure of coal per indicated horse-power was computed at from 8 to 10 lbs. per horse-power per hour, but now it is under 5 lbs. per horse-power per hour in engines that are worked expansively, and even then they are far

[96] Op. cit.
[97] W. Fairbairn, 'On the Expansive Action of Steam, and a new Construction of Expansion Valves for Condensing Steam Engines', *Proc. Inst. Mech. Engineers*, i, July 1849, pp. 21–31; also in *Mechanic's Magazine*, li, and *Artizan*, vii, 1849, pp. 251–4. (My italics.)

below the duty of a well-regulated Cornish engine, with averages from 2¼ to 5 lbs. per horse-power per hour.

. . . One of the chief, if not one of the most important, reasons for the exercise of economy in fuel, is the reduction of profits on articles manufactured by power; under these circumstances, a saving in coal becomes a consideration of some importance: *and to these reductions alone may be traced the powerful stimulus which of late years has been prevalent in that direction.* The low rate of profit in manufacturing operations, and a desire to economize and reduce the cost of production to a minimum, has been of great value in its tendency to improvement in the economy and efficient use of fuel, and also to the use of high pressure steam and its expansive action when applied to the steam engine. In France and most other parts of the continent this system has long been in use; and although its effects as well as its economy have long been known in this country, it was only within the last few years that the benefits arising from it were appreciated.

As described in Chapter 4, the transition from low-pressure to high-pressure steam-engines in Lancashire awaited the development of much neater, direct-acting high-pressure engines, and few of these were to be found before the late 1830s. By the mid-1840s firms such as the Boulton & Watt Company were able to manufacture high-pressure engines at least as cheaply as low-pressure beam engines.[98] On the technical front the changes were admirably summed up by the famous engineer James Nasmyth (inventor of the steam-hammer) in a letter to Factory Inspector Horner, quoted in the latter's Factory Report for 31 October 1852. As a summary of the technological changes this letter is so concise and comprehensive that it is practically worth quoting in full. Short excerpts will have to suffice.

For a great many years after the introduction of steam-power into the mills and manufactories of [Lancashire, Cheshire, and Yorkshire], the velocity at which it was considered proper to work condensing steam engines was about 220 feet per minute of the piston . . .

---

[98] Boulton & Watt in the middle of 1842 quoted J. Ashley £500 for a 10-horsepower high-pressure engine with 13-horsepower boiler, delivered in Liverpool. A few days before, they had given W. Cotton an estimate of £550 for a 10-horsepower independent engine with (10-horse) boiler, delivered. Cf. Farey, *Treatise*, vol. ii, pp. 12–13.

However, at length, either through fortunate ignorance of the 'rule', or by better reasons on the part of some bold innovator, a greater speed was tried, and as the result was highly favourable, others followed the example, by, as it is termed, 'letting the engine away', namely by so modifying the proportions of the first motion wheels of the mill gearing as to permit the engine to run at 300 feet and upwards per minute, while the mill gearing generally was kept at its former speed as best suited to the requirements of the work.

But as in order to obtain increase of power from the same engine by permitting its piston to travel at a higher velocity it is requisite that we supply steam to such an engine in somewhat the same ratio as the increase in the velocity of its piston, it became requisite either to 'fire up harder', that is, burn more coal per hour under the same boiler or employ boilers of greater evaporating capabilities, i.e. greater steam generating powers . . .

It would not be very easy to get an exact return as to the increase of performance or work done by the identical engines to which some or all of these improvements have been applied; I am confident, however, that could we obtain an exact return the result would show, that from the same weight of steam engine machinery we are now obtaining at *least* 50 per cent more duty or work performed on the average, and that, as before said, in many cases, the identical steam engines which in the days of restricted speed of 220 feet per minute, yielded 50 horse power, are now yielding upwards of 100.

The very economical results derived from the employment of high-pressure steam in working condensing steam engines, together with the much higher power required by mill extensions from the same engines, has within the last three years led to the adoption of tubular boilers, in place of the simple waggon-shaped and cylindrical boilers, the tubular boilers yielding a much more economical result than those formerly employed in generating steam for mill engines.[99]

His remarks on new engines are also of some interest: '. . . Most of our mills lately built are worked by the Woolf engines such that the consumption of fuel is at the rate of from $3\frac{1}{2}$ to 4 lbs of coal per horse per hour; while in the engines of the old system the consumption used to be on the average from 12 to 14 lbs per horse per hour.'

His remarks in the penultimate paragraph on the resulting

[99] P.P. 1852–3: XL, pp. 483–7. Fairbairn, *Mills and Millwork*, p. 181, states that this was to regularize high-pressure action.

rise in efficiency may also be compared with those of another of Horner's quoted authorities, an important Lancashire 'millocrat'; '... within a few years (ever since the pressure of steam was raised from an average of 20 lbs to the inch to 35 lbs to the inch), there has been an increase in the power of engines after the rate of 10 horses to every 40.'[100] That is, an increase of 25 per cent—somewhat less than Nasmyth's reckoning, but he is describing only a portion of the increase in pressures that Nasmyth holds out for.

Nasmyth makes it very clear in some of the sections not transcribed here that in the major respects—the lagging of steam pipes and the adoption of higher steam pressures (he mentions ratings of 14 to 70 lb p.s.i.)—the Lancashire engineers were directly borrowing from the Cornish.

A partial exception lay in the gestation of the so-called Lancashire boiler. Patented in 1844 by Fairbairn & Hetherington this boiler directed the heat through its centre by means of cylindrical flues. Trevithick's Cornish boiler, already widely used in Cornwall, adopted the same principle, but with one cylindrical flue rather than two. By this means a much greater amount of heating surface was exposed to the action of the fire than in the old waggon boiler (cf. Chapter 3).

The use of internal-flued boilers such as the Lancashire boiler, and one or two developments of a similarly more 'embodied' kind, may have spread to a greater degree in years of more adequate profits, such as the brief prosperity of the mid-1840s and the more enduring upswing of the early 1850s. Other requirements for the use of high pressures involved minimal outlays of capital (valve adjustments, pipe-lagging, etc.) and one can imagine that even in severe depression there would have been an incentive to carry these out. There were sixty-three engines McNaughted between 1846 and 1851— largely years of depression.[101] No doubt the horrendous casualty rate from boiler explosions over those years (an average of more than one man killed a day in the late 1840s) is partly accounted for by attempts to extract high-pressure operation from low-pressure or at least worn-out boilers.

[100] P.P. 1856: XVIII, p. 236.
[101] *Artizan*, ix, 1851, e.g. p. 111. Cf. also 'Advertisements relating to the Great Exhibition' (B.M.), and Ch. 4 above.

The disembodied or quasi-embodied nature of the improvements, however, all too often makes it difficult to pinpoint the timing of the change in individual firms. If lucky, one obtains the following information:[102]

    1835. Benjamin Hick for Steam Engines... ... ... £1000
    1843. J. Marsland for new engine ... ... ... ... 1240
    1850. Speeding Marsland's engines as per folio 78... 87/2/6½
    1850. ,, Hick's ,, ,, 117/19/4

At the end of the 1850s the Inspector's Reports of the Manchester Steam Users' Association clarify the extent of transition.[103] The report dated 4 January 1859 gives first the distribution o boilers under inspection, according to steam pressures (pressures are measured *above* the atmosphere); this is set out in Table 7.10.

TABLE 7.10

*Distribution of boilers according to type and pressure in Manchester area, 1859*

| Pressure lb p.s.i.: | Under 15 | 15–30 | 30–45 | 45–60 | Over 60 | Total |
|---|---|---|---|---|---|---|
| Type of boiler: | | | | | | |
| Cylindrical, internal flues | 119 | 406 | 326 | 253 | 73 | 1177 |
| Cylindrical, without flues | 36 | 21 | 15 | 16 | 6 | 94 |
| Galloway | 4 | 29 | 36 | 33 | 1 | 103 |
| Multitubular | 3 | 14 | 24 | 43 | 35 | 119 |
| Multiflued | 0 | 8 | 19 | 5 | 1 | 33 |
| Butterley | 24 | 20 | 1 | — | — | 45 |
| Waggon | 7 | — | — | — | — | 7 |
| Total | 193 | 498 | 421 | 350 | 116 | 1578 |

[102] Dugdale MSS. E.
[103] Several of these reports for the late 1850s survive; the one used here appears to be the most comprehensive of those I have unearthed. It comes from Quarry Bank MS. C5/3/2. This association existed for the inspection of steam-boilers in a circuit of 35 miles around Manchester. For the circumstances surrounding its foundation, see W. Dinsdale, *History of Accident Insurance in Great Britain*, London, 1954, pp. 37–40, 133–4. It is admittedly very probable that those employing higher pressures would be more inclined to undergo inspection by the M.S.U.A., not just because they may have been more aware of changes in technique, but also because these boilers and engines were reputed to explode more often. Total indicated horsepower in Lancs cotton in 1861 was 203 567.

It will be noted that the old waggon shape (of which the Butterley Company's boiler is an improved form), which had been almost universal in the mid-1830s, had virtually died out by 1859. The amount of indicated horsepower supplied according to steam pressure and engine type from the same source is shown in Table 7.11.

TABLE 7.11

*Distribution of indicated horsepower according to engine type and pressure in Manchester area, 1859*

| Pressure lb p.s.i.: | Under 15 | 15–30 | 30–45 | 45–60 | Over 60 | Total |
|---|---|---|---|---|---|---|
| Type of engine: | | | | | | |
| Condensing: | | | | | | |
| low expansion | 2297 | 9898 | 3118 | — | — | 15 313 |
| moderate expansion | 680 | 12 370 | 5941 | 614 | — | 19 605 |
| high expansion | — | 4477 | 3731 | — | — | 8208 |
| Non-condensing | — | 30 | 389 | 59 | — | 448 |
| Compound | — | 495 | 9046 | 18 552 | 4189 | 32 282 |
| Total | 2977 | 27 270 | 22 225 | 19 225 | 4189 | 75 856 |

Thus within a decade or so of the extensive introduction of higher pressures there were relatively few of the old, non-expanding, low-pressure condensing engines among those reporting to the Steam Users' Association. Condensing engines were still the largest single type, but 65 per cent of their power came from engines that cut off the steam before one-half of the stroke in the cylinder had been completed; and 85 per cent of the remainder used steam of more than 15 lb p.s.i. above the atmosphere. Engines working compound, such as the McNaughted engines, supplied 42 per cent of the total indicated horsepower and were 46 per cent of the total number of reported engines. Not surprisingly, they dominated at medium and high steam pressures.

SUMMARY

In Volume I of *Capital* Marx spoke of three types of effect that mechanization had for labour. (i) He referred to the exploitative employment of female and child labour. (ii) He noted the tendency towards prolonging the working day. (iii) He stated that, somewhat paradoxically, the advent of machinery had 'intensified' labour, by which he meant that it had the effect in some circumstances of making labourers work harder—working faster in order to keep up with the machine.

In this chapter I have had occasion to refer to each of these, directly or indirectly. The core of my argument exists in the third of Marx's list of consequences. I have formalized a model in which the speed of machinery is explained endogenously. At the margin rapidly rising power costs largely dictate the optimal speed of running the machinery, and thereby the 'intensity' of labour. If all else stays the same, a reduction in the cost of applying 1 horsepower to a mill will promote an increase in machine speeds, and therefore in labour effort.

In Chapter 4 I showed that there was no perceptible fall in the money costs of power from the 1790s up to the mid-1830s. Initially steam-power was mainly confined to particular fields, such as carding and mule spinning. The largest mills, in which one might have expected the greatest economies of scale to be reaped, were still generally water-powered. The exception was provided by some fine spinning mills, so that steam supported the substantial cheapening of finer yarns that had begun in the late 1780s, with noticeable effects on both exports and clothing fashion in Britain. Otherwise the change in kind induced by the introduction of steam into cotton was hardly yet notable.

A slight speed-up occurred in the depression of the late 1830s to early 1840s, resulting in some decline in real value-added per lb woven. A much more startling decline in this variable, however, took place in the next depression, that of the late 1840s, when cotton output fell especially sharply. The industry was able to return to the profitable spinning and manufacturing of cotton at the much reduced level of their gross margins because of the fairly rapid introduction of a new power technology; coupled with greater automation of the industry using machinery such as the power-loom and self-actor.[104] Marx

[104] Success may be gauged by the statement in *The Economist* for 1852: 'Large as this [year's U S.] crop has been, it has been one of the most remarkable facts in connection with the present state of the trade, that it has not proved more than sufficient to supply the extraordinary demand of the year. Stocks, in place of having increased, as would have been expected, had such a crop been anticipated, are everywhere somewhat less than they were a year ago. . . . This remarkable increase of consumption has been effected, probably not so much by an actual

thought that the Ten Hours Act had inspired these developments; my feeling is that it was probably the most important determinant, though there were deeper-seated factors. Note that when the labour force benefited from a shorter working day (in 1847), it did so through even greater 'intensification'.

---

increase of the number of spindles in existence, as by the increased speed applied generally to all existing mills.' (2 Oct. 1852, p. 1092.) According to the figures generated in the Appendix below (for most of which, though not all, I would take responsibility), total factor productivity rose by 64 per cent in spinning and 53 per cent in weaving between 1835 and 1856 (see *Thesis*).

# APPENDIX 7.1

## The growth of the Cotton Industry, 1833–1856

In the latter sections of Chapter 7 I tried to leaven out the specific findings of the earlier sections—as regards power-loom, mule, steam-engine, etc.—with information on what was happening to the industry as a whole. In this way the behaviour of the individual entrepreneur could be gauged in context. The trouble with this broad view is that there is very little agreement in the secondary literature on series for inputs or outputs. This may come as a bit of a shock for an industry which has the reputation of being badly over-exposed in terms of the amount of research devoted to it. In fact, with the exception of Blaug's important article of 1961 there is no serious quantitative study of the industry for the first half of the nineteenth century.

In generating my own series I have aimed to render all the available information mutually consistent. That is, all extractable series are compared with one another, observation by observation—some agreement is taken as a first sign that the orders of magnitude are at least roughly correct. In further checking I have used any outside information I could lay my hands on; including all price series, all import series, contemporaries' assessments of quality, any pertinent engineering relationships, etc. Each additional series to be cross-checked raises the number of calculations factorially, so that the whole procedure was highly labour-intensive. Apart from generating new series on many variables, the method also brought to light inaccuracies in some data that were accepted even in the secondary literature.

Though it might seem slightly illogical, I shall begin by evaluating output before estimating the various inputs; this being the line of least resistance.

(i) *Output*

At first sight there appears little difficulty with estimating the *volume* of output, since all cotton had to be imported. Obviously there are continuous import series on raw cotton, upon which nearly all writers have gratefully relied. After deducting re-exports of raw cotton (available from the same sources), the remaining difficulties standing in the way of a production series, i.e. allowances for inventory movements and wastage of raw cotton in the manufacturing processes, have seemed readily soluble.

## Steam-power and the Cotton Industry

The problem is that we have almost a plethora of series available on most of these variables, even for this period in which statistics are otherwise so scarce. For most of the period 1833–56 there are three published series of lb of raw cotton imported, two of raw cotton re-exported, three of changes in stocks, and three of net consumption. The number of possible permutations is evidently very large, and in addition there are several more alternative observations in some cases obtainable from still other sources.[1] Something therefore needs to be ascertained about the compilation of these series and the possible degree of conflict, and this was summarized in Appendix V.1 of the *Thesis*.

By trial and error I have tried to see which of the many possible estimates gave the most plausible results when worked up. In no case, though, did I make any *ex ante* effort to relate output to fluctuations in inputs.

Deriving the *value* of output of British cotton textiles compounds the above weaknesses in the volume figures with probably greater inadequacies in series for unit values. All such calculations to date that are not just wild guesses begin from the well-known series for the 'declared values' of exports of cloth.[2] The biggest obstacle is that the unit value of cloth sold on the home market differed, and at times widely, from that intended for export. To get the over-all total I simply copied the technique invented by Professor Matthews[3] in his "Study in Trade-Cycle History".

It has been objected[4] that it is expecting too much of Matthews's neat solution to resolve the valuation difficulty over as long a period as, say, 1833 to 1856. In making this criticism, Blaug, by way of an alternative, simply assumed that the quality of production for the home market was *always* one-third above that for export. But this in turn does not square too well with Ellison's assertions about the relative values of domestically consumed and exported

---

[1] The major series of the Board of Trade (worked up from annually published figures by Mann, *Cotton Trade*); Ellison (1886); and R. S. Burn, *Statistics of the Cotton Trade*, London, 1847, are compared in *Thesis*, App. V.1. Up to 1830 J. Marshall, *A Digest of all the Accounts*, London, 1834, has a series overlapping on occasions with these; until 1833 Baines has alternative information (*History of the Cotton Manufacture*, p. 315). The monthly series for the Board of Trade, published from the late 1840s, have minor inconsistencies with their annual versions. Note also A. D. Gayer, W. W. Rostow, and A. J. Schwartz, *The Growth and Fluctuation of the British Economy, 1790–1850*, Oxford, 1953 (esp. microfilm supplement) for other consumption series.

[2] e.g. Blaug (1960–1), pp. 376–8; Mann, op. cit., p. 91; Baines, *History*, p. 431; Ashworth (1858); Ellison (1858), pp. 137–8; W. G. Hoffmann, *British Industry, 1700–1950* (trans. W. O. Henderson and W. H. Chaloner, Oxford, 1955); etc.

[3] Matthews, *Study in Trade-Cycle History*, esp. p. 150.

[4] Blaug (1960–1), p. 379.

cloth.⁵ On Ellison's figures, home-market cloth was 25 per cent more expensive per yard than exported cloth in 1819–21, 14 per cent more in 1829–31, 38 per cent more in 1844–6, and as much as 70 per cent more in 1859–61. Unfortunately, nobody knows how these ratios were obtained.

Now by using the stereotyped contemporary proportion of 5 yd of cloth per lb of yarn,⁶ I can compute the value per lb of both home and exported cotton manufactures implied by my results. The relative quality of home to foreign goods is set out in column (iii) of Table 7.12, following the estimates of total values by Blaug and me in the first two columns. Thus my estimates show that in the early 1830s domestic consumption of cloth is about 16 to 20 per cent higher in quality than exports; in the middle 1840s 30 to 34 per cent higher; and in the middle 1850s up to 55 per cent more. The concordance with Ellison's results is quite remarkable. Regrettably, the calculation performed here is sensitive to the precise conversion ratio between lb of yarn and yards of cloth, so that the case is not entirely clinched.

In columns (iv) and (v) yarn and cloth quality are computed, using Sandberg's estimates of the quality of exports. The particular estimation process makes these series comparatively insensitive to the above-mentioned yarn:cloth conversion ratios; and the picture of substantial stability that emerges will be very important in the later discussion. I may note here that this comes from the decline in quality of exports unearthed by Sandberg being largely offset by rising quality of domestic consumption.

To summarize, I put greater faith in a method of estimation of the value of cotton output which allows the relative quality of goods for home as against overseas markets to vary, especially since the figures of a near-contemporary imply a sizeable variation. I thus work with the series set out in column (ii) of Table 7.12; besides, it is highly compatible with the occasional miscellaneous estimate.⁷

---

[5] L. Sandberg, 'Movements in the Quality of British Cotton Textile Exports, 1815–1913', *J. Econ. Hist.* xxviii, no. 1, Mar. 1968, p. 3; Ellison (1886), p. 60; Blaug, pp. 378–9.

[6] Cf. J. Spencer, 'On the Growth of the Cotton Trade', *Trans. Manchester Stat. Soc.*, 1877; Ellison (1886). R. Burn's contemporary *Commercial Glance* gives as high a figure as 5·79 yd per lb for 1832, averaged over all exported cloths, but these figures have been strongly contested by Baines (*History*, pp. 407–10) and are particularly dubious. For 1844 (the only original copy of the *Commercial Glance* in the B.M.) the mean proportion is 5·16. Blaug uses a constant ratio of 5·47, drawn from results after 1870 (p. 378). Inspector Horner gives 5·15 for 60-reed and 4·31 yd per lb for 72-reed (FR 1855, p. 239). *The Economist* index of six types of cloth, used by Sandberg, Blaug, etc. from 1845, reduces to 4·61.

[7] Esp. Ashworth (1858), who gives £53·9 m. for 1853, cf. my £53·8 m., though this is rather too good to be true. Note that Matthews's procedure requires a somewhat arbitrary initial value—op.cit., pp. 148–9.

TABLE 7.12

*Value and quality of cotton textiles, 1833–1856*

| | Total value of output | | Relative quality Home/Exports | Yarn Quality | Cloth Quality |
| | Blaug £m. (i) | Computed £m. (ii) | (iii) | 1833 = 100 (iv) | 1833 = 100 (v) |
|---|---|---|---|---|---|
| 1833 | 36 | 32·3 | 1·17 | 100·0 | 100·0 |
| 1834 | 38 | 33·7 | 1·19 | 98·3 | 97·7 |
| 1835 | 42 | 36·6 | 1·16 | 93·7 | 94·5 |
| 1836 | 45 | 40·3 | 1·18 | 93·4 | 96·2 |
| 1837 | 43 | 37·5 | 1·18 | 105·4 | 110·5 |
| 1838 | 46 | 41·1 | 1·18 | 93·6 | 95·7 |
| 1839 | 41 | 36·7 | 1·20 | 89·9 | 93·3 |
| 1840 | 47 | 41·5 | 1·19 | 98·4 | 102·0 |
| 1841 | 43 | 38·4 | 1·23 | 98·3 | 100·3 |
| 1842 | 37 | 35·5 | 1·26 | 103·1 | 103·4 |
| 1843 | 42 | 39·2 | 1·29 | 95·4 | 95·9 |
| 1844 | 45 | 41·8 | 1·34 | 97·2 | 98·2 |
| 1845 | 51 | 46·8 | 1·30 | 97·3 | 100·5 |
| 1846 | 48 | 44·5 | 1·33 | 102·0 | 106·3 |
| 1847 | 33 | 33·5 | 1·41 | 103·8 | 109·1 |
| 1848 | 41 | 38·4 | 1·38 | 108·3 | 111·3 |
| 1849 | 44 | 42·0 | 1·43 | 97·5 | 99·6 |
| 1850 | 44 | 41·6 | 1·42 | 88·0 | 91·2 |
| 1851 | 47 | 44·2 | 1·52 | 88·8 | 92·3 |
| 1852 | 55 | 50·8 | 1·41 | 92·8 | 95·6 |
| 1853 | 59 | 53·8 | 1·41 | 91·8 | 94·6 |
| 1854 | 55 | 54·4 | 1·44 | 94·0 | 96·5 |
| 1855 | 57 | 55·1 | 1·55 | 97·6 | 101·2 |
| 1856 | 63 | 59·8 | 1·52 | 93·9 | 97·8 |

*Sources:* Col. (i): Blaug, op.cit., p. 376.
        Col. (ii): See text, and Matthews, op.cit., pp. 148–51.
        Col. (iii): Quantity of cloth exported, in yd, was reduced to lb weight by the ratio of 5 yd per lb of yarn. Lb of yarn consumed in home manufactures was then calculated residually. Since figures of aggregate values are readily available (exports as 'declared values' and total value from column (ii)), it is then straightforward to obtain value per lb of both home and exported goods. The series divides the former by the latter.
        Col. (iv)–(v): Sandberg's Assumption-A estimates of the quality of exports are re-scaled to 1833 = 100; and yarn exports, cloth exports, and domestic cloth are weighted by their proportionate shares in gross output (Sandberg, op.cit., Table 1, p. 8). The re-scaling procedure means that I am not bound to any one ratio of yd of cloth per lb of yarn in these columns.

(ii) *Costs of Production*

The value of output of finished cloth estimated in column (ii) of Table 7.12 can be split up into its constituents, viz., spinning, weaving, and finishing. Firstly, Blaug's thorough price series[8] for standard grades of grey cloth was extracted and used (as he had done) to establish total revenue from 'grey weaving', based on my slightly different output indices. This left value-added in the finishing stages by elimination. Next, the gross margin at the intermediate stage of spinning was estimated by assuming that 45 per cent of the value-added per lb in producing grey cloth arose at this level. This crude algebraic method means that I cannot expect to find any contrasting fortunes of the spinning as compared with the weaving branch, whereas it is known that such swings occurred over the course of the trade cycle. However, I am concerned with trends rather than cycles in this chapter; and in any case the division between spinning and weaving will rarely be needed in further estimates or results.[9] From these considerations I have computed Table 7.13.[10]

With one or two minor exceptions[11] for which allowance can be made where necessary, the results are in mutual accord and agree satisfactorily with extraneous information. The most striking feature of Table 7.13, and really the starting-point of the analysis in Chapter 7, is the rapid decline in the gross margins per lb spun, woven, and finished in the later 1840s (see Fig. 7.4).

---

[8] Blaug (1960–1), p. 376, col. 3. For the definition of grey cloth see ibid., p. 359.

[9] The figures have been checked against price series for yarns, esp. from Mann, op.cit., pp. 91, 95–6; Burn, op. cit.; Ellison (1858), opp. p. 140.

[10] Blaug's price index for grey cloth—an average over six different grades—clearly understates the whole amount of value-added in weaving. The gross wage bill in spinning and weaving in 1833, derived from Wood, *History of Wages*, Table 41, comes to £4·9 m. in factories and £3·2 m. on hand-looms. In spinning-mills wages averaged about 55 per cent of the value-added, and in weaving about 60 per cent. In Bolton in 1840 net wages of hand-loom weavers were some 75 per cent of the gross (clear weekly income was only about $51\frac{1}{2}$ per cent of gross); e.g. P.P. 1840: XXIV, p. 601, and Table 7.5. On these figures, and assuming Wood's hand-looming data are net wages, the value-added ought to have been about £8·7 m. in factories and £4·3 m. on hand-looms, or £13 m. altogether, rather than the £9·3 m. derived from Table 7.13 for aggregate value-added. Printing would have added at most about £$2\frac{1}{2}$ m. in the early 1830s (derived from material in Ellison, Baines, Donnell, etc.).

[11] The final column—value-added in finishing—is probably the simplest check; spinning and weaving look to be underestimated in 1847 and overestimated in 1855. Different averaging procedures are likely to be the most important reason for these irregularities.

TABLE 7.13

*Value-added per lb in cotton, 1833–1856*

|      | Total value, £m. | | Value-added per lb | | |
|------|------|------|------|------|------|
|      | Raw cotton | Cotton cloth | Spinning d. | Weaving d. | Finishing d. |
| 1833 | 10·8 | 20·0 | 4·38 | 5·37 | 15·1 |
| 1834 | 11·4 | 22·0 | 4·88 | 5·96 | 14·0 |
| 1835 | 14·2 | 24·5 | 4·59 | 5·61 | 14·1 |
| 1836 | 14·9 | 26·6 | 4·68 | 5·73 | 14·4 |
| 1837 | 11·1 | 23·0 | 4·68 | 5·73 | 15·2 |
| 1838 | 12·8 | 27·5 | 4·98 | 6·09 | 12·2 |
| 1839 | 13·0 | 25·6 | 4·74 | 5·80 | 11·1 |
| 1840 | 12·0 | 26·7 | 4·52 | 5·53 | 11·9 |
| 1841 | 11·9 | 25·3 | 4·41 | 5·38 | 11·5 |
| 1842 | 10·6 | 21·8 | 3·62 | 4·43 | 12·0 |
| 1843 | 10·5 | 25·7 | 4·15 | 5·07 | 9·7 |
| 1844 | 11·5 | 27·0 | 4·00 | 4·89 | 10·0 |
| 1845 | 11·5 | 29·3 | 4·07 | 4·98 | 10·2 |
| 1846 | 13·0 | 26·9 | 3·18 | 3·89 | 10·6 |
| 1847 | 12·4 | 19·9 | 2·45 | 3·00 | 11·8 |
| 1848 | 10·5 | 22·7 | 2·95 | 3·61 | 9·9 |
| 1849 | 14·0 | 26·7 | 2·80 | 3·42 | 8·7 |
| 1850 | 18·1 | 28·2 | 2·39 | 2·92 | 8·0 |
| 1851 | 16·2 | 30·2 | 2·97 | 3·64 | 7·6 |
| 1852 | 17·0 | 32·8 | 2·92 | 3·57 | 8·3 |
| 1853 | 18·1 | 35·5 | 3·18 | 3·89 | 8·4 |
| 1854 | 18·2 | 34·0 | 2·68 | 3·28 | 8·6 |
| 1855 | 19·6 | 38·6 | 3·10 | 3·78 | 6·7 |
| 1856 | 22·9 | 38·5 | 2·41 | 2·95 | 8·3 |

*Sources:* Col. (i): Volume of net imports, adjusted for stock movements, as described in *Thesis*, App. V. 1. Unit values from Mann, op.cit.

Col. (ii): Adapted from Blaug, op.cit., p. 376, col. (iii); using outputs from Table 7.12, col. (ii).
N.B. Includes yarn exports.

Col. (iii): Quantities of yarn were greater than of cloth, because some yarn was exported. Hence the series in this column is obtained algebraically, as described in the text—by allowing for the yarn exports in the course of comparing cols. (i) and (ii).

Col. (iv): As for col. (iii). Col. (iv) = 0·55 times the sum of cols. (iii) and (iv).

Col. (v): Col. (ii) subtracted from column (ii) of Table 7.12, and divided by output woven in lb.

(iii) *Power*

Far and away the most important primary sources for the estimates of horsepower supplied to the cotton industry are the Reports and censal Returns of the Factory Inspectors, beginning under Act in 1835. The major Returns—for 1835, 1838, 1850, and 1856—have often been quoted by economic historians. Of these the only one to have aroused more scepticism than is customary for any nineteenth-century statistics is that for 1835.[12] In the form in which the Returns were officially published there is one very good reason for this: no horsepower figures are provided for most of the cotton-spinning and -manufacturing area outside Lancashire. Partly for this reason the Lancashire industry is frequently quoted separately in what follows. But the 1835 Returns as they are convey information valuable to this study that cannot be obtained from the otherwise more extensive 1838 Return. Fortunately, Baines summarized the preliminary Returns for 1835,[13] made available to him privately by Inspector Rickards, and these supply much of the missing data. These preliminary horscpower figures are slightly adjusted where required by using the published official Returns on employment.

The major Returns are probably sufficient to carry out comparative static analyses such as in some of the later sections here. However, for a better appraisal of the dynamics of the cotton industry, such as is needed to explain the important discontinuities, there is much more to be gleaned from the half-yearly Reports. The surveys were often carried out at highly unprosperous times for the industry, the urgency of acquiring information or unemployment, etc. making this almost inevitable.

---

[12] Criticisms are summarized by D. T. Jenkins, 'The Validity of the Factory Returns 1833–50', *Textile History*, iv, Oct. 1973, but Jenkins misses a number of points. He claims that the 1835 Return was accepted at the time seemingly without criticism, but this was not the case. Inspector Saunders wrote a few years later '. . . When adverting to the number of persons employed in 1838, I refer to a return printed by the House of Commons in 1839. This return was needlessly complicated, difficult of reference, and collected at a great sacrifice of time and expense. I speak only of what relates to my own district, when I say, that it may on the whole be deemed correct, with the exception of a few clerical errors caused by the publication having been hurried, without the usual re-examination of all the figures, by the original documents. Another retuı n was made in 1835, but I have no confidence in the contents of that volume; I have therefore not made any use of it . . .' (FR for 14 July 1843, p. 372)—though Saunders must have been involved in reporting in 1835 himself. W. R. Greg also had complaints about the discrepancies between 1835 and later Returns, as early as 1838, though his arguments were more casually formulated. Jenkins also fails to bring out the differing definitions of horsepower used (capacity versus utilized, etc.), as noted later in this section.

[13] Baines, *History*, pp. 386–95.

The half-yearly Reports from Inspector Horner do seem to contain enough material to construct an annual series for horsepower in Lancashire cotton from 1838 onwards. The long gap between 1838 and 1850 is broken by two censuses of Horner's area; one in the depths of the 1841–2 depression[14] and another at the peak of a boom late in 1845.

In line with all these Returns, the figures here are in nominal horsepower, i.e. James Watt's measure of physical capacity; this has considerable advantages when it comes to estimating capital. It would be still more useful if one knew how much of the capacity installed was actually at work. The 1835 Returns exhibit the horsepower in actual use in Rickards's district (then mainly Lancashire–Cheshire) alongside the capacity estimates. In 1838 similar adjustments are made for the rest of the country, but Inspector Horner, for all the virtues Karl Marx found to praise in him, for once neglected to do his homework. From the Returns one can find how many mills were not at work in 1838, and such adjustments have been made to the figures of nominal horsepower in Table 7.14.

However, this was not the largest cause of below-capacity working in most years. Quantitatively more significant was the extensive practice of installing a steam-engine in a water-powered mill, to provide relief for the water-wheel in times of drought or flooding (see Chapter 6). Since this correction is not made in Table 7.14; I have called the results 'quasi-capacity'.

The Table disaggregates horsepower in Lancashire into the categories of spinning, integrated spinning and weaving, weaving only, and miscellaneous (mainly smallware manufactures). Little notice can be taken of the last of these, at least in isolation—it seems to suffer from changing classifications. Disaggregation into the remaining three categories could also be branded as over-confidence, with some justification, yet Horner's Reports record year-to-year changes in each section in nearly every case.

National estimates can be generated from the Lancashire totals in any number of ways. The method adopted here was to call in the employment figures in Lancashire, England and Wales, and the United Kingdom, to interpolate between the censal estimates for 1838, 1850, and 1856, (cf. Table 7.18). All that then had to be done

[14] P.P. 1842: XXII (App.); P.P. 1846: XX (App. I). The former compilation provides the one serious embarrassment, since it leads to totals of horsepower and employment *at full capacity* that do not square at all with those for 1838 or 1845, once intervening growth is allowed for. I have so far been unable to reconcile aspects of this particular Return with the picture that emerges from all the others. Happily the discrepancies are very much less for totals of horsepower and employment actually at work (cf. Table 7.14).

TABLE 7.14

*Horsepower by category in Lancashire Cotton Industry, 1839–1856, in thousands*

|      | Spinning | Spinning and Weaving | Weaving | Subtotal | Miscellaneous | Total |
|------|----------|----------------------|---------|----------|---------------|-------|
| 1839 | 15·6 | 18·3 | 0·7 | 34·7 | 0·6 | 35·3 |
| 1840 | 15·7 | 18·6 | 0·8 | 35·1 | 0·6 | 36·2 |
| 1841 | 15·1 | 19·3 | 0·9 | 35·4 | 0·7 | 36·1 |
| 1842 | 13·6 | 19·9 | 1·2 | 34·7 | 0·7 | 35·4 |
| 1843 | 13·7 | 20·0 | 1·2 | 34·9 | 0·7 | 35·6 |
| 1844 | 14·1 | 21·0 | 1·4 | 36·5 | 0·7 | 37·2 |
| 1845 | 15·0 | 22·2 | 1·9 | 39·1 | 0·7 | 39·8 |
| 1846 | 15·8 | 24·1 | 2·2 | 42·1 | 0·7 | 42·8 |
| 1847 | 16·2 | 27·1 | 2·1 | 45·4 | 0·7 | 46·1 |
| 1848 | 15·9 | 27·8 | 2·0 | 45·7 | 0·7 | 46·3 |
| 1849 | 16·9 | 28·5 | 2·3 | 47·7 | 0·7 | 48·4 |
| 1850 | 17·4 | 29·4 | 2·7 | 49·6 | 0·7 | 50·3 |
| 1851 | 18·3 | 31·2 | 3·2 | 52·9 | 0·6 | 53·5 |
| 1852 | 19·3 | 32·6 | 3·6 | 55·5 | 0·5 | 56·0 |
| 1853 | 20·9 | 34·2 | 3·8 | 59·0 | 0·4 | 59·3 |
| 1854 | 22·1 | 35·1 | 4·7 | 61·9 | 0·3 | 62·3 |
| 1855 | 22·8 | 35·2 | 5·0 | 62·9 | 0·2 | 63·1 |
| 1856 | 23·3 | 35·6 | 5·4 | 64·4 | 0·1 | 64·6 |

*Notes and sources:*

All figures are intended to refer to the "quasi-capacity" of horsepower (see text) installed at the *beginning* of each year. Rows may not add exactly to totals because of rounding.

1839: Calculated backwards from 1842, q.v., by adding (algebraically) closures in each year from Horner's 1842 Returns (hereafter FR) and subtracting new mills and additions from ibid., App. 2. Checked forwards from the 1838 FR for the total, and gives a plausible result.

1840: Do.

1841: Do.

1842: From Horner's 1842 FR (but see text). Data were collected Sept.–Dec. 1841.

1843: 1843 FR (1842 Returns plus new mills less mills not at work). Only 196 HP are recorded as the net gain in the year of severe depression, 1842, and this has been distributed among the subdivisions in proportion to their 1842 shares. The figure is too small to be worth more prolonged investigation.

1844: New mills and additions in each category from 1844 FR, added to 1843. Occasional instances of failure to note HP of new mills were treated as 30 HP.

1845: Do., from 1845 FR.

1846: Horner's complete Return for his district in 1846 FR, lowered by 1502 HP to allow for that part of Horner's district outside Lancs. For individual categories, whole increment for 1845 is extrapolated on partial FR for Jan.–Apr. 1845.

1847: Net addition to HP from 1847 FR. Spinning-only and weaving-only columns are taken as 1848 capacities (from 1850 FR, q.v.); residual attributed to combined firms.

1848: Backwards from 1850 FR. 1848 May–Dec. figures multiplied up for 12 months. In 1850 FR new tenants cannot be properly separated from new additions. Since the additions figures are almost certainly understated, by Horner's own admission (FR), my figures suppose that the two conflicting sources of error offset.

1849: Do.

1850: From county data in FR.

1851: Figures obtained backwards from 1856 were modified by averaging against 1850 Jan.–Apr. figures together with mills being erected in mid-year from 1850 FR. Totals thus obtained for spinning–only were identical, while those for combined and weaving–only virtually offset each other.

1852: Backwards from 1856, using FR of relevant year. Miscellaneous column has to be superimposed on a downward trend, others are not adjusted.

1853–5: Do.

1856: From FR, county data. Individual counties checked with 1850 FR for consistency.[15]

was to establish how power-intensive each Lancashire operative was in comparison with the rest of the country, and since this relative power-intensity changed little until after 1850 the rest was straightforward. Other approaches would give similar results, but it would be unwise to read much into the annual changes in U.K. horsepower.

(iv) *Capital*

The customary procedure for estimating capital accumulation in the spinning and weaving branches of the industry has been to extrapolate on the number of spindles and looms in use by means of conventional estimates of costs of mill and equipment per spindle and per loom. Blaug[16] re-used this method in his article on productivity in nineteenth-century cotton. The capital figures are not really crucial to my work here so that instead of fully reworking them I have accepted Blaug's figures but subjected them to my own amendments.

---

[15] Additions to horsepower between 1851 and 1856 are also given in J. R. T. Hughes, *Fluctuations in Trade, Industry and Finance*, Oxford, 1960, p. 78.

[16] Blaug (1960–1), pp. 364, 379–81; Fong, *Triumph of Factory System*, pp. 43–4.

TABLE 7.15
*Horsepower in the cotton industry, 1835–1856*

|  | Lancashire | England and Wales | United Kingdom |
|---|---|---|---|
| 1835 | 23·2 | 36·0 | 42·6 |
| 1838 | 33·0 | 50·3 | 59·8 |
| 1839 | 35·3 | 53·3 | 63·3 |
| 1840 | 36·2 | 54·3 | 64·3 |
| 1841 | 36·1 | 53·8 | 63·6 |
| 1842 | 35·4 | 52·4 | 61·8 |
| 1843 | 35·6 | 52·3 | 61·6 |
| 1844 | 37·2 | 54·4 | 63·1 |
| 1845 | 39·8 | 57·8 | 67·7 |
| 1846 | 42·8 | 61·8 | 72·2 |
| 1847 | 46·1 | 66·2 | 77·1 |
| 1848 | 46·3 | 66·1 | 77·0 |
| 1849 | 48·4 | 68·8 | 79·9 |
| 1850 | 50·3 | 71·1 | 82·6 |
| 1851 | 53·5 | 74·8 | 86·4 |
| 1852 | 56·0 | 77·6 | 89·0 |
| 1853 | 59·3 | 81·4 | 92·9 |
| 1854 | 62·3 | 84·8 | 96·1 |
| 1855 | 63·1 | 85·1 | 96·0 |
| 1856 | 64·6 | 86·4 | 97·1 |

*Notes*:

The 1835 Lancs. figure is from FR. For England and Wales and the British Isles the 1835 preliminary count, published by Baines, has been adopted and adjusted by the no. employed at the final count, as explained in the text.

The main evidence used in extrapolating from Lancs. to the rest of the country is the relative no. employed (from Table 7.18 below). See text and notes to Table 7.18 for details. Interpolations are between 1838, 1850, and 1856.

Blaug's results on the capital:output ratio were quite unexpected, given that he was describing a period of rapid mechanization.

TABLE 7.16
*Capital: Output ratios at constant prices, 1834–1886 (Blaug's results)*

|  | 1834 | 1845 | 1856 | 1886 |
|---|---|---|---|---|
| Total COR, spinning and weaving | 2·0 | — | 1·4 | 1·9 |
| Fixed COR, spinning and weaving | 1·4 | — | 0·9 | 1·3 |
| Total COR, all branches | 1·2 | 1·2 | 1·1 | 1·1 |

Blaug himself did not especially draw attention to the decline in all ratios between 1834 and 1856, since he was more concerned with the subsequent rise. For the 1834 figure Blaug mentions Ure's guesstimate of 12 million spindles. On the assumption that all spindles in the country were mules of the hand-powered type, that the average count was 40s, and that mules could produce 18 hanks per spindle per week in the early 1830s, the number of spindles in the country would be:

| Year | Cotton consumption | Mule spindles |
|---|---|---|
| 1833 | 263·5 m lb | 11·7 m. |
| 1834 | 274·3 ,, ,, | 12·2 ,, |
| 1835 | 286·2 ,, ,, | 12·7 ,, |

This agrees reasonably well with Ure's figure of 12 million spindles in 1835,[17] although he may have calculated similarly. Self-acting spindles do not greatly complicate the picture; Ure states that there were 300 000 to 400 000 of these by December 1834.[18] On the other hand throstle spindles may have constituted as much as a quarter of total spindleage. The higher average productivity of these and self-acting spindles would mean that I have over-estimated spindleage.

For 1856 there is less cause for complacency. According to the Factory Returns, on which Blaug bases his estimate, total U.K. spindleage (excluding that engaged on smallwares) rose from about 20½ m. in 1849–50 to 28 m. in 1855–6 and only 30·4 m. in 1861–2. This 1856 figure gave very odd results when used in all the cross-checking procedures I employed. These led me to believe that it is almost certainly incorrect. The correct figure probably lies between 24 and 25 m. spindles in 1856.[19] By itself, such a correction would accentuate the decline in the capital:output ratio shown in Table 7.16.

However, there are other problems. The most serious is that other forms of spinning than by hand-mule ought to be considered. I have not brought these out here, though they have been mentioned more casually in Chapter 7. What I have done instead is to recruit

[17] Ure, *Cotton Manufacture*, vol. ii, pp. 312–15. Contemporary estimates vary widely from Baines's 9·3 m. in 1832 (*History*, p. 368), to the 13 m. guessed by Andrew (1887) to exist in 1837, to even higher guesses quoted by Andrew. Ure's figure seems plausible in the light of related information including the estimate of 17½ m. made by Messrs. Du Fay & Co., the brokers, in a survey in 1845 (quoted by J. Baynes, *The Cotton Trade: Two Lectures*, Blackburn, 1857). Blaug himself is inconsistent, using 12 m. on p. 371 of his article and 10·8 m. on p. 379.
[18] Ure, *Cotton Manufacture*, vol. ii, p. 155.
[19] Interpolation is hampered by the change from nominal to indicated horsepower in the Factory Returns between 1856 and 1861, with no obvious conversion factor.

the output data generated earlier (Table 7.13) to calculate alternative capital:output ratios for spinning and weaving; these are set out in Table 7.17.

TABLE 7.17
*Capital: Output ratios at constant prices (modified results)*

|  | 1834 | 1856 |
| --- | --- | --- |
| Total COR, spinning and weaving | 2·1 | 1·6 |
| Fixed COR, spinning and weaving | 1·4 | 1·1 |

The net effect of these incomplete adjustments is to leave Blaug's results still in need of explanation.

(v) *Employment*

Everyone undertaking research on the nineteenth-century cotton industry is indebted to G. H. Wood for his compilations of wages, in the course of which he extracted virtually all the information available from secondary sources, arranged it according to region and specific job, and summarized it in index form. He also estimated total employment in spinning and weaving, but these estimates are a much more casual affair than the wage studies. Wood interpolated between the major censal years, using some supporting evidence from leading contemporaries. Occasionally he gets his census years confused; he uses Stanway's 1833 estimate for England and Wales as if it were for the U.K.; and—the most surprising mistake of all—he neglects the minor censal data for 1841, 1845, and 1847. His estimates for the 1840s of operatives in factories show U.K. employment increasing linearly by about 2000 a year to 1847, whereupon there are three large jumps of 18 000 p.a. in the mediocre years up to 1850. It is important for present purposes to establish a pattern which, if anything, is the reverse of Wood's for the 1840s.[20]

Firstly, factory employment in Lancashire is found by taking the numbers employed per horsepower wherever available[21] and inter-

[20] Wood, op.cit., esp. Table 41. Wood's 2000-a-year increase in the middle 1840s seems quite inconsistent with the evidence of labour shortages and rising wages during the boom, e.g. in Horner's words: 'Wherever I have been, I have heard statements of the difficulty of getting hands, of the rise of wages, and of demands for a still further advance.' (P.P. 1846: XX, Report for 26 Nov. 1845.)

[21] The population censuses for 1841 and 1851 were included, but because definitions were not the same as in FR (allowing for non-factory employment, etc.) a dummy variable had to be inserted to adjust them, so that their net influence on the calculations was correspondingly slight.

polating the ratio linearly between. This use of nominal horsepower to forecast employment was precisely what contemporaries did—Horner indeed seems to have regarded this as the most important function of the horsepower statistics he returned annually. There is much further comment in his reports of the appropriate employment:power ratio to be used.[22] One additional refinement beyond Horner used here was to use separate ratios for each category as defined in Table 7.14. For final confirmation the results were checked against other material for 1844 and 1845.[23]

TABLE 7.18
*Employment in cotton mills, 1835–1856, in thousands*

|  | Lancashire | England and Wales | United Kingdom | U.K. (Wood) |
|---|---|---|---|---|
| 1835 | 122 | 183 | 219 | 220 |
| 1838 | 150 | 219 | 259 | 250 |
| 1839 | 161 | 235 | 276 | 259 |
| 1840 | 164 | 236 | 277 | 262 |
| 1841 | 166 | 237 | 278 | 264 |
| 1842 | 164 | 232 | 271 | 267 |
| 1843 | 163 | 230 | 266 | 269 |
| 1844 | 169 | 236 | 273 | 271 |
| 1845 | 179 | 249 | 287 | 273 |
| 1846 | 191 | 264 | 303 | 275 |
| 1847 | 204 | 280 | 320 | 277 |
| 1848 | 202 | 278 | 317 | 295 |
| 1849 | 209 | 285 | 323 | 313 |
| 1850 | 216 | 293 | 331 | 331 |
| 1851 | 227 | 306 | 344 | 339 |
| 1852 | 235 | 315 | 354 | 347 |
| 1853 | 246 | 328 | 366 | 355 |
| 1854 | 255 | 339 | 378 | 363 |
| 1855 | 256 | 338 | 376 | 371 |
| 1856 | 259 | 341 | 378 | 379 |

*Notes*: Similar basis of estimation as Tables 7.14 and 7.15, i.e. all figures are "quasi-capacity" and refer to 1 January of stipulated year. Thus these cannot be used e.g. for estimating factory unemployment, though they are regarded as much more adequate for medium-term fluctuations than the series by Wood (Table 41) in col. (iv).

[22] e.g. P.P. 1842: XXXV (for Stockport); P.P. 1843: XXVIII; P.P. 1857: XVIII.
[23] R. H. Greg et al., 'A Return from 412 Cotton Mills in Manchester and Surrounding Districts', Manchester, 1844; P.P. 1846: XX (Horner). Both had to be geographically adapted to conform here. After doing so, Horner's total came to within 150 of the present estimate, and Greg's was a little over 1000 away.

Secondly, the ratio between Lancashire and total English or U.K. employment was established for all censal years, including 1847. A quadratic time trend was fitted to these ratios; from which the national totals were readily obtained from the Lancashire figures.

Quantitative estimates of the number of hand-loom weavers are embroiled in ideological debate. Bythell[24] argues that figures such as those of Wood are too high for the 1830s—if so, estimates of productivity growth in weaving will be biased downwards. In any case, as hand-loom weavers carry a proportionately small weight per capita in total weaving the deficiencies may not be calamitous.

[24] Bythell, *The Handloom Weavers*, esp. pp. 263–8. Babbage agrees on a decline before 1832 in Stockport.

# 8. Textiles, Steam, and Speed

CAN THE results of Chapter 7 be generalized to apply to other industries? A minimum test would be the significance of power costs for machinery speeds and thus productivity growth in other branches of the textile industry. In the interests of brevity I shall argue by example rather than give comprehensive economic histories of the textiles concerned.

For the woollen industries, broadly defined, it has been observed that the worsted branch took over from pure woollens in the early nineteenth century as the pacemaker for the sector. S.G. Checkland propounds this view as follows:

> The phases of development in the woollen (using the shorter, more easily matted fibres) and the worsted (using the long staple wools) showed important differences, both in comparison with cotton and between themselves... It was the worsted manufacturers who more quickly followed the lead of the cotton men. They were the younger branch of the industry, less bound by the past, more efficient, and in consequence, more highly capitalized. They enjoyed a good market for their yarn outside Yorkshire, in the worsted and hosiery trades of East Anglia... The adoption of the power loom, as we have seen, was delayed even in cotton, so that the thirties saw its almost simultaneous appearance and extension in both industries. Yet the greater initiative of the worsted men over the woollen was still apparent.[1]

## WOOLLEN AND WORSTED SPINNING

From medieval times the technology of spinning evolved along two main paths: the continuous and the discontinuous.[2] Worsted spinning was continuous. The primitive distaff was long employed until the arrival of the 'one-thread wheel' with its flyer. In the Industrial Revolution this gave way directly to Arkwright's water-frame and subsequently the throstle.

---

[1] S. G. Checkland, *The Rise of Industrial Society in England, 1815–1885*, 1971 edn., pp. 117–18.
[2] This is clearly set out by K. G. Ponting, *The Woollen Industry of South-West England*, Bath, 1971, p. 48.

In the woollen branch, however, the shorter staple and especially the mattedness of the fibres prevented ready application of roller drafting, as in Arkwright's frame. Out of the early 'Jersey' wheel, suited by its intermittent spinning and superior twisting to woollens, had arisen Hargreaves's jenny; essentially duplicating the motions for a large number of spindles. Within a few years of its invention the jenny was used for spinning woollen yarns, and as early as 1776 had been recorded as far afield from its native Lancashire as Shepton Mallet.[3] Greater success awaited the important technical improvements made to the jenny by Hargreaves's successors, plus the complementary development of the technically similar slubbing billey. In contrast to cotton, where the jenny had been so quickly overhauled, the woollen industry relied on it for about half a century. Jenny-spinning continued as a domestic occupation in Yorkshire and elsewhere long after other processes of woollens had been integrated into the mill. As such, it constituted an intermediate technology slowing down the introduction of the woollen mule. (Even the type of mule used in woollens, as Catling stated, 'could, perhaps, be more properly designated a self-acting jenny'.[4])

A rough rate-of-return calculation will demonstrate this. All the available material confirms the judgement of H. S. Chapman, in his well-known Parliamentary Report of 1840, that the productivity of a man engaged on mules was four or five times as large as on a jenny.[5] According to Baines,[6] gross wages for mule-spinning in Leeds in 1826 were 40s. a week, including 12s. for the piecers; whereas male jenny-spinners got 20s. 4d. in that year. On this basis, the saving in wages from using 200 mule spindles rather than 200 jenny spindles is cancelled out by the annual costs (in coal, lubrication, and labour) of having to supply artificial power to the mule[7].

---

[3] J. de L. Mann, *Cloth Industry in the West of England*, p. 123. See also Cunningham, *Growth of English Industry and Commerce*, p. 654.
[4] H. Catling, 'The Evolution of Spinning', in J. G. Jenkins, *The Wool Textile Industry*, p. 114.
[5] P.P. 1840: XXIII, p. 430.
[6] K. G. Ponting (ed.), *Baines's Account of the Woollen Manufacture of England*, Newton Abbot, 1970, p. 95.
[7] Adult males usually worked on 80-spindle jennies, so it would take $2\frac{1}{2}$ of them to propel as many spindles as this conventional-size woollen mule. As well as being smaller than cotton mules, woollen ones needed much more power—85

However, mule spindles worked faster than jennies, in line with Chapman's assessment. Large (80-spindle) jennies seem to have been producing about $1\frac{1}{2}$ lb an hour in the early nineteenth century; the 1856 Factory Returns suggest about $2\frac{1}{3}$ lb from the same spindleage.[8] If the latter is taken to represent mules, and the discount rate set at 5 per cent, then a decision by the entrepreneur to scrap jennies and erect mules would have broken even at the end of six years' use.

With an average life-expectancy of $13\frac{1}{3}$ years anticipated, the decision would seem rational, though not overwhelmingly lucrative. There would, though, have been another possibility: to substitute female jenny-spinners for males. At 8s. a week for females—a figure indicated by the 1833 Parliamentary Enquiry[9]—there would be no wage saving on a mule (one adult male plus piecers) compared to the five jennies needed as the alternative. The greater output of mules would be sufficient to cause the mule to break even in the course of its twelfth year of installation. Since many jenny-spinners were, in practice, females, it is little wonder that the conversion to mules was languid. Reasonably generous allowances for error in the data would still give much the same results.

Power costs strategically affect these conclusions. The purchase and erection of the (portion of) steam-engine needed to drive the 200-spindle mule amounts to some 40 per cent of the set-up costs. If power were to be assumed costless, the conversion to mules would cover expenses out of increased revenues within a year and a half, instead of the six years of the former example.

It was thus the jenny that brought about the biggest fall in costs of producing woollens, accounting for about 6d. a lb reduction for average superfine woollens (woollens that had cost 21d. to 27d. a lb for all processing in the 1780's).[10] The introduction of scribbling and carding machinery in the 1790s

---

[8] Jenny output for Yorkshire from Gott MSS. by interpolating on productivity of scribbling machines; for West Country from P.P. 1840: XXIV. 1856 figure for mules from Baines's estimate of aggregate wool consumption, in Ponting (1970), and spindles from P.P. 1857 (I): XIV.
[9] P.P. 1834: XIX (answers of mill-owners to questionnaires).
[10] *Thesis*, Tables VI.3, VI.9.

---

spindles per horsepower being the average for Yorkshire in 1850. Coal was costed at 6s. a ton; otherwise power costs were as in Table 4.11.

lowered costs by another 1d. The significance of these early advances may be seen in Table 8.1, where some of the more reliable cost estimates are broken down between spinning (including preparation), weaving, and finishing.

TABLE 8.1

*Percentage of labour costs of woollens in each process*[11]

|  | 1683? | 1724 | 1781–1796 | 1798 | 1796–1805 | 1820–1827 | 1830 |
|---|---|---|---|---|---|---|---|
| Spinning, etc. | 39·2 | 44·4 | 41·2 | 41·8 | 21·4 | 20·6 | 22·0 |
| Weaving, etc. | 18·1 | 25·8 | 33·2 | 29·0 | 44·1 | 48·9 | 40·1 |
| Finishing | 26·7 | 29·7 | 20·7 | 29·2 | 25·1 | 16·8 | 31·3 |

The hiatus between the first four columns and the last three is immediately apparent—and this is explicitly stated to coincide with the advent of the jenny. Rougher assertions support the conclusion: Heaton states that spinning was twice the cost of weaving in the late eighteenth century, but by 1840 H. S. Chapman could note that the value of spinning was one-third to one-half that of weaving.[12]

The throstle redressed the advantage the jenny had given woollens when it came to be introduced into worsteds in the Bradford area after about 1805. As in cottons, the worsted throstle was probably first water-powered, with steam soon utilized when water sites became congested.[13] These heavy demands on power were sometimes augmented, after 1831, by borrowing the American 'dead-spindle' system for finer yarns.

### WOOLLEN AND WORSTED WEAVING

Though flying shuttles were known in Huddersfield and elsewhere in the 1740s, their substantial adoption in Yorkshire dates from the 1780s like that of the jenny. The worsted sector was still slower to adopt them, but the positions were reversed in the 1830s when worsteds led the transition from hand-loom to power-loom. By 1856 there were 35 398 power-looms for worsteds in Yorkshire alone, compared to only 6275 for woollens.[14] Herein

[11] Ibid., Table VI.4.
[12] Heaton, 'Benjamin Gott and the Industrial Revolution in Yorkshire', *Econ. Hist. Rev.* iii, 1931; P.P. 1840: XXIII.
[13] Noted by M. T. Wild, in J. G. Jenkins, op.cit., pp. 213 ff.
[14] *Thesis*, Table VI.1.

lies one of the major indictments of the nineteenth-century woollen branch.

At John Foster & Sons (Black Dyke Mills) in 1849, worsted power-looms wove twice as many pieces a week as hand-looms.[15] This ratio is much lower than the conventional 3:1 for cottons (also supported in Chapter 7)—a potentially important factor in the lag of worsteds behind cotton in power-weaving. In woollens the ratio was evidently much lower still, for the following technical reasons:

(i) Power-looms were less economical (or technically efficient) in coping with higher qualities of cloth, and by and large woollen goods such as the West Country superfines were, *mutatis mutandis*, consistently finer than average Yorkshire worsted stuffs;

(ii) the process of woollen manufacture required a very loosely spun yarn compared to that for worsteds, since this looseness allowed it to be 'felted' in the finishing processes, but this itself made the application of the power-loom especially difficult—there was a tendency for anything less than a delicate touch to pull it to shreds;

(iii) the woollen loom was generally much broader than its worsted (or cotton) equivalent. In 1901 Clapham noted that most woollen looms were 5 to 8 feet in width (less than one-eighth were under 5 feet).[16] Nineteenth-century broadlooms could be 100 inches wide in the West Country.

These factors together meant that the customary speed at which woollen power-looms were run in the 1830s was only 40 to 48 picks per minute, compared to 80 or 90 for worsteds.[17] Indeed even hand-looms in woollens could be worked at close to 40 picks per minute, so that the incentive to mechanize on these grounds was exceptionally weak. To have attained a rate of output the same as that for worsteds would have implied, amongst other things, a disproportionate increase in power requirements and costs. This took place in the 1850s: the number of power-looms in woollens roughly doubled between 1850 and 1856, whilst at the same time the number of woollen power-looms per horsepower fell from $6\frac{1}{2}$ or $7\frac{1}{2}$ in 1850 to 3 or

[15] Taken from Sigsworth, *Black Dyke Mills*, pp. 192, 205–6.
[16] J. H. Clapham, *The Woollen and Worsted Industries*.
[17] Ponting (1970), pp. 70–1.

$3\frac{1}{2}$ in 1856.[18] By contrast, 1 horsepower would drive 8 to 10 worsted power-looms on average, and about the same number of cotton looms.

Reduced power costs and rising power productivity help account for these changes in woollens. John James, the author of the classic *History of the Worsted Manufacture in England*, made the same claim for worsted when he wrote in 1857:

> Certainly the present production of yarn and pieces is, in proportion to either the horse power, the number of spindles, or the looms, very much greater than it was a few years past. This, as regards the horsepower, arises from the very many improvements made of late in the steam engine, whereby its capabilities, compared with that [*sic*] of former days, are twofold augmented. . . . Then as to the spinning frames; they are now made with all the nicety of clockwork, and where formerly it was thought to be a capital performance for the cylinder to make 250 revolutions in a minute, now 360 is a common velocity. Since the increased use of what is termed the 'dead spindle', and which is adapted for some descriptions of yarn, the quantity of thread spun per spindle is very much larger. In power-looms 80 or 90 picks a minute were considered as a good speed, now the shuttle will traverse the web 160 or 180 times . . .[19]

Similarly, Henry Forbes, in a lecture to the Society of Arts in 1851,[20] declared that the number of picks per minute on a 6/4 loom had risen from 90 in 1839 to 180 or 200 by 1850. To maintain these speeds at lowest cost required a more efficient steam-engine.

In sum, the woollen branch stood to gain most in mechanical terms from falling costs of steam-power: (i) because in spinning the cost advantage of the mule over the jenny was rather indecisive in the 1830s, at least by comparison with that of the throstle over the one-thread wheel; (ii) because in weaving so much more power was needed to give woollen power-looms an advantage over hand-looms than for worsteds. Despite the 'power-intensive' throstle in worsteds, the woollen industry had twice as much steam-horsepower (alone) across the British

---

[18] From the Factory Returns. The respective higher figures are for Yorkshire alone.
[19] J. James, *History of the Worsted Manufacture in England*, London, 1857, pp. 355–6.
[20] H. Forbes, *The Rise, Progress and Present State of the Worsted, Alpaca and Mohair Manufactures of England*, London, 1852.

Isles as worsteds in 1838—a greater share in steam-power than in mill employment or total output.[21]

### FLAX-SPINNING

In the early nineteenth century the flax-spinning industry of England became concentrated in Leeds and that of Scotland in Dundee. The linen industry of Ulster rose to eclipse both, especially in the 1850s. Table 8.2 compares spindleage per horsepower, region by region.

TABLE 8.2
*Spindles per horsepower in 1850 and 1856*[22]

|      | England | Scotland | Ireland | British Isles |
|------|---------|----------|---------|---------------|
| 1850 | 88·9*   | 49·5     | 116·1   | 79·0          |
| 1856 | 115·3   | 53·0     | 81·2    | 80·4          |

The model of Chapter 7 would predict that, other things equal, lower spindleage per horsepower would result from cheaper power. Coals were cheaper in Leeds than in Dundee: 8*s*. a ton for the flax-spinners in Leeds in 1821, compared with 12*s*. in Dundee. Yet Table 8.2 shows that the predictions of the model are controverted, and this was equally so in 1821. Conversely, output per spindle was higher in Dundee than in Leeds: William Brown[23] estimated daily production at 950 'leas' (an English measure of length) per horsepower in Dundee but only 722 in Leeds. However, Leeds was producing the

---

[21] The woollen branch had 66 per cent of the steam-power, or 74 per cent of total power, in mills in 1838. By contrast, woollen-mills employed 63 per cent of total mill hands in the industry at this date. Figures for total output are highly dubious; data from the Select Committee of the House of Lords in 1828 indicate that some 54 per cent of total wool inputs (allowing for imports) were short wool (see P.P. 1828: VIII). For the few dates that we have relatively reasonable data this seems fairly close to the average (for 1850 61 per cent, see *Thesis*, Table VI.21).

[22] From P.P. 1850: XLII and P.P. 1857 (I): XIV. * = includes the additional 100 000 spindles for Yorkshire first suggested by J. G. Marshall ('Sketch of the History of Flax Spinning in England, especially as developed in the town of Leeds', British Assoc. for the Advancement of Science, 1858), also Fong, *Triumph of Factory System*, p. 117. This figure is especially dubious.

[23] W. Brown, 'Information regarding Flax Spinning at Leeds', 1821 (Leeds Reference Library), p. 1.

finer yarn. The low output and high spindleage per horsepower in Leeds, contrary to the spirit of the earlier model, can be explained by spinning these finer grades—other things are not equal.

Flax-spinning was carried out on a modified Arkwright frame. As with worsteds, there was a preparatory phase in which the long flax fibres ('line' or 'lint') were sorted out from the inferior short ones ('tow'), the tow being carded and spun into coarser yarns usually in the same mill. The analogous process to combing in worsteds was called 'heckling' or 'hackling'. After heckling the wetting of the line on the drawing rollers (in cold water) had been practised in mills as early as 1789 to permit easier drawing and the production of rather higher-quality yarn, imitating the use of saliva in hand-spinning. Wetting during preparation in factories, however, fell out of favour as it not only involved costs of its own but extra costs for drying, none of it adding up to the gains achieved. In the 1820s it was found that passing the line through warm water (90–150° F.) in the actual spinning process was vastly more effective. In such conditions the long flax fibre, held together in its natural state by a gummy substance (pectose), was briefly split up into what Rimmer calls its 'ultimate fibres', which could then be spun much finer.[24] The improvement was evidently of the anonymous learning-by-doing kind. However, to carry it effectively into practice required associated changes which were not at all costless.

(i) The old dry spinning frames needed fairly expensive conversion, because wet spinning called for a much shorter 'draught' between the front and back rollers of the frame—often the dry frames were simply scrapped.[25]

(ii) To produce the *same* size of yarn (line) a wet frame required a finer roving than did a dry frame. The key was an advance of far greater technical skill than the wet spinning frame: the use of 'gills' in the preparatory frames. Gills were

[24] W. G. Rimmer, *Marshalls of Leeds, Flax-Spinners, 1788–1886*, Cambridge, 1960, p. 169 ff.
[25] Marshalls sometimes broke up their old frames and rebuilt them, e.g. at their Shrewsbury mill in 1829. Cf. also Marshalls MS. 16/27: 'Reach of old spinning line—14 to 24 [inches]; do. tow, 4 to 12; do. of Wet spinning line and tow, $1\frac{1}{4}$ to $3\frac{1}{2}$ . . .'

really mobile heckles, mounted on 'fallers', travelling endlessly from the front (drawing) rollers to the back (receiving) rollers in the drawing and roving frames, moving a little faster than the front rollers so as to split the sliver and render the line increasingly regular. The most practical advance was Peter Fairbairn's screw gill, i.e. mounting the gills on a kind of corkscrew propulsion.[26]

(iii) Improvements in heckling the flax even before it went to the preparing frames were needed if heckling were not to become a production bottleneck. Cylindrical heckling machines to replace the skilled hand labourers were patented by Matthew Murray in 1805, though early machines saved little labour (except in substituting children for adults), were costly in machinery (requiring four to five cylinders of heckles), and produced excessive amounts of tow.[27] Pendulum heckling machines appeared in 1825, cutting labour and capital costs by reducing the action to one machine rather than four to five. Sheet heckling and improved circular heckling followed. According to the famous Leeds firm of Marshalls, these all contributed to the eventual success of wet spinning by furnishing a better and finer product to the gill drawing and roving frames.

The wet spinning itself involved ancillary costs, e.g. drying the line afterwards. Particularly relevant here were the additional charges upon power: wet bobbins were about half as heavy again as dry ones. H. C. Marshall reckoned the spinning frames to take up three-quarters of the total power of the Marshalls's factory even before going over to wet spinning.[28] This they did in the severe contraction following the boom of 1825.

In isolation one might expect these factors to have lowered the number of spindles per horsepower. In practice, as Table 8.2 showed, the ratio was twice as great in Leeds as in Dundee. Given the greater focus on wet spinning and high quality in

[26] D. Bremner, *The Industries of Scotland*, Edinburgh, 1869; Marshalls MSS.
[27] As early as 1810 John Marshall thought that heckling machines had reduced wage bills by a third, were cutting the line better than by hand, and were giving a greater yield of line than tow. (Marshalls MS. 15, and 'General State of the Business'.) See also *Thesis*, Fig. VII.2, for falling unit costs of heckling in the mid-1820s.
[28] Marshalls MS. 32, dated 1821–5.

the former, this must imply consequent shifts in the cost curves of Fig. 7.2 dictating considerably lower speeds. I have noted that Dundee speeds were already higher in 1821. At Marshalls the average number of 'leas' of line per spindle per day in 1824—before wet spinning—was 12·42, in 1846 only 8·12; for tow the respective figures were 11·00 and 7·7.[29] In 1834 their situation was summed up tersely: 'We produced the same number of dozens [yet another measure of length] per week in 1827 from $\frac{3}{4}$ the spindles spinning and twisting as we now have. There are no doubt many reasons for this: thread $\frac{1}{3}$ finer; time reduced from 72 to 66 hours per week; wet twisting introduced.'[30]

The Leeds industry thus persisted with laborious preparation, slow driving, and hard twisting, in the interests of higher qualities. Dundee spinners, on the other hand, were well known for their single-minded attention to faster rates of output, even at the cost of high wastage. Their machinery was modified with this in mind, e.g. inserting fine steel spindles and lightweight flys and bobbins.[31] Dundee's concentration on coarse flax evolved towards its more famous jute industry from 1836.[32]

The model is therefore inadequate for explaining the choice of yarn or cloth quality. Determinants of quality change must remain quite speculative in this work, but the need to investigate them further is heavily underlined. In all probability the demand schedules for varying grades of flax and linen shifted in the early nineteenth century under the influence of cheapening cotton and other competitors. Choosing the lowest-cost speed and spindleage for middling qualities of flax in Leeds would probably have meant minimizing losses rather than making any profits—in fact, I was able to show in the *Thesis* that if Marshalls had produced the same composition of output according to quality in 1830 as for 1824, they would have run up a loss for the year of several hundred pounds, as compared with the gross profits of £23 000 in 1824.[33]

[29] Compiled from their Annual Check Accounts. [30] Marshalls MS. 37.
[31] E. Gauldie (ed.), *The Dundee Textile Industry, 1790–1885*, Edinburgh, 1969.
[32] D. Chapman, 'The Establishment of the Jute Industry: A Problem in Location Theory?', *Rev. Econ. Studies*, vi, 1938–9. This article is a fascinating precursor of the 'New Economic History'.
[33] *Thesis*, Table VII.3.

For any given quality, though, the specification seems valid. A similar situation arose in the worsted sector in the 1850s, where despite the rise in output per spindle, etc., noted by James in the quotation on page 246, the average spindleage per horsepower rose between 1850 and 1856. From the later 1830s, and especially from 1847, the products of the worsted industry altered sharply. Comparing the first half of the 1850s with the latter half of the 1840s, James noted, somewhat to his own surprise, that: '. . . the manufacture of mixed goods had been extended, but not that of pure worsted stuffs, so that the increase is to be found in goods fabricated with cotton or silk warps, combined with Alpaca, and the finer description of worsted weft'.[34]

In the same vein, Baines said a year later: 'I am informed by a Bradford merchant of great knowledge, that "out of 100 pieces of worsted goods manufactured, at least 95 are made with cotton warps . . ."'[35] These trends, too, must owe something to the pronounced decline of cotton prices (Fig. 7.4), on both supply and demand sides.

The results obtained for the cotton industry in the last chapter can thus be generalized in certain directions. Specific machines in other textile branches owed their adoption in part to reductions of the costs of power. In the various weaving departments power-looms were resorted to most quickly in industries where a high ratio of power-looms to horsepower gave a productivity advantage over hand-looms.[36] At the same time other incidental charges (including the power costs themselves) would have to be covered, and this sometimes implied a *fall* in the number of power-looms driven by 1 horsepower when adoption became economic. In both spinning and weaving the relationships depended, often critically, on the quality of yarn or cloth under consideration, so much so that the simple predictions could be reversed.

[34] James, op.cit., p. 513.   [35] Ponting (1970), pp. 74–5.
[36] This also fits the case of linen-weaving.

# 9. Cornish Mining and the Cornish Engine

THE PREVIOUS two chapters have given pride of place to the introduction of the high-pressure engine, a process which has gone all but unrecognized in economic history. Available information relating to the diffusion of the high-pressure engine to the various textile and metal manufacturing regions of Britain was summarized in Chapter 4, but it was so fragmentary that the findings were perforce inconclusive. In this chapter I shall conduct a much more microscopic, but comprehensive, study of the origins of the stationary high-pressure engine, by examining its diffusion within Cornwall.

In the social savings calculations for the Watt engine (Table 6.5) it emerged that one-half of the fuel savings on reciprocating engines, and one-third of the savings over all, were accounted for by just one county—Cornwall. My estimate of just over £60 000 is closely in accord with W. J. Rowe's figure[1] of annual savings of £40 000 to the Cornish adventurers from Watt's engine, once slightly different time periods are allowed for. The significance of the Cornish copper- and tin-mines in the spread of best-practice technology in the eighteenth century cannot be exaggerated by these ratios. In invention as well the Devonians Savery and Newcomen early in the century are usually thought to have been stimulated by the needs of the mining districts of the south-west. Later in the century both Smeaton and Watt were to have great impact upon Cornish mining practice; and both men, Watt especially, were profoundly influenced in the nature of their improvements to the steam-engine by the demands of the Cornish miners.[2]

Boulton & Watt had made use of expansion of the steam inside the cylinder before condensation as early as 1785 in some of their engines in Cornwall. But so long as the steam from the boiler was kept at or little above atmospheric pressure, expan-

[1] W. J. Rowe, *Cornwall in the Age of the Industrial Revolution*, Liverpool, 1953.
[2] Watt not just by his engines but also by his participation in Cornish copper-mining ventures (see H. Hamilton, *The English Brass and Copper Industries to 1800*, London, 1926).

## Cornish Mining and the Cornish Engine 253

sion could not be taken very far. Jonathan Hornblower had earlier attempted to develop the use of expansion by constructing the first practical compound engine, but the advantages of the Hornblower engines *vis-à-vis* the Watt type were never clearly demonstrated, since low-pressure steam continued to be used. It was left to Richard Trevithick to be the first Englishman to make use of high-pressure steam, originally without condensation as in his locomotive engines. His engine at Wheal Prosper in 1812 is usually regarded as the first true Cornish engine, because it combined high-pressure steam with its subsequent condensation once its pressure had been allowed to fall to near atmospheric levels in the cylinder. Arthur Woolf was, however, of much greater significance in economic terms in Cornwall (and arguably deserves just as much attention in engineering terms[3]). Woolf's engines combined Trevithick's use of high pressures with Hornblower's of compounding in two cylinders. At this stage, however, compounding did not prove to have any very great advantage over carrying out the expansion and condensation phases in a single cylinder, and even Woolf himself reverted to one-cylinder engines. James Sims revived compounding in the 1840s, presaging the multiple compounding that was to characterize the most efficient steam-engines of the later nineteenth century.

Study of the engine in Cornwall is possible at a level unmatched elsewhere in this book (and indeed rarely in any literature covering the Industrial Revolution) because detailed reports were issued monthly from August 1811 by the Lean family. For each engine investigated (at the rate of one guinea a visit) they published its performance (load lifted, length of stroke, number of strokes per minute, etc.) as well as general remarks on the pumps, state of the boilers, cylinder size, and so on.[4] One column of the reports give 'lbs raised one foot high per bushel of coals consumed', or as it was called by contemporary Cornish engineers, the 'duty' of the engine.

As noted in Chapter 4, duty is a measure of the largest variable input into steam-engines, so that it can be directly

---

[3] e.g. for his water-tube boiler, cf. *Thesis*, Ch. I, and T. R. Harris, *Arthur Woolf: The Cornish Engineer, 1766–1837*, Truro, 1966; also R. Jenkins, 'A Cornish Engineer: Arthur Woolf'.

[4] A nearly complete set of the *Engine Reporter* was tracked down in Redruth Public Library.

put to use to construct an index of diffusion of a new technology. There is every indication of a qualitative kind that the goal of Cornish engineers, and frequently the criterion by which their efficiency was judged, was maximizing the duty of the engines they were entrusted with. In some cases the managers paid their engineers 'million money'[5] as a bonus for raising duty. The very publication of Lean's *Engine Reporter* was both a symptom of and a stimulus to the competitive spirit prevailing between rival engineers over duty. Accordingly, a body of data is available on precisely the kind of improvement that these engineers most desired to make.

The reliability of the Lean reports was extensively examined in an earlier article.[6] It seems reasonable to conclude that as a measure of what might be termed 'effective horsepower' they were subject to as many downward sources of bias as upward ones. If anything, they probably gave a less favourable picture of the duty attainments of Cornish engines compared to other types, e.g. the Watt engine. A much greater problem for my purposes than errors of measurement was estimating how representative the sample that appeared in the report was. Intuitively one might expect only the better engines to be submitted for reporting. Certainly the miscellaneous winding and stamping engines found at many mines, and some of the smaller pumping engines as well, would barely be worth the monthly fee as their aggregate coal consumption (and hence opportunity for absolute fuel economy) was so low. The neglect of these smaller engines limits discussion of the diffusion of best-practice techniques to the more powerful pumping engines, those for which economy of fuel was important in determining the over-all profitability of the mine. The Leans themselves commented upon the 'scale effects' of larger engines attaining higher duties, with economies generally being obtained up to 80-inch cylinders, then diseconomies (of duty but not necessarily of physical scale) for still larger engines.[7]

[5] W. J. Henwood, 'Presidential Address' to the Royal Institution of Cornwall, 1871. The reason for the term will soon become evident.
[6] N. von Tunzelmann, 'Technological Diffusion during the Industrial Revolution: the Case of the Cornish Pumping Engine', in R. M. Hartwell (ed.), *The Industrial Revolution*, Oxford, 1970.
[7] T. and J. Lean, *Historical Statement of the Improvements made in the Duty Performed by the Steam Engine in Cornwall*, London and Camborne, 1839.

## Cornish Mining and the Cornish Engine 255

The life-history of a typical engine can be sketched in briefly. In the first few months of operation its duty normally rose. The most important reasons for this were as follows: (i) time was required to find any leaks in the pumping equipment and to stop them; (ii) experience had to be gained in managing the fires and dampers optimally; (iii) the size of fire grates and number of boilers actually used could be varied; (iv) the most suitable way to regulate feed water entering the boilers had to be discovered; (v) in this period of rather primitive lubricants, any irregularities in rubbing surfaces had to be allowed to wear themselves smooth; (vi) the strength at which the steam was best supplied to the cylinder, and the length to which its expansion could best be carried, had to be found. There was therefore an initial period of learning about every engine, for each had its individual quirks. Thus some of the above relate to deficiencies in the construction of the engine itself. Others have reference to the surroundings, e.g. in the long run the extent to which expansion was carried depended on the weight of pit-work and balances to be lifted, because if these were too light, very high expansion would involve large fluctuations during each stroke in the rate at which power was transmitted, subjecting the machinery to great strain. (Perhaps the main reason why scale effects were obtained in duty was because larger engines were usually loaded less heavily p.s.i. of the piston.[8] Certainly many of them did not have to be over-exerted by unwatering requirements in the first year or two of their life, as the pits would still not be deep enough.)

Gains in duty through the learning process increased at a diminishing rate and eventually were outweighed by physical deterioration. Cornish engines were run twenty-four hours a day, so that this decline could be quite rapid, and overhauls soon called for. Any stoppage in the pumping action usually meant not just a zero advance but positive retrogression, with the mine 'going out of fork' and filling up with water. Important sources of physical decline were (i) the engines, pumps, and pit-work simply wearing out; (ii) leakages from the boilers into the fires becoming considerable; (iii) increased loads imposed

[8] Farey, *Treatise* ii. 254–60.

on the engine by new pumps as the mine deepened, especially if the new shaft were not vertical but inclined, as often occurred in the 1840s and subsequently; (iv) even without increased loads the plates of the boilers wore thin and the pressure of the steam had to be reduced for safe operation.

After the early success of compounding and high-pressure steam in the wake of the activities of Trevithick and Woolf in the second decade of the nineteenth century the performances achieved by Cornish engines lapsed somewhat. There were few outstanding technical advances during the early 1820s, though the opportunity was taken to strengthen pit-work and pumping equipment. These did not improve duty ratings—rather the reverse—but did make mechanically bearable the more extended use of expansion by engineers such as Samuel Grose in the later 1820s. Nothing structural had to be done to the engine itself, but every chance of conserving fuel in the whole operation was seized upon. All surfaces such as steam pipes that could be cooled by the atmosphere were thickly lagged to limit heat loss between boiler and cylinder, etc.

. . . It was essential to keep the cylinder as near the heat of the entering steam as possible and to this end it was customary to steam jacket the working cylinder occasionally, by enclosing it within an older worn-out one of larger diameter. To prevent further heat loss and, what was worse, condensation, the jacket was surrounded with brickwork and/or a wooden casing containing an insulating medium such as sawdust, ashes, or felt. This whole was then finished off with a polished mahogany or other wood cladding, banded with brass.[9]

The boiler was also covered under several feet of ashes. Using the combination of these relatively simple methods Grose and others were soon recording monthly duties over 70 millions, and in the middle 1830s some newer engines reached as much as 90 millions plus. They were thus three to four times as efficient thermodynamically speaking as Watt's own engines.[10]

The nature of the innovation makes a piecemeal engine-by-engine dating of the adoption of, say, greater expansion extremely difficult. And rather than resort to selecting a small number of engines for which the data might be adequate in

[9] Barton, *The Cornish Beam Engine*, p. 97.
[10] Cf. P. Mathias (ed.), *Science and Society, 1600–1900*, Cambridge, 1972, p. 73.

the hope that the sample may be representative, I continue with the use of duty, the productivity index, as a surrogate for the introduction of new methods. In the March observations being used here[11] no Watt engine in pristine state burning highest-quality coals is ever known to have performed at as high a level as 35 millions. The percentage of engines attaining that duty or more is thus a cautious proxy for studying the dissemination of expansive working when coupled with steam-jacketing and superheating.[12]

The diffusion of a new technology has frequently been suggested to be best approximated by the logistic curve.[13] This curve is asymptotic to zero and to the chosen ceiling value, and is symmetric around its centre, i.e. it takes on a leaning S-shape. A heuristic explanation might go as follows. Initially the new technique is in an experimental stage. The scope for its application can be known only at some cost. There may be considerable resistance perhaps out of ignorance but also because its cost in relation to the advantage it brings is high. This resistance will be especially strong if to adopt the new method requires a substantial diversion of resources. If its adoption does, however, lead to significant economic gain, other firms in different cost situations or less favourably disposed to innovation will be induced to follow suit. The technique has now blossomed out of the experimental stage; its scope and perhaps its flexibility widen. More and more firms introduce it, until diminishing returns start to set in. Gradually a ceiling is approached at which all those firms that can benefit will have done so.

This explanation may over-specify or under-specify the requirements for the curve as actually applicable to the Cornish engine. It is not possible to verify whether any or all factors

[11] The March figure was used as a compromise between light summer and heavy winter water demands; occasionally months either side were taken if the March figure was missing (von Tunzelmann, op.cit., p. 87).
[12] Some superheating was rather accidentally introduced by the drying-out of the steam through lagging, etc. Additional economy from drying the steam can be dated back as far as the Cornish engineers, Budge and Wise, before the 1780s (Harris, op.cit., p. 17 n.).
[13] *Locus classicus* in economic history is Zvi Griliches, 'Hybrid Corn: An Exploration in the Economics of Technological Change', *Econometrica*, 1957. The general derivation used here is taken from A. Lotka, *Elements of Physical Biology*, Baltimore, 1925.

operated. A more systematic derivation of the logistic trend will, however, be qualified below.

In this example the ceiling value is established at 91 per cent. The logistic curve can then be formulated algebraically as:

$$P = 91(1 + e^{m-nt})^{-1} \qquad (9.1)$$

where $m$ and $n$ are the parameters to be estimated, P is the percentage of engines of 35 millions in duty or over, and $t$ is time (zero in 1811).[14]

For regression purposes, eqn 9.1 is transferred to:

$$\frac{P}{91 - P} = e^{-m+nt}$$

Natural logarithms were taken and the curve fitted from 1816 (when P first became non-zero) to 1838, to give:

$$\log_e \frac{P}{91 - P} = -4 \cdot 402 + 0 \cdot 2516t \qquad R^2 = 0 \cdot 93$$

The fitted curve, along with actual observations, is plotted in Figure 9.1.

The high $R^2$ gave some strength to the belief that the logistic was appropriate description of the trend, but, as Griliches has warned, it measures the fit of the transform and not P (as in eqn 9.1) itself. The estimated values of P, call them P*, were evaluated by substituting in the fitted curve for $t$ and solving. 'Residuals' were then calculated as P* − P. A chart of these residuals showed, as indeed is obvious from Figure 9.1, that the logistic curve underrated the rate at which the improved practices were diffused during the central years 1825 to 1832.

It would be incorrect to suppose that because the steam-engine was a large fixed item of capital investment technical progress could be achieved only by being incorporated in new engines. Practically all the advances mentioned required at most the replacement of a small portion of the engine. Watt engines could be converted to high-pressure operation by lengthening the boiler feed-pipe, widening the steam-valves,

[14] There was a typographical error in the equations quoted in von Tunzelmann, op.cit., p. 92. This was kindly pointed out to me by Professor G. R. Hawke. The results are unaffected.

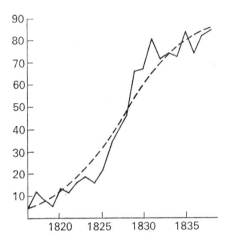

Fig. 9.1. Percentage of all reported engines performing at duties over 35 millions, 1816–1838.
(Solid line: actual percentages (March of each year). Dashed line: percentages estimated from the logistic curve).

and strengthening the steam-case and boilers (initially by extra internal stays).[15] The adoption of cylindrical boilers, etc. could come in due course. This piecemeal fashion of advance is still more marked in the late 1820s—lagging the pipes, etc. involved no necessary interference with the machinery. If embodied technical progress is defined as by Johansen, where by implication the productivity of each machine (disregarding depreciation) is date-stamped on that machine at the time of its construction until demolition, then these new techniques did not have to be embodied in new engines.[16] Older vintages of engines did not have to be dismantled to make way.

In Figure 9.2 the performances of three important engines are charted for each month from August 1827 to August 1829,

[15] Farey, ii. 182–3, 280–1; Fairbairn, *Mills and Millwork*, p. 180.
[16] L. Johansen, 'Substitution versus Fixed Production Coefficients in the Theory of Economic Growth: A Synthesis', *Econometrica*, Apr. 1959. Cf. also Salter, *Productivity and Technical Change*, and R. M. Solow, 'Investment and Technical Progress', in K. J. Arrow *et al.*, *Mathematical Methods in the Social Sciences*, Stanford, 1960.

during which each engine experienced a sharp but pronounced rise in duty of around 20 millions. It will be noticed that there is no discernible break in reportings while the rapid increase takes place; though if the engine were out of work for much less than one month its duty might still be reported.[17]

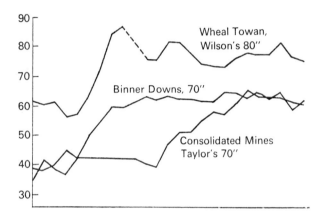

Fig. 9.2. Monthly duties of three major engines, August 1827 to August 1829.

Other definitions of embodiment are less exacting. If disembodied change consists solely of organizational change, then pure disembodiment cannot account for a dominating proportion of the later improvements. In a paper read in 1824 Joseph Carne described at least eleven 'economical improvements which have nothing to do with the construction of the engine...' yet which bore upon its performance, ranging from altering the management of the fires (not building the fire too deep, etc.) to better cleaning of the engine.[18] These could be even more widely diffused and perfected in the late 1820s, but their relative influence must have been waning. Yet the progress achieved and graphed in Figure 9.1 must qualify in most respects for inclusion in Landes's class of 'anonymous technical change'. The outlays involved were

[17] Data from Farey, ii. 217–18.
[18] In *Transactions of the Royal Geological Society of Cornwall*, iii, 1828.

small—in 1817 the engineers Jeffery & Gribble advertised that they could effect improvements to a 63-inch Watt engine at a cost of under £150[19]—while the rises in efficiency were disproportionately great.

The hypothesis being suggested is that the relatively inconspicuous and piecemeal nature, or even disembodiment, of the technical advances helps to account for their diffusion being unduly rapid.[20] Along the fitted logistic curve the rate at which high-duty engines increase is governed by the number of engines already performing at high duty (to a quadratic approximation). In this case the adoption of extensive expansion, superheating, and the like proceeded at a rate unencumbered by past history.

Alternative explanations can be conjectured for the fast rate of diffusion. It might have been simply the construction and reporting of an unprecedented number of large engines, since these consistently averaged higher duties. In fact the proportion of engines over 60 inches (cylinder diameter) appearing in the *Reporter* was exactly the same in 1832 as in 1825, while the ratio of engines under 40 inches to those of 40 to 60 inches doubled over the same years. Nor did my earlier work throw out any suggestions that trends in either price or quality of engine coals accounted for the especially rapid emergence of high-duty engines after 1825-6.[21]

Were profits the key stimulus? It is virtually impossible to derive accurate figures on profits even for *one* Cornish mine, because of the cost-book system of accounting, which failed to distinguish profits from capital expenditures in a way that modern theorists would demand. Still less, therefore, is it possible to derive a profits series for all or even a fair sample of Cornish mines. With a host of obvious reservations the most

---

[19] Barton, op.cit., p. 39 n.
[20] Even the more embodied changes associated with raising the horsepower of the engine could take place in a piecemeal manner. For example, it was common to replace cylinders by new ones under the existing 'superstructure': see Barton, op. cit., p. 127. Paul David, in his study of what he calls the 'Horndal effect', in New England textile-mills of this period, uses a definition of disembodiment which includes small components (*Technical Choice*, p. 178).
[21] Reasons for the possible contribution of these factors, and dismissal of them as operative in our period, are given in von Tunzelmann, op.cit., and *Thesis*, Ch. III. The coal-price series are for Cornwall, and are mainly drawn from the records of Harvey & Co. of Hayle, stored in the Cornwall County Record Office.

straightforward procedure is to take product prices as an index of profitability. Rowe takes the steady fall in copper and tin prices to indicate reasonably satisfactory cost reductions in the longer run. In the short run I shall suppose them to be adverse for profits, with increases in duty one of the more common eventual responses in the medium term.

Table 9.1 gives the zero-order correlation matrix for first differences of duty and of both copper and tin prices, the latter being given both concurrently with duty changes and lagged one year. The much higher correlations between changes in duty and lagged changes in prices (rather than concurrent changes) indicate that causation ran from fluctuations in the prices of copper and tin to changes in engine duty, and not the other way round.

TABLE 9.1

*Simple correlations between changes in average duty of Cornish engines and changes in copper and tin prices, 1813–1837*[22]

|  | $\Delta D$ | $\Delta P_c$ | $\Delta P_{c-1}$ | $\Delta P_t$ | $\Delta P_{t-1}$ |
|---|---|---|---|---|---|
| $\Delta D$ | 1·0 | −0·030 | −0·535 | −0·164 | −0·405 |
| $\Delta P_c$ |  | 1·0 | −0·028 | +0·372 | −0·233 |
| $\Delta P_{c-1}$ |  |  | 1·0 | +0·054 | +0·375 |
| $\Delta P_t$ |  |  |  | 1·0 | +0·239 |
| $\Delta P_{t-1}$ |  |  |  |  | 1·0 |

After some further experimentation, eqn 9.2 was accepted as a description of the relationships under scrutiny:

$$\Delta D = 1\cdot08 - 1\cdot18\Delta P_{c-1} - 3\cdot47\Delta P_{t-1} \quad R^2 = 0\cdot343 \quad (9.2)$$
$$\phantom{\Delta D = 1\cdot08 - 1}(0\cdot49)\phantom{\Delta P_{c-1} - 1}(2\cdot84)$$

The relative high standard errors shown in parentheses for each variable are partly the result of high intercorrelation

---

[22] Sources: $\Delta D$ (changes in average duty of reported engines)—from Sir Charles Lemon, 'The Statistics of the Copper Mines of Cornwall', *J. Stat. Soc.* i, 1838 (Farey has a series which differs in each year, but leads to the same substantive results in the regressions). $\Delta P_c$ (changes in prices of Cornish copper ore)—derived from ibid. by dividing the value of ore by tonnage. $\Delta P_t$ (changes in prices of tin ore)—from Mitchell and Deane, *Abstract*, p. 154 (a series for the price of common tin in London, 1818–37, given by J. Carne, 'Statistics of the Tin Mines in Cornwall and of the Consumption of Tin in Great Britain', *J. Stat. Soc.* ii, 1839, gave virtually identical results). The subscript -1 indicates a one-year lag.

between them, as shown in Table 9.1—presumably because of their vulnerability to similar shifts in demand. Even disregarding this complication the equation suggests that a negative relationship between duty changes and lagged ore price changes can be accepted at the 95 per cent level of confidence in the case of copper and 75 per cent level for tin. In other words, if my earlier, admittedly questionable, assumption about profitability being positively related to prices in the short run is correct, falling profits would give rise to greater technical advance than rising profits.

This is the opposite of Griliches's conclusion.[23] In line with my explanation of the rapidity of technological diffusion I take this to be the result of mostly inexpensive or at-hand methods for raising duties.

SUMMARY

Apart from rank disbelievers, most authorities regarded the Cornish engine as the highest level of technological accomplishment in its field between, say, 1815 and 1840.[24] In thermodynamic terms this view is powerfully supported by the evidence mustered here. Such performances were achieved in a succession of remarkably sharp bursts. By detailed investigation of the last of these bursts, from the mid-1820s to the early 1830s, I concluded that this was possible chiefly because of the 'disembodied' character of the developments—they involved little or no change to the outward appearance of the engine. As such it needed little in the way of capital investment to incorporate the advances, with the consequence that their diffusion was especially rapid in times of depressed prices for tin and copper. In this respect there is something of a parallel with the later extension of some of these ideas to the cotton industry around the time of the 1847 depression.

In many other respects, though, the parallels ought not to be taken too far. The different circumstances of coal prices and capital requirements meant more emphasis on compact

[23] Op.cit., and also his articles in *Science* and the *Journal of Political Economy*, reprinted in N. Rosenberg (ed.), *The Economics of Technological Change*, Harmondsworth, 1971, Chs. 9, 11.
[24] From the late 1830s the performances of the engines within Cornwall fell off. For suggested explanations see *Thesis*, Ch. III.

engines with a smooth transmission in the textile-manufacturing districts, as explained at the end of Chapter 4. But in France and on the Continent generally coal prices were higher even than in Cornwall, and the response was an unusually early adoption of the Cornish engine in Europe.[25] This will be demonstrated in the case of Belgium in the next chapter.

[25] French theorists were quick to advocate high pressures in the interests of optimal fuel economy, e.g. E. P. C. Dupin, *Rapport fait à l'Institut de France*, Paris, 1823; A. Morin, *Leçons de mécanique pratique*, ii, Paris, 1850; J. M. Hoëne-Wroński, *Machines à vapeur; Aperçu de leur état actuel*, Paris, 1829, p. 16.

# PART IV. PERSPECTIVES

## 10. The International Context in the Early Nineteenth Century

IN EXAMINING the forward linkages from steam to cotton textiles in Chapter 7 I adopted a counterfactual approach to costs, in showing that production costs for Lancashire spinning-mills would have been higher if in the 1830s they had run their spindles much faster. In Chapter 8 though, it emerged that the model was weakened by having to take account of changes in quality of product. Is my attribution a false one? It would be helpful to know whether the model can clarify diffusion patterns not only through time in the U.K. but across countries at certain points of time. Professor Redlich[1] has asserted, for instance, that counterfactual explanations lie outside the realm of traditional historiographical practice, and that a preferable alternative is 'comparative history', viewing the impact of economic change in one country in the light of what was happening in others. This looks suspiciously like a false dichotomy, since the 'comparative' approach simply replaces one kind of counterfactual with another: the behaviour observed in other countries is assumed to act as 'what might have been'. In some cases it may be a much poorer counterfactual than devising hypothetical situations for a given country, since too many factors may alter from one country to the next; and even if complex estimation procedures are recruited, many of them may escape attention: the 'control' gets out of control. This in fact happens, in certain respects, in comparing the U.K.

[1] F. Redlich, ' "New" and Traditional Approaches to Economic History and their Interdependence', *J. Econ. Hist.* xxv, 1965.

on the one hand and the U.S.A. and Belgium on the other. Nevertheless, the counterfactual employed in this 'comparative' model does have the defence of being a situation that has, for whatever reasons, actually come to pass. Arguments of the sort that, regardless of its theoretical merits, the hypothesized alternative rests on specious historical reasoning thus cannot apply. In terms of the limitations of 'comparative history', the main problem is that costs other than that of power vary among nations. A complete study has to take account of labour alongside capital, and of heterogeneous labour (especially in composition according to skills) alongside heterogeneous capital.[2] For instance, Wallis in 1854 pointed out that Americans tended to run their power-looms slower than did the English, because by so doing the employee could look after a greater number of looms (three or four) at once.[3] The smallish loss in extra power costs was outweighed by the savings on labour, in a country where traditionally labour scarcity has been allowed such an important role in explaining the evolution of technology.

### THE U.S.A.

Did productivity differ in the United States? If so, can the model adumbrated in Chapter 7 help to explain why? This might be seen as a minimum test of its practicality. More ambitiously, can the model further our understanding of the nature of those techniques likely to be more rapidly diffused?

On the issue of relative productivities it follows from the model that if power costs are lower, speeds of operation will be faster, and the amount of machinery per horsepower greater, other things being equal. David J. Jeremy among others has recently demonstrated that Americans worked their machines more quickly than the British.[4] He explains this as just one of nine factors exhibiting the response of American entrepreneurs

---

[2] For an approach to the matter of labour-force composition, see D. L. Brito and J. G. Williamson, 'Skilled Labor and Nineteenth Century Anglo-American Managerial Behavior', *Expl. Econ. Hist.* 2nd Ser. x, 1973.
[3] N. Rosenberg (ed.), *The American System of Manufactures*, Edinburgh, 1969, p. 216. For a similar situation in Rhode Island in 1842 see P. McGouldrick, *New England Textiles in the Nineteenth Century: Profits and Investment*, Cambridge, Mass., 1968, p. 23.
[4] D. J. Jeremy, 'Innovation in American Textile Technology during the Early 19th Century', *Technology and Culture*, xiv, Jan. 1973.

## International Context in the Early Nineteenth Century 267

on the supply side to British competition and American tastes. Jeremy gives two reasons for higher speeds in the U.S. textile industry: (i) the saving of labour, explicitly in the Habakkuk mould, and (ii) the introduction of power-looms in the U.S.A.

The first point appears to be mistaken. The usual interpretation of Habakkuk is one of American technology being *relatively* labour-saving, i.e. in relation to the saving of capital, natural resources, or whatever. Here and elsewhere in his article Jeremy misinterprets the argument as one for *absolute* labour-saving (a reduction in labour input per unit of output)—a very weak version of Habakkuk's thesis. In *relative* terms, as my argument in Chapter 7 showed in passing, an increase of speeds is likely to have augmented the amount of labour used per machine, in order to cope with additional breakages of the yarn or cloth (there may also have been some rise in machine depreciation, but insufficient to offset this trend). The traditional Habakkuk argument should therefore have led to higher speeds in Britain than in the U.S.[5] Jeremy's second explanatory factor is even less clear. He presumably does not mean a change in the type or quality of yarn in order to suit the newly introduced power-looms. Most probably he is resorting to something like the 'challenge-and-response' view of textile technology, with improvements in weaving stimulating improvements in spinning, and conversely. If so, then the argument is even more under-specified than the orthodox one for eighteenth-century England. Why not more spindles instead of faster spindles as a reaction to the power-loom, especially as faster spindles tended to reduce the capital: labour ratio?

My intention is therefore to argue that differing power costs were important in explaining the faster spindle speeds adopted in the U.S. The obvious departure point for the discussion is to ask what differences in costs of power there were. In Appendix 6.1 I transcribed Temin's data on costs of steam- and water-power (taken from C. T. James[6]) and expressed them in English currency at effective exchange rates. As such they bear out the proposition starkly: American power, whether

---

[5] Not, of course, that Habakkuk himself would have made this contention (see below).
[6] 'Justitia' (C. T. James), *Strictures on Montgomery on the Cotton Manufactures of Great Britain and America.*

steam or water, was barely half as expensive as that in the textile centres of Britain. If anything, the American estimate maybe more inclusive of supplementary costs such as steam-heating than the British figure, thus underrating the real differences.

Regrettably, the American figures cannot be accepted as they stand. C. T. James spoke impassionedly in favour of steam-power, but his attempt to prove that it was competitive with water-power remains unconvincing. He selected two Massachusetts towns for his comparison: Lowell—the hub of the *ante-bellum* American industry—for water-power, and Newburyport for steam. The $1\frac{1}{3}$ tons of coal he supposed were consumed daily by the 75-horsepower engine at the Bartlett no. 1 mill in Newburyport are equivalent to about $3\frac{1}{3}$ lb per horsepower per hour, *including* heating.[7] One of his newspaper correspondents pointed out in addition that the consumption per horse he had estimated for the steam-powered Bartlett no. 2 mill (not yet in operation when James was writing) was actually less than the water-powered Boott Mills at Lowell needed simply for heating, according to James's calculation for water-power. James stood unabashed and talked vaguely of technical improvements in the steam-engine, noting the Cornish engine especially. If consumption of fuel by the steam-engine were the same at Newburyport as in Lancashire—say $14\frac{1}{2}$ lb per horsepower per hour, including heating—its annual cost would be about $150 rather than the meagre $35 per horsepower recorded by Temin.

Corroborative detail on the costs of American steam-power is lacking. The figure just quoted is indefensible in a different way from Temin's: it assumes the same sort of low-pressure condensing engines as those in Lancashire mills, whereas in fact the American steam technology had evolved for the most part along rather different lines. In the wake of Oliver Evans the characteristic American version was the high-pressure non-condensing engine. Without compounding or the use of exceptionally high pressures it could not be very effective

[7] Ibid., pp. 30–1. In his Appendix James remains unfazed by these criticisms, which were clearly also levelled at him by his contemporaries. Temin also takes them in his stride ('Steam and Waterpower . . .', p. 198 n.), though he quotes $5\frac{1}{2}$ lb per horsepower per hour for the *1850s*, and obfuscates the question of including heating.

in economizing upon coal.[8] Before the advent of the Corliss engine compounding was not especially popular in the U.S.A., and though pressures were no doubt higher than on, say, the Cornish engines studied in Chapter 9, the proportional increase had to be very great indeed to reap the kind of economies that Temin appears complacent about.

From my earlier discussion, steam- and water-power costs ought to be the same in equilibrium, and Temin in fact made this point explicitly in his article. If this presumption is not to be scrapped, then several possibilities emerge to explain these results. First, fuels cheaper than coal (e.g. cordwood[9]) may have been burned in the engines. Second, it may be that steam-powered mills for textiles were not viable in Massachusetts at this date.[10] Some of the 317 engines recorded as being at work in 1838 could have been employed on non-competing products, as I have shown occurred in Britain. Third, the water-power cost figures may also be awry. One is alarmed by the 6 per cent discount rate allowed by Temin on all capital equipment; an upward revision to provide more generously for interest, depreciation, wear and tear, etc. would have a larger impact on water-power with its high capital costs.

Further research is needed before it will be possible to appraise satisfactorily whether these, or other, revisions are warranted. The work of Temin and his source can however serve to buttress my argument in other ways; for in these fields, more important in the over-all context of this book, they are supported by both theoretical presupposition and other contemporary sources. For this purpose I plan to extract a consensus on ratios among spindles, horsepower, and speeds in the U.S.A., then try to explain them.

At Lowell power was let off in units referred to as 'mill

[8] Temin (op.cit.) notes a Philadelphia trial that yielded similar coal consumptions for Evans and for low-pressure Boulton & Watt engines, and also that other researchers have put the American consumption higher.
[9] See N. Rosenberg, *Technology and American Economic Growth*, New York, 1972, for use of wood by Midwestern steamboats.
[10] Except where the water resources were less than fully adequate, e.g. at Newton. Note especially McGouldrick's verdicts: 'Until the 1850's, water power was much more economic than steam power, and this restricted mill building to sites where the power supply had a natural limitation' (op.cit., p. 136)—a conclusion apparently based on the James mill at Newburyport (p. 288 n.)—and, 'Peter Temin's conclusion that a high proportion of new cotton mills erected after 1835 had steam power is mistaken' (p. 297 n.).

270  *Perspectives*

powers', representing 3584 throstle spindles, plus all the necessary machinery to convert the raw cotton into cloth. Montgomery[11] showed that the effective power supplied was about $54\frac{1}{2}$ horsepower (together with about 4 acres of land); making about 66 spindles-plus-looms per horsepower. His sometime critic, C. T. James, ranked the power required for mule spindles at one-quarter less than that for throstles.[12]

Outside verification and clarification are hampered by having to deal statistically with the integrated 'Waltham system' mill that characterized best American practice after 1814. Despite the comments of Wallis quoted above, Montgomery noted that American power-looms worked at 120 picks per minute in the 1830s, compared to 95 on equivalent qualities in Scotland.[13] I therefore take alternatives of 6 and

TABLE 10.1
*Spindleage per horsepower in some American mills in the 1830s*[14]

| Source: | Montgomery | James | Temin |
|---|---|---|---|
| Spindles | 4992 | 6336 | 29 248 |
| Looms | 128 | 144 | 830 |
| Horsepower | 80 | 75 | 439 |
| Spindles-cum-looms per HP | 62·4 | 84·5 | 66·7 |
| Spindles per HP (6 looms/HP) | 85 | 124 (93*) | 97 |
| Spindles per HP (10 looms/HP) | 74 | 105 (79*) | 82 |

* James's figures are for Bartlett's no. 1 mill, for mules spinning 40s; they have been converted to throstles per horsepower in parentheses at his estimate of relative power consumption.

[11] *A Practical Detail*, p. 215.
[12] Op.cit., p. 23. This seems a cautious estimate in view of the fact that in England mule spindles needed little more than a third of the power of throstles, but this may be further evidence of the very high marginal power component involved in running mule spindles at higher speeds.
[13] *A Practical Detail*, pp. 117 ff. Until this date the two countries were probably using power-looms of much the same type. Thereafter, according to Copeland, the English went over to the over-pick loom, invented by Dickinson at Blackburn in 1828, whereas the Americans stayed with the older under-pick form. The over-pick loom was generally run at 20 to 40 picks per minute faster than the under-pick one, though at the cost of greater attendance and therefore lower output per labourer (M. T. Copeland, *The Cotton Manufacutring Industry of the United States*, Cambridge, Mass., 1912, pp. 89–90).
[14] Montgomery, *A Practical Detail;* Justitia, op.cit., p. 26; Temin, p. 197. At Lowell, Montgomery reckoned 77 throstle spindles (with preparation) per horsepower or 8·5 looms (with dressing) per horsepower in the spinning and weaving of 14s, its standard count. On 40s the respective figures were 127 and 11 (*A Practical Detail*, p. 108).

10 looms per horsepower, to cover the likely range of power requirements in American weaving.

These estimates all fall within the range of 80 to 100 spindles per horsepower usually quoted as applicable to the dead-spindle system,[15] and seem adequately consistent in view of the differing qualities under consideration.

Any calculation of output per spindle from secondary sources runs into the same problem of vertical integration. I have taken Davis and Stettler's results for Massachusetts and for the U.S.A. as a whole in the 1830s and 1840s, and applied to their productivity figures Montgomery's assumptions about lb of cotton spun per yd of cloth, plus C. T. James's about the length of the working year (310 days). Given the

TABLE 10.2

*Hanks of 18s per spindle per day in Massachusetts and the U.S.A., 1831–1849*[16]

|      | U.S.A. | Massachusetts |
|------|--------|---------------|
| 1831 | 3·73   | 4·70          |
| 1837 | —      | 4·52          |
| 1845 | —      | 4·33          |
| 1849 | 4·38   | 4·68          |

very rough nature of the assumptions, these results agree quite comfortably with James's assertion that the output of a much finer product—40s, spun on mules—was 4 skeins (hanks) per spindle a day in Massachusetts.[17]

It seems apparent that, even granted all these difficulties with the data, one finds the Americans at a different point on the contour relating output per spindle to spindles per horsepower. According to the model, the greater output per spindle and

[15] Montgomery, p. 72; the live-spindle system usually encountered in Britain operated at 100 spindles per horsepower and above.
[16] Sources as in text, and L. E. Davis and H. L. Stettler III, 'The New England Textile Industry, 1825–60: Trends and Fluctuations', in National Bureau of Economic Research, *Output, Employment, and Productivity in the United States after 1800* (Studies in Income and Wealth, xxx), New York, 1966, pp. 228–31.
[17] *Justitia*, p. 22.

smaller spindleage per horsepower in the States should have been related either to higher plant and labour costs per lb or to lower power costs per lb, quality differences aside. Both of the conditions would appear to hold good. The higher level of real wages earned in the U.S. relative to those in Britain is, of course, a favourite observation both of contemporaries and of present-day historians.[18]

Jeffrey Williamson[19] recently constructed an argument about optimal replacement of textile machinery in New England as compared with Britain partly on the assumption that there was little or no difference in the price of textile machinery between the two. This assumption was largely unsupported by historical evidence, and in fact appears quite unwarranted.[20] Williamson's argument rested on the assertion that the tariff would have influenced consumption-goods prices more than investment-goods prices. But the tariff directly influenced the price of materials out of which the machinery was built.[21] In 1820 it was considered that prices of textile machinery were about twice as high in the States as in Britain.[22] Despite a steep fall in the price of mills and machinery from 1815 to 1830[23] the relative situation in the 1830s does not seem to have been very different. The cost of equipment per throstle spindle at the Boott no. 1 mill in 1836 was $9·45, or £1·94 at official exchange rates.[24] This compares with the conventional £2 per throstle spindle for machinery *plus mill* charged in England from the end of the eighteenth century on

[18] Habakkuk, op.cit., etc. But note Copeland's remark: 'The situation in 1860 was summed up by Mr. Samuel Batchelder, a competent critic, who said: "The advantage of manufacturing in England on account of wages is much less than we have generally supposed . . ."' (op.cit., p. 11).
[19] J. G. Williamson, 'Optimal Replacement of Capital Goods: The Early New England and British Textile Firm', *J. Pol. Econ.* 79, Nov.–Dec. 1971, esp. pp. 1,329 ff.
[20] Williamson based his historical observation substantially on Zachariah Allen's comparison of wage rates for skilled and unskilled labour in each country in 1825; as interpreted by Nathan Rosenberg, 'Anglo-American Wage Differences in the 1820s', *J. Econ. Hist.* xxvii, 1967.
[21] See S. L. Engerman, 'The American Tariff, British Exports, and American Iron Production 1840–1860', in McCloskey, *Essays on a Mature Economy;* and cf. Baines, *History of the Cotton Manufacture,* p. 509; etc.
[22] G. S. Gibb, *The Saco-Lowell Shops: Textile Machinery Building in New England, 1813–1849,* Cambridge, Mass., 1950, p. 42.
[23] McGouldrick, op.cit., p. 19, estimates the decline at 40 per cent or more.
[24] Gibb, op.cit., p. 632; exchange rates from Davis and Hughes (1960–1).

to the 1830s.²⁵ According to the Secretary of the Treasury, Levi Woodbury, New England mills cost from $28 to $44 per spindle in the early 1830s; McGouldrick's annual data for his sample of mills indicate the lower figure—or roughly £6 per throstle spindle.²⁶ Costs of mule-equipped U.S. firms are unknown at this date, but in 1866, when throstle firms cost roughly the same in money terms, mule firms ran at about £5 per spindle.²⁷ This compares with the standard £1 per spindle (or £1. 4s. for self-acting mules) in the U.K. If Williamson were correct, my argument could be much stronger. Alas, he cannot be upheld.

Exactly how much these higher plant and equipment costs contributed to higher American speeds cannot be estimated without more information than appears in the secondary sources. But one has to give due credit also to the effects of having cheaper power. McGouldrick referred to '. . . the low cost of power in New England, which made it economic to produce very durable fabrics made of wiry, highly twisted yarn',²⁸ while Montgomery himself noted that: 'Driving machinery at high speed does not always meet with the most favourable regard of practical men in Great Britain; because in that country where power costs so much, whatever tends to exhaust that power is a matter of some consideration but in this country [the U.S.], where water power is so extensively employed, it is of much less consequence.'²⁹ Habakkuk, following similar quotations from Montgomery, stated:

Montgomery seems to be referring to spinning machinery, though he does not explicitly limit his observation to this operation, and elsewhere (p. 162) he concluded that 'The factories of Lowell produce a greater quantity of yarn and cloth from each spindle and loom (in a given time) than is produced in any other factories without exception in the world'. It was certainly the general impression later in the century that the Americans ran their spinning machinery faster. Mule for mule, said Young, New England produced more than Lancashire. One reason for the higher American speeds in spinning was the low marginal cost of power in the New England mills. But this was not the only reason. Spinning was a

---

[25] See *Thesis*, Ch. IV. The ratio was still effective in the case of the Quarry Bank Mill in the 1830s.
[26] McGouldrick, pp. 19, 240–1.     [27] Ibid., p. 243.
[28] Ibid., p. 31.     [29] *A Practical Detail*, p. 71.

branch of the industry where American equipment was more capital-intensive than the English, and the Americans therefore had an incentive to run their machines at higher speeds, so long as by so doing they did not add disproportionately to labour costs per unit of output . . .[30]

In an important sense, therefore, much of my work in this sphere is in the nature of an extended gloss on Habakkuk's footnote. The argument may be extended to note that the influence of power costs was not simply to cause the Americans to run the same type of machinery at faster speeds. Thus far I have made only a rather minimal test of the model. On a grander scale, the effect of cheap power (cheap, that is, in the everyday sense) was also to help dictate the very technology used. Two breakthroughs patented in 1828 were the key. Firstly, ring spinning was patented in that year by Thorp, as well as by a number of others over the next few years. This technique has of course dominated production in the late nineteenth and twentieth centuries, especially in coarse yarns. However, even in the U.S. it was slow to catch on before the 1850s, while in Britain its success was deferred much later still.[31] Secondly, Danforth patented his throstle in 1828, to be known subsequently as the cap frame. The Danforth throstle was patented in England in 1829, and introduced into Yorkshire worsteds in 1831 and to Lancashire cottons about the same time.[32] In the latter it was capable of churning out $5\frac{1}{2}$ to $7\frac{1}{4}$ hanks of 36s per day, compared to only $3\frac{1}{2}$ on the common throstle.[33] Even so it did not prove a permanent success in Britain. It spun a soft and spongy yarn, suitable for some branches of calicoes but not for velveteens and the like which required the wirier, smoother yarn of the common throstle. The combination of such factors with the demands on power meant that what was technically best-practice was not long after laid aside in the U.K.

There were equivalent variations between the countries in their preparatory machinery. Even where the English equipment was very similar to the American, e.g. the fly frame, which

---

[30] Op.cit., p. 54 n.
[31] e.g. L. Sandberg, 'American Rings and English Mules: The Role of Economic Rationality', *Quart. J. Econ.* 83, 1969.
[32] W. English, *The Textile Industry*, London, 1969, p. 169.  [33] Ure (1861), vol. ii.

paralleled the American double speeder, the former worked more slowly (under 800 rev/min, cf. 900–1000 for the latter).[34] The English was also much more economical of power—Montgomery estimated a 20-spindle double speeder to require double the amount of power for a 48-spindle British fly frame. Thus in the 1820s and 1830s the Americans had a whole range of preparatory machines such as extensors and speeders whose names meant little in Lancashire.

The isocost contours used in explaining the differences in speed between similar or not-so-similar machines across the two countries assume no differences in quality. There is, however, abundant evidence that U.S. yarn and cloth were of coarser quality than the average British product. Fine spinning in the British sense was virtually absent in the U.S.A. where the finest yarn counts found in the 1830s were 60s and the quantity produced finer than 40s was rather minimal. This continues to fit the pattern though, because it has already been established in Chapter 7 that coarse yarns and fabrics are relatively power-intensive. R. H. Greg, of the famous Cheshire cotton family, clearly considered relative power costs as an important factor when questioned in 1846.

Q. 7566: Do you see any reason why the Americans in the U.S. should not make just as fine goods as we do here?
A: Yes, I certainly see a reason; because capital will always be cheaper in this country, and machinery and labour is lower. There we have natural advantages.
Q. 7567: Is labour lower in this country than in the U.S. in factories?
A: Yes, it is rather lower. America has another natural advantage of unlimited water power, besides the raw material. In the coarser goods, there is a large quantity of power employed; and in the water power, and the raw material consumed in the coarse goods, she has great natural advantages . . .
Q. 7590: Would not the greater abundance of water power in the U.S. affect the manufacture of fine goods as well as of coarse?
A: It affects them all of course; but the power that is taken in coarse goods is much greater than that taken in fine. Perhaps, with an equal amount of capital, where 50 horse-power would be wanted for the fine, 200 would be wanted for the coarse.[35]

[34] Montgomery, op.cit.    [35] P.P. 1846: VI(I).

As the quotation suggests, there are a large number of reasons that have been brought forward to explain why Britain on average spun finer yarns than the U.S. Greg here notes particularly the effects of freight costs of raw material prices; one could also mention climatic factors, demand factors (mass consumption in the U.S.A.), differential incidence of tariffs, and above all labour scarcity à la Habakkuk. It is beyond my abilities and material to disentangle these causes here; in any case I am simply trying to reinforce rather than rebut the earlier argument.

Thus the costs of power are shown to be important in determining the drift of best-practice technology in the U.S.A. to the extent that it parted ways with that in Britain. Even when the machines were the same in each country they were operated faster (at least in spinning) in the U.S. because of its abundant resources of accessible water-power. More interestingly, perhaps, inventors were spurred along towards machines that would optimally work at much higher speeds than those that suited the British structure of production. Complex interrelations of technology transfer await anybody researching this field, providing they possess sufficient technical insight. The cap frame, invented in the U.S., was soon tried in Britain, but took twenty years or so to find wide use, despite being half or more as fast again as its rivals. Conversely, the British fly frame, though slightly slower than the American double speeders, had other advantages that enabled it to supersede the latter in the U.S. in the 1840s, and to find further development along lines best suited to American conditions.

Finally, the quality of product differs between the two countries in a way which power-cost differentials alone would predict. But in this case a large number of other factors must be awarded the main credit. The experience of Continental Europe shows that high power costs will not be enough to concentrate production in the finer qualities.

## BELGIUM

Belgium is often taken to be the first Continental country to have imitated industrialization on the British pattern. Its economic historians have frequently blamed the Belgian lag

## International Context in the Early Nineteenth Century 277

behind Britain four-square upon the lack of steam-engines.[36] However, the first Newcomen engine was being built in (and exported from) Liège province as early as 1720, and the deeper collieries of the Borinage district supported wider diffusion during the middle years of the eighteenth century.[37] Manufacturers in this period included Dorzée near Mons, Lejeune & Billard at Jemappes, Braconnier at Basse-Ransy, and Wasseige at Liège.[38] Though Watt engines were available, e.g. by 1785 at Jemappes, the Newcomen engine was generally preferred in the Borinage, at least until the adoption of double-action and high pressures. Thus lack of information about steam-engines would hardly seem to have been the case.

Undoubtedly, though, there were economic objections to high rates of investment in steam-engines. John Cockerill, founder of the metal works at Seraing that in their heyday were regarded as the finest in Europe, charged 1250 francs per horsepower for low-pressure engines of up to 20 horses, and 1100 francs per horsepower for those of larger size, in the late 1830s.[39] At prevailing exchange rates these amount to £50 per horsepower for smaller engines and £44 for larger—and these were considered to be lowest prices. Through higher costs of metals and relative scarcities of skills, plus any monopoly profits that may have been going, Cockerill was charging about half as much again for his low-pressure engines than the high-cost firm of Boulton & Watt could attain in Britain. However, high-pressure engines—presumably non-condensing —were available for use as a cheaper substitute, at £36 per horsepower for larger and £40 for smaller engines. Their coal consumption was put at 12 lb per horsepower per hour.[40] It has long been recognized that, largely for technical reasons such as the necessity of extraction from narrow, inclined coal seams, coal was expensive in France. In the 1830s the average

---

[36] e.g. R. Demoulin, *Guillaume Ier et la transformation économique des provinces belges, 1815–1830*, Liège, 1938. Cf. J. Mokyr, 'The Industrial Revolution in the Low Countries in the First Half of the Nineteenth Century', *J. Econ. Hist.* xxiv, no. 2, June 1974, pp. 371–3.
[37] J. Dhondt and M. Bruwier, *The Industrial Revolution in Belgium and Holland, 1700–1914*, London, 1970, pp. 8–10, 27–8. See also J. Lewinski, *L'Évolution industrielle de la Belgique*, Brussels, 1911.
[38] Demoulin, op.cit., p. 262.    [39] P.P. 1839: XLII, pp. 686, 690.
[40] Ibid., p. 685.

price of coal in manufacturing districts of France was stated to be 36s. per ton, and nowhere less than 28s.[41] compared to the 6s. in Bolton and 9s. in Manchester used in calculations in Chapter 7. Extraction does not appear to have imposed the same difficulties in Belgium as in France in the eighteenth century, but after the expiration of French rule prices rose sharply. Lowest prices were at Mons:[42] 7 to 8 frs. a (metric) ton there in 1836 and 12 to 13 frs. in 1837 (at this date a Belgian franc was almost exactly equal to 10d.). At Charleroi, in the heart of the metallurgical area, coal prices were 6 frs. a ton in 1820, 9 frs. in 1830, 13 to 14 frs. in 1836, 18 to 19 in 1837, and 15·60 frs. in 1839. In the textile areas Ghent, the capital of the Belgian cotton industry, paid about 20 frs. a ton in 1839, Verviers, headquarters for wool, 25 frs., and Brussels, 30 frs.[43]

In a mill near Liège spinning coarse cotton yarns (20s), coals were estimated at more than a third the wage cost per lb, and nearly 11 per cent of the value-added alone. The owner complained 'Le charbon que nous employons pour nos machines à vapeur, coûte à Liège 1 fr. 44 c. les 100 kilogs., qui, avec 1 fr. 20 c. de transport, nous revient ici a 2 fr. 64 c. les 100 kilogs.; c'est environ quartre [sic] fois autant qu'il coûte au fabricant de Manchester ou de Leeds!'[44]

These limitations were compounded by having a stock of machinery whose productivity was low by even British standards. In the cotton industry a Charleroi engineer of quintessentially English origin named Grenville Withers wrote in 1839:

... The spinning and weaving is done on the system of Manchester 20 years ago, but carried out very slovenly. There is little or no throstle spinning, and their mule jennies, as well as all their other machines, are very badly made. I am quite sure, notwithstanding the low price of labour, that the spinning of a lb. of cotton, costs 40% less at Manchester than at Ghent. The advantages of the Manchester spinner are,—
1. Cheapness of coals . . .
2. Superior machinery and cheapness of do.
3. Greater speed of do., turning off 30% more work.
4. Greater ability of work-people, making less waste.
5. Better market for the raw materials.[45]

[41] P.P. 1841: VII.
[42] P.P. 1839: XLII, p. 671.
[43] Ibid., pp. 676, 680, 696.
[44] Ibid., p. 695.
[45] Ibid., pp. 78–9.

## International Context in the Early Nineteenth Century 279

Not only was the machinery less efficient, it was also more expensive than that of Manchester. Cockerill's cotton mule jennies were sold for 16 to 22 frs. per spindle, i.e. over half as much again as a self-acting mule spindle in Manchester. Power-looms from the same firm sold for £19 to £27 in 1839. For the small worsted sector, throstles sold at 28s. a spindle, compared with 15s. for best equipment in Bradford.[46] Of course labour was cheaper to hire for any given period of time, but it is clear that its productivity was more than commensurately low; so that cotton may well have been 40 per cent more expensive to spin (as suggested in the quotation above), while worsteds, behind tariff barriers, were relatively even more costly.

From the model I have used it is straightforward to explain the slow rate of turnoff noted above in terms of operating the machinery more slowly in order to conserve power. From the parameters evaluated in Chapter 7 it follows that if all other costs of a 30-horsepower mill were the same between Belgium and England, the speed of Belgian machinery ought to have been only about 3000 rev/min or a little more, rather than the 4200 rev/min achieved in Manchester. This agrees well with the average of 30 per cent slower speeds stressed by Withers. Other costs *were* different, but apparently the effect of higher mill outlays and lower labour charges, coupled with what might be termed the quality-structure, were all self-neutralizing. Even on Roberts's self-acting mules, where differences in, say, skills can hardly have been of importance, the rate of output in Belgium in 1842 was put at less than two-thirds that in Manchester.[47] In addition it was said that

Mules are almost universally preferred to throstles (of which I saw but few abroad), owing to the much less power they require. They calculate that one horse-power drives 500 mule spindles but only 200 throstle spindles; and as the object with them is the greatest possible produce, quality being a secondary consideration, they seldom adopt throstles at all.[48]

[46] Ibid., p. 690. Cf. also P. Lebrun, *L'Industrie de la laine à Verviers*, Liège, 1948, for earlier prices at Cockerill's (pp. 242–4), and P.P. 1841 (Sess. I): VII, pp. 50–2.
[47] P.P. 1841: VII, qs. 1083–4.
[48] P.P. 1839: XLII, p. 669. Throstles produced more expensive yarns within the low-count range, as discussed in Ch. 7. The author of this quotation, J. C. Symons, went on from this statement that the object in Belgium was 'the greatest possible

These impressions can be confirmed for earlier periods by direct observation. A survey taken in Ghent in 1808 shows that the average firm produced about ¾ hank per spindle per day on coarse grades from 20s to 26s (individual firms ranged from 1¼ hanks down to ½ hank).[49] More regular information is available on Ghent and surrounding districts from 1810 to 1813,[50] and it bears out the 1808 results.

These data relate to the initial 'hot-house' period of Belgian cotton-spinning, evidenced in the years immediately after the cessation of war in 1815 by steam-engines being said to sell for 3000 to 3500 francs per horsepower,[51] i.e. for the extraordinary sum of £125 to £145 (assuming similar exchange rates). It is thus not surprising to find relative speeds so inferior to those in Britain.

Under these conditions it is less easy to see the steam-engine as the ultimate salvation of the Belgian textile industry, even when improved by Watt. Early on, water-power was preferred in the rather hillier region occupied by the woollen industry between Verviers and Aachen.[52] Given the high price of grain during the war years, engines were no doubt still cheaper than animals, but there cannot have been a great deal in it. The leading French advocate of Watt's design, and the first to adopt such an engine in France, J. C. Perier, wrote in 1810 (probably about the double-acting engine):

Il résulte de ces calculs et d'expériences bien constatées que le travail de la journée d'un cheval fait par une machine à vapeur consomme 66 lbs de charbon de terre qui coûtent dans ce moment à Paris £1/12/6d. Cette dépense est moindre que la nourriture d'un cheval, et la machine ne coûte rien les fêtes et dimanches qu'elle ne travaille pas; son mouvement est infiniment plus régulier.[53]

[49] Derived from data in F. de Bondt, 'La Dimension du progrès technique dans le textile', *Recherches économiques de Louvain*, Oct. 1970. See also J. Dhondt, 'The Cotton Industry at Ghent during the French Regime', in F. Crouzet, W. H. Chaloner, and W. M. Stern (eds.), *Essays in European Economic History, 1789–1914*, London, 1969.
[50] Provided on half-yearly and quarterly bases by Dhondt, p. 35.
[51] Demoulin, p. 268. The effective exchange rate is uncertain.
[52] P.P. 1839: XLII, pp. 676–7.
[53] J.-C. Perier, 'Sur les machines à vapeur', *Bulletin de la Société d'Encouragement pour l'Industrie Nationale*, no. lxxiii, July, 1810, p. 169.

produce', to argue in alarmist fashion that the rate of output was actually higher than in Britain. Withers's material in the same report shows that this was not so, as does the empirical information given here.

## International Context in the Early Nineteenth Century 281

The adoption of the high-pressure engine offered another avenue for further cost savings. The non-condensing engine was widely adopted at an early stage for functions such as colliery winding, where its comparatively low first cost helped offset its rather high coal consumption. If, however, stress was to be placed upon the latter, with relative disregard for purchase price, then the Cornish engine was a possible alternative. Humphrey Edwards had introduced this engine into France before 1815.[54] Grenville Withers, who had recently been in Manchester inquiring after a 30-horsepower engine, was asked by the Select Committee on the Exportation of Machinery in 1841 about power sources in Belgium.

Q. Do you find generally throughout Belgium that English steam engines are getting into disrepute? A: They are getting into very great disrepute, in general, because they consume so much coal; English manufacturers having coal extremely cheap, pay very little attention to how much the engine consumes.

Q: That applies to the low-pressure engines? A: It applies to the nature of the engine you employ; for example, in France and Belgium, manufacturers find that Wolf's [sic] principle of steam engine, which is both high and low pressure, is on a much more economical system, that is, it burns much less coal; we calculate, in England, that a low-pressure engine will burn about 10 lbs. per horse-power per hour, whereas the manufacturers abroad pretend that Wolf's engine burns something under 8, and they go down to 6.[55]

### SUMMARY

The 'comparative history' of this chapter is inconclusive by itself; extended proof of the hypotheses would need more research into the economic histories of both textiles and energy in the U.S.A. and in Belgium. Even then, complex statistical procedures will probably be required to analyse the various factors involved. At best I hope I have pressed the case set out in several of the earlier chapters of this book.

Manufacturers were constrained in the first place by the price they had to pay for coal, which I have taken to be

[54] J. Payen, *Capital et machine à vapeur au XVIIIe siècle*, Paris, 1969, pp. 225–8; W. O. Henderson, *Britain and Industrial Europe 1750–1870*, 3rd edn., Leicester, 1972, p. 58.
[55] P.P. 1841: VII, qs. 640–1.

influenced primarily by geological and geographical factors. In both Massachusetts and Ghent textile firms would have had to pay about two and a half times as much for coal per ton as in Manchester in the 1830s. The problem was alleviated, and indeed actually reversed, in the American case by having available abundant water-power—at the heart of the *antebellum* U.S. industry in Lowell power costs were scarcely half those charged in Manchester. It is unlikely that the steam-engine offered this cheap water-power much competition, especially in view of the concentration on high-pressure non-condensing engines in the U.S.A. before the middle of the century. In Belgium topography seems to have limited water-power to hillier regions where the woollen industry was located, rather than to centres for cotton. By this time, manufacturers as well as mine-owners were trying to reduce power costs by adopting the compound engine of the Woolf type—several years in advance of their counterparts in Lancashire.

From these observations one would expect on the basis of my earlier reasoning to find (i) the ratio of spindles to horse-power being lower in the U.S. and higher in Belgium than in England; (ii) productivity per spindle thus being higher in the States and lower in Belgium. These expectations are fulfilled. Particularly interesting is the incentive offered to develop a new technology permitting still higher rates of output in the U.S., and embodied in the diffusion of ring spinning from the late 1840s.

The other components of the model must, however, be allowed their due. (i) Plant and equipment costs were higher in both countries than in Britain—this by itself should have raised their optimal operating speeds. (ii) Labour costs were probably higher in America and lower in Belgium—in isolation this would have had the effects actually observed, of higher speeds in the former and lower in the latter. (iii) Quality of product was lower in both than in the U.K.—this, like plant costs, should have raised their relative operating speeds. These aspects, along with the horde of climatic factors, demand, etc. therefore remain to be disentangled at the international level.

## 11. Conclusions

THE AIM of the book has been to combine economics, engineering and history to reassess the contribution of the steam-engine to British economic growth during the Industrial Revolution. In the text these elements have been intermingled; here I shall be reconsidering them in turn. The conclusions point to further avenues for investigation.

### HISTORICAL METHOD

It was my intention to amalgamate the strengths of the British antiquarian approach to economic history with the strengths of the American statistical-cum-theoretical approach. At its worst, the British school can indulge in heaping fact upon fact, higgledy-piggledy, possibly in the hope that the sheer weight of paper bearing on the matter in hand will be convincing. This often becomes counter-productive: the reader is left with no confidence whatever in the choice (or representation) of facts that happen to appear; he has no guidance other than his own imagination about how to weigh up pros and cons.

For its part, the newer American school all too often conveys the impression that history has been ridden over roughshod. Long periods are casually spanned by the econometrics, with at best a time trend to convey all the year-by-year variations in outside conditions that so beset the British historian. Contradictory evidence is sometimes airily dismissed as non-quantitative, without consideration of any deeper virtues it may have. At its extreme, the theoretic schema adopted (no doubt necessarily) for the sake of the argument is transformed into 'real history'.

For all these weaknesses on either side, there are many gains accruing from them both, and a reconciliation is overdue. The price I have paid for attempting such a synthesis in this book is that my methodological offerings new to one or other school are distinctly modest.

It is, for instance, standard procedure in current American writings on the historical contribution of specific technological advances to employ the concept of 'social savings'. However, in progressing from the savings on Boulton & Watt engines alone to those on all steam-engines later in Chapter 6, I did make some effort to improve upon existing method. The advent of the steam-engine almost certainly did not lower the costs of power to nascent British industry in the medium run. What it did was to convert the power situation from one giving rise to what Boulding would call a 'spaceship economy', in which resources are finite and costs would be driven up through time, to one giving rise to what he terms a 'cowboy economy', where expansion could go ahead, largely unchecked by power limitations, at more or less constant costs.[1] In reverse one therefore ought to take account of the 'spaceship' features of a world without Newcomen and Watt. This involves estimation of the parameters describing the (upward-sloping) cost curve. Ideally, statistical calculation would require a period in which the cost curve for steam-power shifted sharply but the demand curve for all power stayed fixed. Figures on aggregate steam-horsepower are elusive enough, those on aggregate water-horsepower non-existent. The elasticities selected could best be described as merely consistent with the observed behaviour of power costs in the cotton industry in the 1850s (when the above constraints on schedule shifts were reasonably well satisfied for the first time). But in the present example even a vertical cost curve would not greatly alter the results computed, because when costs rose another 50 per cent or so above the assumed average for steam-power in 1800, other technologies than water-power that happened to have the property of fairly constant costs would become viable alternatives to steam. (The calculations of social savings on railroads by Fogel, Hawke, and others, ought—in theory—to be modified in the same manner, though the empirical difficulties might be even greater than in my case.)

Important defects remain in my own calculation of 'social savings'. The neglect of external diseconomies of both water- and steam-power makes the use of the word 'social' rather

---

[1] K. Boulding, 'The Economics of the Coming Spaceship Earth', in H. Jarrett (ed.), *Environmental Quality in a Growing Economy*, Baltimore, 1966; also W. Nordhaus, 'Resources as a Constraint on Growth'. *Amer. Econ. Rev., Papers and Proceedings*, lxiv, 1974.

capricious. Casual observation suggests that in comparing steam- and water-power the social diseconomies are much greater for steam than for water (greater smoke and noise pollution, greater cost in human life from boiler explosions, etc.), so that the computation continues to be an upper bound.

In assessing the forward linkages the method has been more that of the traditional British economic historian from some points of view, but also more radical. I retain a deep-rooted scepticism about the worth of any statistics relating to the early nineteenth century or before. No figure is accepted, if I can help it, without being verified independently. The process of verification nearly always involves a chain of mutually supporting relationships. One must have faith that substantial errors will eventually cancel out. It remains true that many of the quantitative results are 'manufactured', so that to attempt econometric model-building on the processed data involves the risk that the relationships tested may have been guaranteed *a priori* by assumptions underlying the method of compilation. Even if that were not so, it is hazardous to work with data containing errors of measurement, as is well known.[2] The plain fact here is that econometrics have been employed, sometimes in quite sophisticated fashion, to data on the nineteenth-century cotton industry which are hopelessly astray.[3]

### RESUMÉ OF HISTORICAL FINDINGS

The mutual sustenance of the steam-engine (especially Watt's engine) and the iron industry in the late eighteenth century has been a favourite hypothesis of historians. In Chapter 5 I investigated only one of the two possible connections, the feedback from the construction of engines to the growth of the iron industry. It is well known that output of pig iron doubled every eight years or so from the late 1780s up to the early 1800s. But the steam-engine cannot be held substantially responsible for this. Even in the late 1790s, when the production and sale of Boulton & Watt engines rose to a peak, their consumption of

[2] See G. N. von Tunzelmann, 'The "New Economic History": An Econometric Approach', in R. Andreano, *The New Economic History: Recent Papers in Methodology*, New York, 1970.
[3] e.g. Wright (1971); E. Asher, 'Industrial Efficiency and Biased Technical Change in American and British Manufacturing: The Case of Textiles in the Nineteenth Century', *J. Econ. Hist.* xxxii, no. 2, June 1972.

iron would have amounted to under one-quarter of 1 per cent of annual output. Moreover, this tiny demand was mostly for cast iron, a sector which, in contrast to wrought iron, was not hovering on the brink of a major technical breakthrough. For much the same reason Hyde has shown that the advances in the steam-engine bore little relationship to improvements in the smelting of iron, though developments in iron-refining are a more open question. The linkages back to coal, however, cannot be neglected in the same way as those to iron, though again much of the demand was for coal of an especially low added value (slack, etc.). The difficulty here is that most economic historians have considered the technical development of coal to have taken place before the Industrial Revolution.

In textiles, horse gins were giving way to water and steam in the last quarter of the eighteenth century, so that by the 1790s they were to be found only in very small mills. The rising ratio of horse-feed to coal prices spelled their end, especially as the costs of steam-power (in real terms) declined somewhat late in the decade.

However, the resourceful entrepreneur could still acquire water-power sites at which total costs worked out cheaper than for steam (certainly if imputed rentals are excluded). The low marginal costs of water-power made it particularly desirable for power-intensive activities, including the spinning of warps on frames such as the throstle, though external economies, e.g. in heating, could encourage the adoption of steam for weft and fine spinning. As noted in the last section, the achievement of steam-power in the early nineteenth century was to restrain the kind of pressure on energy costs that would have taken place if resort had had to be made to successively undesirable water sites.

With these considerations in mind it is not surprising to find that the social saving estimated for 1800 is very low even by the normal standards of such reckonings. For Boulton & Watt engines alone (including their pirates) the social savings over atmospheric engines can be put at about 0·11 per cent of national income in 1800. If total real output was then growing at its average rate for the take-off years, the level of national income reached on 1 January 1801 would not have been attained much before 1 February 1801 without James Watt.

## Conclusions 287

If all steam-engines, Watt and atmospheric alike, were hypothetically replaced with other means of motive power (a combination of water and wind would be optimal), the setback would have been about two months. These are upward-biased figures.

The result is comparatively insensitive to minor changes in dates once the parameters such as coal prices and labour costs are altered appropriately. There is a strong likelihood that the figures employed for horsepower embodied in atmospheric engines will prove demonstrably incorrect, but in view of the high running costs of these engines in many areas a drastic upward revision of the social savings is unlikely.

On the other hand, it seems highly probable that the ratio of social savings to national income would be higher in 1830, say, or 1850. From the scattered evidence on rates of growth of horsepower in leading industrial cities summarized in Chapter 2 it follows that steam-power grew much faster than national income through these years. To be more exact, we need to know vastly more about both aggregate horsepower and its geographical location in these later years. Unless one is prepared to take estimates of horsepower such as Mulhall's on trust, to my mind the only feasible approach is to work backwards from output and power-productivity data, as is done for cotton textiles in App. 7.1. It must be confessed that while the textile industries are highly important users of steam-power, they are also one of the easiest sectors to check.

Even if one could settle on a likely range for steam-horsepower, and adjust for locational shifts (to or from regions with expensive coal, labour, etc.), one would still be obliged to reassess the relevant costs for 1830. Speaking in money terms, I have shown that the supply price of steam-power did not fall from 1800 to the 1830s and early 1840s, whilst it is likely that the supply curve for other forms of power shifted downwards (i.e. demand for water-power at the going charge would be larger). Valuation of resource costs raises a further problem. There are indications in the 1790s and early 1800s of excess demand; of an inflationary gap opening up as resources were strained in order to put the economy on a war footing. But this was unusual. There seems no reason to depart from the conventional view that there were long periods of underemployment (and sometimes open unemployment) over the years 1815 to 1845 or 1850. If it is assumed: (i) that the relevant wage rates are based on subsistence considerations (institutionally defined); (ii) that other factor inputs are in excess supply in a 'glut'; (iii) that leisure has zero value—then the social savings will tend towards zero, since sectors other than power will be surfeited with low-cost resources. The contribution of steam-power would then be entirely one of putting otherwise unemployed or under-employed factors to work. The translation from direct cost savings to enlarged incomes for society in general breaks down under these conditions, because prices and marginal costs will diverge.

This may or may not be a valid interpretation of the later Industrial Revolution period, and it is out of the question to conduct formal tests for validity here. But if it is roughly correct, it encourages a very different assessment of the economic effects of Watt and others from that implied by the 'energy-crisis' notion. The same would be true if it were subsequently to be found that Watt's major gift to early nineteenth-century English society was in reducing the level of smoke pollution below that created by the atmospheric engines that I have assumed his to have replaced. Note that if external economies and diseconomies, such as pollution here, or leisure in the preceding paragraph, were to be accounted for, then the denominator as well as the numerator of the social savings ought to be adjusted: national income as conventionally measured is too narrow for the concept of national welfare that we have in mind. Whether the latter would be bigger or smaller than national income in 1800 (or after) is uncertain.

On other grounds as well it may be objected that national income is an inappropriate deflator for the social savings, for instance, because steam-power was at this stage being used only on a narrow industrial front. Nevertheless, there were no obvious technical reasons why agriculture, for example, could not then gain access to steam-engines. What restricted diffusion in agriculture was chiefly the high price of coal in most rural areas. The calculation of social savings for 1800 mirrors this very limited diffusion. Naturally the impact on certain selected industries may have been quite profound, and this makes it important to complement the over-all figure with studies of individual industries, as was done in Part III of this book.

The reduction in cost of a typical unit of power was based less on Watt's efforts than on those of the Cornish engineers (see Chapter 9) and the diffusion of their high-pressure condensing engine to the Midlands and North-West (Chapter 4) The technical requirements of pumping water out of mines allowed the Cornishmen to economize on fuel to a far greater degree than was possible in manufacturing. One of the consequences of their procedure was, for instance, enormous irregularity of action. Even in Cornwall the engine was frequently surrounded by a mass of balance beams. Adoption there and elsewhere involved relatively large outlays of fixed capital, not for the engine itself so much as for smoother transmission, for higher ratios of power capacity to power delivery, for larger boilers, for thicker boiler plates to maintain safety standards, etc. These were justified by the relative expense of coal in Cornwall; whereas in the northern manufacturing districts there was no offsetting cheapness of fixed capital. Compromise solutions such as the McNaughted engine were sought.

This forms the background to the study of the introduction of more automatic machinery in the cotton industry in Chapter 7. Such forward linkages came late in the day. Virtually all the famous 'wave of gadgets' were developed for either human, animal, or water-power (and usually the first two of these). Apart from certain preparatory functions such as carding, steam-power was late in being recruited for cotton manufacture, and as mentioned above water-power remained long pre-eminent in some power-intensive sectors. Moreover, most of the really large early mills were water-powered; even if one believes that the very diffusion of the factory was among the most important contemporary advances (and this looks highly dubious on economic if not on social grounds), one has as much right to see this as generated by water-power as by steam.

The failure of power costs to fall by very much delayed the diffusion of more automated equipment, with its relatively greater power consumption. Importance here is not demonstrated simply by the proportion of power costs in total expenses of installing and operating such machines. Viewing the problem as a historical rather than a technological matter, one ought to take account of the flexibility of each cost item. In this light

I estimated the advantages of scrapping hand-mules and hand-looms in favour of self-acting mules and power-looms respectively, around the mid-1830s. In both cases the economically feasible range was largely governed by the quality of product (yarn or cloth). For example, on 25s yarn the self-actor installed in Manchester at this time would have earned almost 9 per cent p.a. greater return than an equivalent investment in hand-mules (cruder data suggest a clear return on the latter of about $4\frac{1}{2}$ per cent, implying approximately 5 per cent for self-actors); while on 36s wefts, other information suggests that the hand-mule would still have been economically superior. Similarly, a power-loom in Bolton working on 50-reed cambrics would have netted a return of another $3\frac{1}{2}$ per cent p.a. or more if installed in place of a hand-loom; but on 60-reed cambrics the hand-loom continued to be cheaper, at the level of wages which hand-loom weavers seemed resigned to accepting by this stage.

Cheaper forms of steam-power in the later 1840s and 1850s should thus have permitted mill-owners to adopt self-acting mules for finer yarns or power-looms for finer or fancier cloths, even disregarding technical progress in the machines themselves. Until the mid-1840s the spinning of average yarns on self-actors and the weaving of above-average cloths on power-looms do not emerge as highly lucrative uses of capital resources.

The repercussions so far as the sector as a whole are concerned can be traced. First, there seems to have been little change in the quality of cotton yarn or cloth produced over the years 1833 to 1856, the decline in quality of exports that has been recognized by earlier authors being offset by rising quality of production for the home market. Second, aggregate figures indicate that the most notable feature of the behaviour of value-added in cotton manufacture over the period was a squeeze on gross margins per lb spun, woven, and finished, beginning in 1845–7. In the short run most manufacturers met this squeeze by minimizing losses, so long as they could cover variable costs. The severity of the decline in the short term and the apparent permanence of reduced gross margins over a longer duration encouraged the search for improved techniques that would allow profitable production at these margins. The adoption of a substantially cheaper power source was central

## Conclusions 291

to this reduction of costs. At the same time, it is obvious from the quite low ratio of power to total cotton costs (13 to 15 per cent in spinning, at a fairly generous estimate) that the effects were largely indirect. Lower costs of power meant that optimal speeds of operating machinery rose, assuming unchanged quality. High rates of turnoff would actually have made little difference to the power constituent of costs, since most of the gains would have been taken out in reduced plant and labour cost per lb.

Techniques in the worsted industry were similar to those in the coarse branch of the cotton industry, based on continuous spinning by means of throstles. Both spinning and weaving branches of this sector speeded up over the same years as for cotton. In worsteds the competitive response to the reduction in costs in cotton was given particular impetus by the adoption of cotton (and other miscellaneous) warps in the course of weaving. Although this seems to have raised labour costs in weaving and dyeing, it greatly reduced the price of the finished article below that of pure worsteds.[4] Such mixed cloths had quite a long history, but proliferated after 1837 and dominated the expansion of the sector after the 1847 slump.

Woollens have customarily been regarded as having lagged far behind worsteds in the first half of the nineteenth century. The evidence of Chapter 8 revises the belief that entrepreneurial shortcomings were fundamentally to blame for whatever lag occurred. In spinning, the jenny gave woollens an advantage in costs that the worsted branch did not make up until fully mechanized. Where coal was quite expensive, as in the West Country, woollen-spinners were unlikely to earn greatly enlarged profits from going over from jennies to mules at this stage. On the weaving side, the slowness of handling woollens, arising partly out of heavy power requirements and partly out of the intractability of the material, made widespread mechanization too costly until the late 1840s.

In one notable branch of the linen industry—the fine flax-spinning of Leeds—the response to cheaper power was actually

---

[4] K. V. Pankhurst, 'Investment in the West Riding Wool Textile Industry in the Nineteenth Century', *Yorks. Bull. of Econ. & Soc. Research*, vii, 1955, shows a dramatic reduction in the importance of labour costs in a certain worsted-mill in the 1850s. His own explanation of mechanized combing does not fit his pattern.

one of reducing speeds. Slower driving of the machinery permitted the spinning of much higher qualities, a move apparently forced on firms like Marshalls of Leeds by developments in flax-spinning elsewhere (Dundee and Belfast) and especially by the very cheapening of cotton that reduced power costs facilitated.

Thus steam-power did eventually have important forward linkages into the textile industries. But what is significant for the Rostow thesis is that they took place roughly over the years 1847 to 1860, with only a few minor foreshadowings. Yet in his chart Rostow places the point of maturity of the British economy at about 1850. So if we accept his chronology we have to see the forward linkages as carrying the economy from the drive to maturity into a subsequent stage of development, rather than from take-off to the drive to maturity as he himself envisages.

### THE ECONOMIC HISTORY OF TECHNOLOGY

In what respects can historians of engineering benefit from carefully investigating the economics of their studies? In regard to invention the question of patenting raises interesting problems of allocation and of private versus social profitability, as both contemporaries such as John Farey and present-day economists like Arrow and Nordhaus have pointed out.[5] In his apparently decisive submissions to the Select Committee on Patents in 1829, Farey in fact used the relative contributions of Watt and Woolf to the development of the steam-engine as the crux of his practical argument.[6] Two years earlier, in the first volume of his *Treatise*, and in spite of his seeming adulation of Watt the man, he had come a little regretfully to conclude that the extension of Watt's patents for twenty-five years offered in 1775 had been too indulgent.

Mr. Watt's original specification was necessarily very imperfect, because it was made at the time when he had not perfected his engine; but when parliament granted him an extension of the term, it ought to have been on condition that he should instruct the

---

[5] K. J. Arrow, 'Economic Welfare and the Allocation of Resources for Invention', in N.B.E.R., *The Rate and Direction of Inventive Activity: Economic and Social Factors*, Princeton, 1962; W. D. Nordhaus, *Invention, Growth, and Welfare*, Cambridge, Mass., 1969.
[6] P.P. 1829: III, e.g. pp. 445–6.

public as far as he was then able to do. According to the ordinary practice of the courts of law in other cases, Mr. Watt's patent ought to have been annulled, for the insufficiency of the specification, which is a series of principles of action, without any description of the means of carrying them into effect.[7]

The same point is made by Professor T. S. Ashton:

James Watt deserved well of his fellows; but when, in 1775, Parliament extended the patent, granted him in 1769, for a further 25 years, it gave great power to a man whose ideas were, before long, to become rigid. Watt refused applications for licences to make engines under his patent; he discouraged experiments by Murdoch with locomotive models; he was hostile to the use of steam at high pressure; and the authority he wielded was such as to clog engineering enterprise for more than a generation. If his monopoly had been allowed to expire in 1783 England might have had railways earlier.[8]

Ashton's judgement is strongly confirmed by the present book. Since it is possible to transnumerate from social savings to social rates of return, a cost-benefit study of Watt's patents could conceivably be undertaken, using earlier results, though it would be inadequate for fully testing Ashton's counterfactual in the last sentence quoted. The social savings summarized in the last section assumed that the technology of power supply other than that developed by James Watt was 'frozen' at the levels of 1769. However, Watt's long-lived patents virtually prohibited anyone else from improving upon the condensing engine, for the simple reason that any improvement beyond Watt required the separate condenser (particularly if it was going to be defined as embracing such disguises as Jonathan Hornblower's second cylinder). Those who tried to pirate Watt's engine in the 1780s and 1790s have gone down in the literature as rascals and scoundrels, partly through the influence of Watt's own descriptions of them. Yet their behaviour hardly seems more immoral than that of Watt's and Boulton's sons, when they came to dominate the company's policies in the 1790s, and in any case it is difficult to see how else they could have competed. Some of them did genuinely have new ideas

[7] Farey, *Treatise* i. 649.
[8] T. S. Ashton, *An Economic History of England: The Eighteenth Century*, London, 1964 edn., p. 107.

to offer; but whether they would ever have developed the separate condenser (or something similar) independently must remain to some extent an article of faith. If it were true, then the social costs of the patent would be grossly understated by my figure for social savings.

That science made a sizeable contribution to technical progress during the Industrial Revolution is a view recently revived by Musson and Robinson, although with reservations. Nothing in this study leads me to dissent from L. J. Henderson's dictum that 'science owed vastly more to the steam engine than the steam engine ever owed to science'.[9] The Cornish engine and its duty ratings were what inspired Sadi Carnot to devise the laws of thermodynamics. Carnot's work was quite ignored by British contemporaries, who had no respect for French theoreticians. Joule's experiments were conducted after the high-pressure engine reached the Midlands and the North, and in any case practising engineers knew little of them until Rankine publicized them in his *Manual*—but this work was not published until 1859, by which time (as we have seen) the transformation to high-steam pressures in the manufacturing areas was everywhere in full swing.

The case for a 'heroic age' of inventors in the seventeenth and eighteenth centuries could still be argued for the steam-engine. The traditional datings of, say, 1712 and 1769 could nevertheless be revised if one were to take an economic rather than a mechanical view of progress. One could hold that the Newcomen engine 'took off' in the early 1730s (except where coal was cheap) and the Watt engine in the late 1790s (except where coal was dear)—both periods when the patents were soon to expire. In the nineteenth century invention became widely diffused and frequent, its direction no longer linear but closely tailored to a wide variety of local factor-price situations. The course of invention cannot be understood without reference to these economic interactions.

The same seems true for the machinery driven by power. Rarely have I unearthed cost reductions from steam-powered inventions in textiles on the scale often intuitively supposed. A possible exception was the application of the mule to the

[9] Quoted by C. C. Gillispie, in A. E. Musson (ed.), *Science, Technology and Economic Growth in the Eighteenth Century*, London, 1972, p. 123.

spinning of fine cotton yarns (see Table 7.1)—though with coarse yarns it was not a great deal cheaper than the jenny. Note, in this connection, that fine yarns long remained uneconomic for weaving into high-quality cloths on the power-loom. In this example, the traditional challenge-and-response explanation of the great textile inventions warrants critical re-examination. In the woollen and worsted industry the greatest single cost reduction came from the improvement of hand-powered methods: the conversion from the distaff to the spinning-wheel in worsteds and especially from the wheel to the jenny in woollens. For the most part the application of steam-power was belated and of little structural importance.

Above all one has to beware of equating the decisive breakthrough with the application of inanimate power. Modern authors have criticized William Radcliffe's claims to have invented power-weaving in Britain; his actual contributions were the independent dressing machine and the devising of a take-up motion for the cloth beam. Yet there is every reason to believe that these called for a higher degree of imaginative insight than simply recruiting artificial power essentially to duplicate the actions of the human body (in fairness, Edmund Cartwright did perceive these and other difficulties, albeit after early disappointments). Aside from the engineering problems the application of power was far from costless in many circumstances: I have attempted to show how the diffusion of the self-acting mule and the various power-looms was constrained by this. The human frame was probably still the cheapest method of supplying very fractionalized power (say under $\frac{1}{4}$ horsepower) before the coming of electric motors. Even in the cotton industry it is likely that human beings supplied more motive power than steam-engines up to the 1820s.

Like most economic historians, I have paid far more attention to the imitation and the diffusion of a technique than to its original invention. Precise quantitative information is extremely scarce on any invention before 1860; the exception in this study is the Cornish engine, and even there it is inferential. In that example, the rate of diffusion seems in inverse ratio to the degree of engineering complexity. Seemingly straightforward advances of the kind often described as 'learning' were more rapidly diffused in periods of economic adversity, a result that contravenes Griliches's for hybrid corn in the Midwestern United States from the 1930s.[10]

Twentieth-century economists have been unable to agree on whether innovation and/or diffusion is more frequent when conditions are improving or deteriorating. Mansfield, in an empirical study of the American petroleum, coal, and iron and steel industries, concluded that a point somewhere between high prosperity and deepest depression produced highest rates of process innovation, varying from two-thirds of full capacity in steel to three-quarters of capacity in coal.[11] He found no statistically significant difference according to whether economic conditions were rising or falling, and no significant relationship for product innovation. Among received opinions, he quotes Keirstead as representing what might be called the optimist case:

... when the economy is approaching the top of the cycle, entrepreneurs 'will welcome cost-reducing or output-expanding inventions. But when the economy turns down ... entrepreneurs will be chary of taking up inventions, *except such as offer great cost reductions at a small increase in fixed costs*'.[12]

The italicized proviso makes this view fairly reconcilable with the more 'pessimistic' viewpoint in this book. The conversion of low-pressure engines to high-pressure action would qualify as an example of cheap or largely disembodied innovation (with some reservations about boilers); so would much of the power-driven machinery. But the diffusion of the engines themselves (i.e. the cylinder, boilers, pumps, and works) might be better favoured by the easier credit conditions and higher profit expectations of the upswing. Matthews's concept of semi-reserve capacity in which managers invested 'excessively' in mills and power in booms and filled them up with machinery in the ensuing slump bears witness to such an oscillation.

It has commonly been assumed that the lag between invention and its diffusion has steadily been shortening up to

---

[10] See above, pp. 257 ff.
[11] E. Mansfield, *Industrial Research and Technological Innovation*, London, 1968, p. 117.
[12] Ibid., p. 111, quoting B. Keirstead, *The Theory of Economic Change*, New York, 1948, p. 145. My italics.

the twentieth century, and in some cases detailed sociological predictions have rested on this premiss.[13] The evidence in its favour, however, is not very strong. If invention is taken as the moment of applying for a patent, and innovation that of first commercial introduction, as has been customary in the few extant empirical studies, then the lag for steam-engines and the machinery they drove was rarely much more than five or six years, and sometimes 'negative' (as with Hargreaves's jenny).[14] The main exception was Watt's engine, for which the gap between conception and practical execution (neglecting the Kinneil engine) was about ten years—though the point Farey made about Watt patenting a vague notion ought to be remembered. These figures are all much lower than for inventions which might be supposed to have been of approximately similar 'intensity' in the late nineteenth and twentieth centuries. One explanation is no doubt that inventors and innovators were almost always the same people; this is less true though not unusual nowadays.

It is difficult to avoid being arbitrary in dating diffusion processes. Granted that, there is no powerful evidence that entrepreneurs were especially slow to innovate in the early Industrial Revolution period. Existing methods of communication were quite fast enough within the country to promote a dissemination of new techniques that was rapid by any reasonable standards: one does not think in terms of terse telephone conversations as being essential for 'rapidity' even in the 1970s.[15] Even in the countryside, the diffusion of the spinning-jenny in the cotton industry of the 1760s or the woollen industry of the 1780s seems to be respectably quick by modern standards. In Chapter 7 I examined two nineteenth-century cotton innovations whose diffusion is conventionally regarded as slow: the self-acting mule and the power-loom. The gradualness of their introduction was shown to be a function partly of technical limitations—which were re-expressed

---

[13] e.g. D. Schon, *Beyond the Stable State*, London. 1971.
[14] This also emerges from the table presented by Mansfield (*Industrial Research*, p. 110). He quotes Enos's conclusions: 'Mechanical innovations appear to require the shortest time interval, with chemical and pharmaceutical innovations next. Electronic innovation took the most time.' (pp. 110–11.)
[15] This point is made by W. Parker, 'Economic Development in Historical Perspective', *Econ. Devpt. & Cult. Change*, Oct. 1961.

as economic phenomena—and partly of outside economic circumstances, such as the level of power costs.

At no point can I definitely refute the view that such gradualism represented rational behaviour by entrepreneurs, given the constraints. At the same time, it would be going too far to claim that the businessmen of this period always and everywhere reacted instantaneously to changing economic and technical circumstances. Much more wide-ranging evidence would be needed to establish such a proposition. My belief is that it could not be supported. Take the diffusion of the high-pressure rotative engine. The delay in its adoption in manufacturing (long after my information hints that adoption was justified), and the eventual bunching of its diffusion in the years following the sharp slump of 1847, suggest that higher productivity in energy could have been attained in the 1830s and 1840s from known production possibilities.

The existence of X-inefficiency of this kind is a practical obstacle to using neo-classical production theory in this study. On the theoretical side, the difficulties erected by the heterogeneity of capital goods have been noted in Chapter 1. Relationships between the engine and the machinery it drove were central to my analysis of the forward linkages. The conventional microeconomic approach using isoquants would give a solution, but not an especially instructive one. I have preferred instead to adapt engineering concepts allowing stress on the vertical alignment of the capital goods used for production. I consider that this more accurately registers the choices presented to nineteenth-century 'millocrats' in the ordinary course of events. To be sure, the present model is hardly more effective than the conventional one in theoretical analysis of changes in quality.

The final point is just that. Industrialization even in one country cannot be satisfactorily appraised without reference to changes in quality of inputs and outputs. Still more is this so in comparing the pattern of industrialization across countries. If there is one major area of theoretical and empirical weakness in economics highlighted by this book, it is the neglect of qualitative change.

# Select Bibliography

Material to which no allusion was made in *Thesis* is excluded, as are general items that do not explicitly relate to the historical context of this work.

(A) MSS. COLLECTIONS

Additional MSS. (British Museum)
Ashworths MSS. (John Rylands Library)
Aubrey MSS. (Bodleian Library)
Barton MSS. (John Rylands Library)
Black Dyke Mills MSS. (Queensbury; and Brotherton Library)
Boulton & Watt MSS. (Birmingham Reference Library)
Bradford Reference Library
Brotherton Library (University of Leeds)
Colquhoun MS. (Baker Library, Cambridge, Mass.)
Cusworth Hall Museum Library (Doncaster)
Dugdale MSS. (John Rylands Library)
*Engine Reporter* (Redruth Public Library)
Egerton MSS. (British Museum)
Farey typescript (National Reference Library of Science and Invention, Holborn Division)
Goodrich MSS. (Science Museum Library)
Gott MSS. (Brotherton Library)
Halifax Public Library
Harvey & Co. MSS. (Cornwall County Record Office)
House of Lords Record Office
McConnel & Kennedy MSS. (University of Manchester Library)
Marshalls MSS. (Brotherton Library)
Public Record Office (Board of Trade papers; Home Office papers)
Quarry Bank MSS. (Manchester Reference Library)

(B) THESES

BECKINSALE, R. P., 'A Geographical Survey of the Textile Industries of the West of England' (D. Phil., Oxford, 1949).
GRIBBLE, I. A., 'Labour Supply and Industrialization: Two Case Studies' (Ph.D., Sheffield, 1971).
HARTWELL, R. M., 'The Yorkshire Woollen and Worsted Industry, 1800–1850 (D. Phil., Oxford, 1955).

HYDE, C. K., 'Technological Change and the Development of the British Iron Industry, 1700–1870' (Ph.D., Wisconsin, 1971).

JENKINS, D. T., 'The West Riding Wool Textile Industry, 1780–1835' (D. Phil., York, 1969).

LANGTON, J., 'The Geography of the South-west Lancashire Mining Industry, 1590–1799' (Ph.D., Wales, 1970).

LAZENBY, W., 'The Social and Economic History of Styal' (M.A., Manchester, 1949).

LEE, C. H., 'McConnel & Kennedy, Fine Cotton Spinners: A Study in Business Enterprise in the Cotton Industry, 1795–1840' (M.Litt., Cambridge, 1967).

TAYLOR, A. J., 'Concentration and Localisation of the British Cotton Industry, 1825–1850' (Ph.D., Manchester, 1947).

TURNER, T., 'History of Fenton, Murray & Wood' (M. Tech. Sc., U.M.I.S.T., 1966).

(C) BOOKS, EIGHTEENTH AND NINETEENTH CENTURIES

ABRAM, W. A., *A History of Blackburn, Town and Parish* (Blackburn, 1877).

AIKIN, J., *Description of the Country from Thirty to Forty Miles Round Manchester* (London, 1795).

ALBAN, E., *The High Pressure Steam Engine Investigated*, trans. W. Pole, (London, 1847–8).

ALDERSON, M. A., *An Essay on the Nature and Application of Steam* (London, 1834).

ANSTIE, J., *Observations on the Importance and Necessity of Introducing Improved Machinery into the Woollen Manufactory* (London, 1803).

APPLETON, N., *Introduction of the Power Loom, and Origin of Lowell* (Lowell, 1858).

ARMSTRONG, R., *An Essay on the Boilers of Steam Engines* (London, 1839).

—— *The Modern Practice of Boiler Engineering* (London, 1856).

ASHWORTH, H., *Cotton: Its Cultivation, Manufacture and Uses* (Manchester, 1858).

BABBAGE, C., *On the Economy of Machinery and Manufactures* (London, 1832).

BAINES, E., sr., *History, Directory, and Gazetteer of the County Palatine of Lancaster* (Liverpool, 1824–5).

BAINES, E., jr., *History of the Cotton Manufacture in Great Britain* (London, 1835; 2nd edn., ed. W. H. Chaloner, Frank Cass, London, 1966).

BAINES, T., *Yorkshire Past and Present* (Liverpool, 1871).

## Select Bibliography

BANFIELD, T. C., *Industries of the Rhine*, Series i: *Agriculture* (London, 1846).
BANKS, J., *A Treatise on Mills* (London and Kendal, 1795).
—— *On the Power of Machines* (Kendal, 1803).
BARLOW, A., *The History and Principles of Weaving by Hand and by Power* (London, 1878).
BARLOW, P., 'A Treatise on the Manufactures and Machinery of Great Britain' (*Encyclopaedia Metropolitana*, Mixed Sciences, vol. vi, London, 1836).
BATCHELDER, S., *Introduction and Early Progress of the Cotton Manufacture in the United States* (Boston, 1863).
BAYNES, J., *The Cotton Trade: Two Lectures* (Blackburn, 1857).
BEATSON, R., *An Essay on the Comparative Advantages of Vertical and Horizontal Windmills* (London, 1798).
BELIDOR, B. F. DE, *Architecture hydraulique* (Paris, 1737–53).
BEVAN, G. P., (ed.), *British Manufacturing Industries*, vol. v (London, 1876).
BIRKBECK, G., and ADCOCK, H. and J., *The Steam Engine Theoretically and Practically Displayed* (London, 1827).
BISCHOFF, J., *A Comprehensive History of the Woollen and Worsted Manufacture* (London, 1842; new edn., Frank Cass, London, 1968).
BOURNE, J., *A Treatise on the Steam Engine* (London, 1846).
—— *A Catechism of the Steam Engine* (London, 1847).
—— *A Popular Description of the Steam Engine* (London, 1856).
—— *Recent Improvements in the Steam Engine* (London, 1856).
BREMNER, D., *The Industries of Scotland* (Edinburgh, 1869; new edn., ed. J. Butt and I. L. Donnachie, David & Charles, Newton Abbot, 1969).
—— *Great Industries of Great Britain* (London, 1876–80).
BREWSTER, D., *Ferguson's Lectures on Select Subjects in Mechanics* (Edinburgh, 1805).
BROWN, J., *The Basis of Mr. Samuel Crompton's Claims* (1825; reprinted 1868).
BRUNTON, R., *A Compendium of Mechanics* (Glasgow, 1824).
BUCHANAN, R., *Practical and Descriptive Essays on the Economy of Fuel and the Management of Heat* (Glasgow, 1810).
—— *Practical Essays on Millwork and other Machinery* (2nd edn., ed. T. Tredgold, London, 1823).
BURN, R. S., *Statistics of the Cotton Trade* (London, 1847).
BURNLEY, J., *The History of Wool and Woolcombing* (London, 1889).
BUTTERWORTH, E., *Historical Account of the Towns of Ashton-under-Lyne, Stalybridge, and Dukinfield* (Ashton, 1842).

CHARLEY, W. T., *Flax and its Products in Ireland* (London, 1862).
CLAUSSEN, P., *The Flax Movement: Its National Importance and Advantages* (London, 1851).
CLELAND, J., *Description of the City of Glasgow* (2nd edn., Glasgow, 1840).
COLLINSON, E., *The History of the Worsted Trade, and Historic Sketch of Bradford* (Bradford, 1854).
COLQUHOUN, P., *An Important Crisis in the Calico and Muslin Manufactory in Great Britain Explained* (London, 1788).
COMBE, J., *Remarks on the State and Prospects of the Flax Yarn Manufacture* (Leeds and London, 1849).
— — *Suggestions for Promoting the Prosperity of the Leeds Linen Trade* (Leeds, 1865).
CORIOLIS, G. C., *Du calcul de l'effet des machines* (Paris, 1829).
CURR, J., *The Coal Viewer and Engine Builder's Practical Companion* (Sheffield, 1797; 2nd edn., Frank Cass, London, 1970).

DEMAN, E. F., *The Flax Industry* (London, 1852).
DESAGULIERS, J. T., *A Course of Experimental Philosophy* (London, 1744).
DICKSON, J. H., *A Series of Letters on the Improved Mode in the Cultivation and Management of Flax* (London, 1846).
DODD, G., *History and Explanatory Dissertation on Steam Engines and Steam Packets* (London, 1818).
— — *The Textile Manufactures of Great Britain* (London, 1844).
DONNELL, E. J., *Chronological and Statistical History of Cotton* (New York, 1872).
DUPIN, E. P. C., *Rapport fait à l'Institut de France* (Paris, 1823).
DUVAL, E., *Des machines à vapeur aux États-Unis d'Amérique* (Paris, 1842).
DYMOCK, J., *The Manufacturer's Assistant* (Glasgow, 1798).

ELLISON, T., *A Handbook of the Cotton Trade* (London, 1858).
— — *The Cotton Trade of Great Britain* (London, 1886).
EMERSON, W., *Mechanics: or the Doctrine of Motion* (London, 1769).
*The Engineer and Machinist's Assistant* (Glasgow, 1850).
EVANS, O., *The Abortion of the Young Steam Engineer's Guide* (Philadelphia, 1805).
— — *The Young Mill-Wright and Miller's Guide* (5th edn., Philadelphia, 1826).
EWBANK, T., *A Descriptive and Historical Account of Hydraulic and other Machines for Raising Water, Ancient and Modern* (New York, 1842).
*Exposition Universelle de 1851: Travaux de la Commission Française*, II$^e$ Groupe (Paris, 1857).

FAIRBAIRN, W., *Treatise on Mills and Millwork* (London, 1861–3).
FAREY, J., jr., *Treatise on the Steam Engine* (vol. i, London, 1827; reprinted with vol. ii, David & Charles, Newton Abbot, 1971).
FORBES, H., *The Rise, Progress and Present State of the Worsted, Alpaca and Mohair Manufactures of England* (London, 1852).
FOSTER, T. C., *Letters on the Condition of the People of Ireland* (London, 1846).
FOX BOURNE, H. R., *English Merchants* (London, 1866).
FRENCH, G. J., *The Life and Times of Samuel Crompton* (London, 1859; 2nd edn., ed. S. D. Chapman, Adams & Dart, Bath, 1970).

GALLOWAY, E., *History and Progress of the Steam Engine* (London, 1830).
GALLOWAY, R. L., *Annals of Coal Mining and the Coal Trade* (London, 1898–1904; new edn., ed. B. F. Duckham, David & Charles, Newton Abbot, 1971).
—— *The Steam Engine and its Inventors* (London, 1881).
GASKELL, P., *Artisans and Machinery* (London, 1836; new edn., Frank Cass, London, 1968).
GAUDRY, J., *Traité élémentaire et pratique de l'installation, de la conduite, et de l'entretien des machines à vapeur* (Paris, 1856).
GLYNN, J., *Rudimentary Treatise on the Power of Water, as Applied to Drive Flour Mills* (London, 1853).
GRAY, A., *A Treatise on Spinning Machinery* (Edinburgh, 1819).
GREGORY, O., *A Treatise of Mechanics* (London, 1806; 4th edn., 1826).
GRIER, W., *The Mechanic's Calculator, or Workman's Memorial Book* (Glasgow, 1832).
—— *The Mechanic's Pocket Dictionary* (Glasgow, 1837).
GROUVELLE, P., and JAUNEZ, A., *Guide du chauffeur et du propriétaire des machines à vapeur* (4ᵉ édn., Paris, 1858–9).
GUEST, R., *A Compendious History of the Cotton Manufacture* (Manchester, 1823; new edn., Frank Cass, London, 1968).
GUYONNEAU DE PAMBOUR, F. M., *The Theory of the Steam Engine* (London, 1839).

HACHETTE, J. N. P., *Histoire des machines à vapeur* (Paris, 1830).
HANN, J., *A Short Treatise on the Steam Engine* (London, 1847).
HEAD, G., *A Home Tour through the Manufacturing Districts of England, in the summer of 1835* (London, 1836; 2nd edn., ed. W. H. Chaloner, Frank Cass, London, 1968).
HENWOOD, W. J., *The Metalliferous Deposits of Cornwall and Devon* (Penzance and London, 1843).

HIRST, W., *History of the Woollen Trade, for the Last Sixty Years* (Leeds, 1844).
HODGE, P. R., *Analytical Principles of the Expansive Steam Engine* (London, 1849).
HODGSON, J., *Textile Manufacturers and other Industries in Keighley* (Keighley, 1869).
HOËNE-WRONSKI, J. M., *Machines à vapeur: Aperçu de leur état actuel* (Paris, 1829).
HYDE, J., *The Science of Cotton Spinning Practically Arranged and Simplified* (Manchester, n.d. [1867].)

INGLIS, H. D., *Ireland in 1834* (London, 1834).
*Invention de la filature mécanique du lin* (Paris, 1850).

JACKSON, R., *The History of the Town and Township of Barnsley* (London, 1858).
JAMES, C. T., ('Justitia'), *Strictures on Montgomery on the Cotton Manufactures of Great Britain and America* (Newburyport, 1841).
JAMES, J., *History and Topography of Bradford* (Bradford, 1841; new edn., 1866).
—— *History of the Worsted Manufacture in England* (London, 1857).
JUBB, S., *The History of the Shoddy Trade* (London and Batley, 1860).
Juries of the Great Exhibition, *Reports* (London, 1852).

KANE, R., *The Industrial Resources of Ireland* (Dublin, 1844).

LARDNER, D., *The Steam Engine Familiarly Explained and Illustrated* (6th edn., London, 1836).
—— *A Rudimentary Treatise on the Steam Engine* (London, 1848).
LAWSON, J., *Letters to the Young on Progress in Pudsey during the Last Sixty Years* (Stanningley, 1887).
LEAN, T. and J., *Historical Statement of the Improvements made in the Duty Performed by the Steam Engines in Cornwall* (London and Camborne, 1839).
LEIFCHILD, J., *Cornwall: Its Mines and Miners* (London, 1857; 2nd edn., Frank Cass, London, 1968).
LUCCOCK, J., *On the Nature and Properties of Wool* (Leeds, 1805).

MCCULLOCH, J. R., *A Descriptive and Statistical Account of the British Empire* (3rd edn., London, 1847; 4th edn., 1854).
MAILLARD, S. DE, *Théories des machines mues par la force de la vapeur de l'eau* (Vienna and Strasburg, 1784).
MANN, J. A., *The Cotton Trade of Great Britain* (London, 1860).

## Select Bibliography

MARESTIER, M., *Mémoire sur les bateaux à vapeur des États-Unis d'Amérique* (Paris, 1824).
MARSDEN, R., *Cotton Spinning: Its Development, Principles and Practice* (London, 1884).
MARSHALL, J., *A Digest of all the Accounts* (London, 1834).
MARSHALL, L. C., *The Practical Flax Spinner* (Belfast, 1885).
MARX, K., *Capital* (English edn., ed. F. Engels, London, 1887; Moscow, 1965).
(MEIKLEHAM), R. STUART, *A Descriptive History of the Steam Engine* (2nd edn., London, 1824).
—— *Historical and Descriptive Anecdotes of Steam Engines* (London, 1829).
MILLINGTON, J., *An Epitome of the Elementary Principles of Natural and Experimental Philosophy* (London, 1823).
MONTGOMERY, J., *The Theory and Practice of Cotton Spinning* (2nd edn., Glasgow, 1833).
—— *A Practical Detail of the Cotton Manufacture of the U.S.A.* (Glasgow, 1840).
MORIN, A. J., *Expériences sur les roues hydrauliques à axe vertical appelées turbines* (Paris, 1838).
—— *Aide-mémoire de mécanique pratique* ($3^e$ édn., Paris, 1843).
—— *Leçons de mécanique pratique* (Paris, 1846–50).
MORIN, A. J., and TRESCA, H., *Mécanique pratique: des machines à vapeur* (Paris, 1863).
MULHALL, M., *Mulhall's Dictionary of Statistics* (London, 1884; 3rd edn., 1892).
—— *Industries and Wealth of Nations* (London, 1896).

NICHOLSON, P., *The Mechanic's Companion* (Oxford, 1825).

OGDEN, J., *A Description of Manchester* (Manchester, 1783).
OWEN, R., *The Life of Robert Owen, Written by Himself* (1857; new edn., Charles Knight & Co., London, 1971).

PARSONS, E., *The Civil, Ecclesiastical . . . and Miscellaneous History of Leeds, Bradford, Wakefield, Dewsbury, Otley* (Leeds, 1834).
PARTINGTON, C. F., *An Historical and Descriptive Account of the Steam Engine* (London, 1822).
—— *A Course of Lectures on the Steam Engine* (London, 1826).
PECQUEUR, C., *Économie sociale* (Paris, 1839).
PERKINS, J., *On the Explosion of Steam Boilers* (London, 1853).
PERRY, J., *An Elementary Treatise on Steam* (London, 1874).
PHILLIPS, J. A., and DARLINGTON, J., *Records of Mining and Metallurgy* (London, 1857).

POLE, W., *A Treatise on the Cornish Pumping Engine* (London, 1844).
—— *The Life of Sir William Fairbairn, Bart.* (London, 1877; 2nd edn., ed. A. E. Musson, David & Charles, Newton Abbot, 1970).
PONCELET, J. V., *Introduction à la mécanique industrielle, physique ou expérimentale* (3ᵉ édn, Paris, 1870).
PORTER, G. R., *The Progress of the Nation* (2nd edn., London, 1847; rev. edn., ed. F. W. Hirst, Methuen, London, 1912).

RADCLIFFE, W., *Origin of the New System of Manufacture, commonly called 'Power Loom Weaving'* (Stockport, 1828).
RANKINE, W. J. M., *A Manual of the Steam Engine and other Prime Movers* (London and Glasgow, 1859).
REES, A. (ed.), *The Cyclopaedia* (London, 1819; selections ed. N. Cossons, David & Charles, Newton Abbot, 1972).
REGNAULT, H. V., *On the Latent Heat of Steam at Different Pressures* (London, 1848).
REID, H., *The Steam Engine* (Edinburgh, 1838).
RENWICK, J., *Treatise on the Steam Engine* (New York, 1830).
RICHE DE PRONY, G. C. F. M., *Nouvelle architecture hydraulique* (Paris, 1790–6).
ROBISON, J., *A System of Mechanical Philosophy* (Edinburgh, 1822).
ROSS, J., *A Treatise on Navigation by Steam, comprising a History of the Steam Engine* (London, 1828).
RUSSELL, J. S., *A Treatise on the Steam Engine* (Encyclopaedia Britannica, 7th edn., 1841).

SAMUEL BROS., *Wool and Woollen Manufactures of Great Britain* (London, 1859).
SAVERY, T., *The Miner's Friend; or, An Engine to Raise Water by Fire* (London, 1702).
SCHULZE-GAEVERNITZ, G. VON, *The Cotton Trade in England and on the Continent* (London and Manchester, 1895).
SCOTT, D., and JAMIESON, A., *The Engineer and Machinist's Assistant* (Glasgow, 1844–7).
SMEATON, J., *An Experimental Enquiry concerning the Natural Powers of Water and Wind to Turn Mills* (London, 1794).
—— *Reports of the Late John Smeaton* (London, 1812–14).
SMITH, J., *Chronicon Rusticum Commerciale, or Memoirs of Wool* (2nd edn., London, 1757).
*The Steam Engine Explained* (Glasgow, 1848).
STEPHENSON, S. M., *On the Linen and Hempen Manufactures of the Province of Ulster* (Belfast Literary Society, 1808).
STEWART, W., *Causes of the Explosion of Steam Boilers Explained* (London, 1848).

Select Bibliography 307

SWITZER, S., *An Introduction to the General System of Hydrostaticks and Hydraulicks* (London, 1729).


TATE, T., *Exercises on Mechanics and Natural Philosophy* (London, 1846).
TAYLOR, J., *Records of Mining*, Part I (London, 1829).
TAYLOR, W. C., *Notes on a Tour through the Manufacturing Districts of Lancashire* (London, 1842; 3rd edn., ed. W. H. Chaloner, Frank Cass, London, 1968).
—— *The Handbook of the Silk, Cotton and Woollen Manufactures* (London, 1843).
TOOKE, T., *A History of Prices and of the State of the Circulation from 1793* (London, 1838–57).
TOYNBEE, A., *Lectures on the Industrial Revolution in England* (London, 1884; new edn., ed. T. S. Ashton, David & Charles, Newton Abbot, 1969).
TREDGOLD, T., *The Steam Engine* (London, 1827).
TREVITHICK, F., *Life of Richard Trevithick, with an Account of his Inventions* (London, 1872).


URE, A., *The Philosophy of Manufactures* (London, 1835; new edn., Frank Cass, London and Haarlem, 1967).
—— *A Dictionary of Arts, Manufactures and Mines* (London, 1839; 4th edn., London, 1853).
—— *The Cotton Manufacture of Great Britain Systematically Investigated* (2nd edn., ed. P. L. Simmonds, London, 1861).


WARDEN, A. J., *The Linen Trade, Ancient and Modern* (London, 1864; 3rd edn., Frank Cass, London, 1967).
WATSON, J. Y., *A Compendium of British Mining* (London, 1843).
WHITE, G., *A Practical Treatise on Weaving, by Hand and Power Looms* (Glasgow, 1846).
WHITE, G. S., *Memoir of Samuel Slater* (Philadelphia, 1876).
WICKSTEED, T., *An Experimental Inquiry concerning Cornish and Boulton & Watt Engines* (London, 1841).
—— *Further Elucidations of the Useful Effects of Cornish Engines* (London, 1859).
WILKINS, C., *The South Wales Coal Trade and its Allied Industries* (Cardiff, 1888).


YOUNG, T., *A Course of Lectures on Natural Philosophy and the Mechanical Arts* (London, 1807).

(D) BOOKS, TWENTIETH CENTURY

ASHMORE, O., *The Industrial Archaeology of Lancashire* (David & Charles, Newton Abbot, 1969).

ASHTON, T. S., and SYKES, J., *The Coal Industry of the Eighteenth Century* (Manchester U.P., Manchester, 1929).

ASPIN, C., and CHAPMAN, S. D., *James Hargreaves and the Spinning Jenny* (Helmshore Local History Society, Preston, 1964).

BALLOT, C., *L'Introduction du machinisme dans l'industrie française* (Paris, 1923).

BARTON, D. B., *The Cornish Beam Engine* (D. Bradford Barton, Truro, 1965).

BIENEFELD, M. A., *Working Hours in British Industry: An Economic History* (L. S. E./Weidenfeld & Nicolson, London, 1972).

BIRCH, A., *The Economic History of the British Iron and Steel Industry, 1784–1879* (Frank Cass, London, 1967).

BOYSON, R., *The Ashworth Cotton Enterprise: The Rise and Fall of a Family Firm, 1818–1880* (Clarendon Press, Oxford, 1970).

BUTT, J. (ed.), *Robert Owen: Prince of Cotton Spinners* (David & Charles, Newton Abbot, 1971).

BYTHELL, D., *The Handloom Weavers* (Cambridge U.P., Cambridge, 1969).

CAMPBELL, R. H., *Carron Company* (Oliver & Boyd, Edinburgh and London, 1961).

—— *Scotland since 1707: The Rise of an Industrial Society* (Basil Blackwell, Oxford, 1965).

CARDWELL, D. S. L., *Steam Power in the Eighteenth Century* (Sheed & Ward, London and New York, 1963).

—— *From Watt to Clausius: The Rise of Thermodynamics in the Early Industrial Age* (Heinemann, London, 1971).

CATLING, H. J., *The Spinning Mule* (David & Charles, Newton Abbot, 1970).

CHAPMAN, S. D., *The Early Factory Masters* (David & Charles, Newton Abbot, 1967).

—— *The Cotton Industry in the Industrial Revolution* (Macmillan Studies in Economic History, London, 1972).

CHAPMAN, S. J., *The Lancashire Cotton Industry* (Manchester U.P., Manchester, 1904).

—— *The Cotton Industry and Trade* (Methuen, London, 1905).

CHECKLAND, S. G., *The Rise of Industrial Society in England, 1815–1885* (Longmans, London, 1964).

CLAPHAM, J. H., *The Woollen and Worsted Industries* (Methuen, London, 1907).
—— *An Economic History of Modern Britain*, vol. i (Cambridge U.P., Cambridge, 1930).
CLARK, V. S., *History of Manufactures in the United States, 1607–1860* (Carnegie Institution, Washington, 1916).
COLLIER, F., *The Family Economy of the Working Classes in the Cotton Industry, 1784–1833* (ed. R. S. Fitton, Manchester U.P., Manchester, 1965).
COPELAND, M. T., *The Cotton Manufacturing Industry of the United States* (Harvard U.P., Cambridge, Mass., 1912).
COURT, W. H. B., *The Rise of the Midland Industries 1600–1838* (Oxford U.P., London, 1938).
CROUZET, F., CHALONER, W. H., and STERN, W. M. (eds.), *Essays in European Economic History, 1789–1914* (Edward Arnold, London, 1969).
CRUMP, W. B. (ed.), *The Leeds Woollen Industry, 1780–1820* (Thoresby Society, Leeds, 1931).
CRUMP, W. B., and GHORBAL, S. G., *History of the Huddersfield Woollen Industry* (Alfred Jubb & Son, Huddersfield, 1935).
CUNNINGHAM, W., *The Growth of English Industry and Commerce in Modern Times*, Part II: *Laissez Faire* (Cambridge U.P., Cambridge, 1913).

DANIELS, G. W., *The Early English Cotton Industry* (Manchester U.P., Manchester, 1920).
DARBY, H. C., *The Draining of the Fens* (Cambridge U.P., Cambridge, 1940).
DAUMAS, M., *Histoire générale des techniques*, tome III (Presses Universitaires de France, Paris, 1968).
DEANE, P., and COLE, W. A., *British Economic Growth, 1688–1959: Trends and Structure* (2nd edn., Cambridge U.P., Cambridge, 1967).
DEMOULIN, R., *Guillaume I$^{er}$ et la transformation économique des provinces Belges, 1815–1830* (Université de Liège, Liège, 1938).
DERRY, T. K., and WILLIAMS, T. I., *A Short History of Technology from the Earliest Times to A.D. 1900* (Clarendon Press, Oxford, 1960).
DHONDT, J., and BRUWIER, M., *The Industrial Revolution in Belgium and Holland, 1700–1914* (Fontana Economic History of Europe, vol. iv, London, 1970).
DICKINSON, H. W., *James Watt: Craftsman and Engineer* (Cambridge U.P., Cambridge, 1936).
—— *A Short History of the Steam Engine* (Cambridge U.P., Cambridge, 1939).

DICKINSON, H. W., and JENKINS, R., *James Watt and the Steam Engine* (Clarendon Press, Oxford, 1927).
DICKINSON, H. W., and TITLEY, A., *Richard Trevithick: the Engineer and the Man* (Cambridge U.P., Cambridge, 1934).
DOBSON, B. P., *The Story of the Evolution of the Spinning Machine* (Marsden & Co., Manchester, 1910).
DODD, A. H., *The Industrial Revolution in North Wales* (Wales U.P., Cardiff, 1933).
DUCKHAM, B. F., *A History of the Scottish Coal Industry*, vol. i (David & Charles, Newton Abbot, 1970).

EDWARDS, M. M., *The Growth of the British Cotton Trade, 1780–1815* (Manchester U.P., Manchester, 1967).
ENGLISH, W., *The Textile Industry* (Longmans, London, 1969).

FINCH, W. C., *Watermills and Windmills* (C. W. Daniel, London, 1933).
FITTON, R. S., and WADSWORTH, A. P., *The Strutts and the Arkwrights, 1758–1830* (Manchester U.P., Manchester, 1958).
FOGEL, R. W., and ENGERMAN, S. (eds.), *The Reinterpretation of American Economic History* (Harper & Row, New York, 1971).
FONG, H. D., *Triumph of Factory System in England* (Nankai U.P., Tientsin, 1930).

GARNETT, W. O., *Wainstalls Mills: the History of I. & I. Calvert Ltd. 1821–1951* (I. & I. Calvert Ltd., Halifax, 1952).
GAULDIE, E. (ed.), *The Dundee Textile Industry, 1790–1885* (Scottish History Society, Edinburgh, 1969).
GAYER, A. D., ROSTOW, W. W., and SCHWARTZ, A. J., *The Growth and Fluctuation of the British Economy, 1790–1850* (Clarendon Press, Oxford, 1953).
GIBB, G. S., *The Saco-Lowell Shops: Textile Machinery Building in New England, 1813–1849* (Harvard U.P., Cambridge, Mass., 1950).
GILL, C., *The Rise of the Irish Linen Industry* (Clarendon Press, Oxford, 1925).
GREEN, E. R. R., *The Lagan Valley, 1800–1850* (Faber & Faber, London, 1949).

HABAKKUK, H. J., *American and British Technology in the 19th Century: The Search for Labour-Saving Inventions* (Cambridge U.P., Cambridge, 1962).

## Select Bibliography

HAMILTON, H., *The English Brass and Copper Industries to 1800* (Longmans, London, 1926).
— — *The Industrial Revolution in Scotland* (Clarendon Press, Oxford, 1932).
— — *An Economic History of Scotland in the Eighteenth Century* (Clarendon Press, Oxford, 1963).
HAMMOND, J. L. and B., *The Rise of Modern Industry* (Methuen, London, 1925; 9th edn., ed. R. M. Hartwell, 1966).
HARRIS, T. R., *Arthur Woolf: The Cornish Engineer, 1766–1837* (D. Bradford Barton, Truro, 1966).
HARTE, N. B., and PONTING, K. (eds.), *Textile History and Economic History: Essays in Honour of Miss Julia de Lacy Mann* (Manchester U.P., Manchester, 1973).
HARTWELL, R. M. (ed.), *The Industrial Revolution* (Basil Blackwell, Oxford, 1970).
HAWKE, G. R., *Railways and Economic Growth in England and Wales, 1840–1870* (Clarendon Press, Oxford, 1970).
HEATON, H., *The Yorkshire Woollen and Worsted Industry, from the Earliest Times up to the Industrial Revolution* (Clarendon Press, Oxford, 1920; new edn., 1965).
HENDERSON, W. O., *Britain and Industrial Europe 1750–1870* (3rd edn., Leicester U.P., Leicester, 1972).
HIGGINS, J. P. P., and POLLARD, S. (eds.), *Aspects of Capital Investment in Great Britain, 1750–1850: A Preliminary Survey* (Methuen, London, 1971).
HILLS, R. L., *Machines, Mills, and Uncountable Costly Necessities: A Short History of the Drainage of the Fens* (Goose & Son, Norwich, 1967).
— — *Power in the Industrial Revolution* (Manchester U.P., Manchester, 1970).
HOFFMANN, W. G., *British Industry, 1700–1950* (trans. W. O. Henderson and W. H. Chaloner, Basil Blackwell, Oxford, 1955).
HOOPER, F., *Statistics relating to the City of Bradford and the Woollen and Worsted Trade* (W. Byles & Sons, Bradford, 1904).
HUGHES, J. R. T., *Fluctuations in Trade, Industry and Finance* (Clarendon Press, Oxford, 1960).

JENKINS, J. G. (ed.), *The Wool Textile Industry in Great Britain* (Routledge & Kegan Paul, London, 1972).
JEVONS, W. S., *The Coal Question* (3rd edn., ed. A. W. Flux, Macmillan, London, 1906).

KOLIN, I., *The Evolution of the Heat Engine* (Longmans, London, 1974).

LANDES, D. S., *The Unbound Prometheus: Technological Change 1750 to the Present* (Cambridge U.P., Cambridge, 1969).
LEBRUN, P., *L'Industrie de la laine à Verviers* (Liège, 1948).
LEE, C. H., *A Cotton Enterprise, 1795–1840: A History of McConnel & Kennedy, Fine Cotton Spinners* (Manchester U.P., Manchester, 1972).
LÉVY-LEBOYER, M., *Les Banques européennes et l'industrialisation internationale* (Presses Universitaires de France, Paris, 1964).
LEWINSKI, J., *L'Évolution industrielle de la Belgique* (Brussels, 1911).
LIPSON, E., *History of the Woollen and Worsted Industries* (A. & C. Black, London, 1921; new edn., Frank Cass, London, 1965).
—— *A History of Wool and Wool Manufacture* (Heinemann, London, 1953).
LORD, J., *Capital and Steam Power, 1750–1800* (P. S. King & Son, London, 1923; 2nd edn., ed. W. H. Chaloner, Frank Cass, London, 1966).

McGOULDRICK, P. F., *New England Textiles in the Nineteenth Century: Profits and Investments* (Harvard U.P., Cambridge, Mass., 1968).
MANN, J. DE L., *The Cloth Industry in the West of England from 1640 to 1880* (Clarendon Press, Oxford, 1971).
MANTOUX, P., *The Industrial Revolution in the Eighteenth Century* (Jonathan Cape, London, 1928; rev. edn., ed. T. S. Ashton, Methuen, London, 1966).
MARSHALL, A., *Industry and Trade: A Study of Industrial Technique and Business Organisation* (Macmillan, London, 1919).
MARSHALL, J. D., and DAVIES-SHIEL, M., *The Industrial Archaeology of the Lake Counties* (David & Charles, Newton Abbot, 1969).
MATHIAS, P., *The Brewing Industry in England, 1700–1830* (Cambridge U.P., Cambridge, 1959).
—— (ed.), *Science and Society, 1600–1900* (Cambridge U.P., Cambridge, 1972).
MATSCHOSS, C., *Die Entwicklung der Dampfmaschine* (vol. i, Julius Springer, Berlin, 1908).
MATTHEWS, R. C. O., *A Study in Trade-Cycle History: Economic Fluctuations in Great Britain, 1833–1842* (Cambridge U.P., Cambridge, 1954).
MIDDLETON, T., *The History of Hyde and its Neighbourhood* (Higham Press, Hyde, 1932).
MITCHELL, B. R., with DEANE, P., *Abstract of British Historical Statistics* (Cambridge U.P., Cambridge, 1962).
MILWARD, A., and SAUL, S. B., *The Economic Development of Continental Europe, 1780–1870* (Allen & Unwin, London, 1973).
MOORE, A. S., *Linen: from the Raw Material to the Finished Product* (London, 1914).

Select Bibliography 313

Musson, A. E. (ed.), *Science, Technology and Economic Growth in the Eighteenth Century* (Methuen, London, 1972).
Musson, A. E., and Robinson, E., *Science and Technology in the Industrial Revolution* (Manchester U.P., Manchester, 1969).

National Bureau of Economic Research, *Output, Employment, and Productivity in the United States after 1800* (Studies in Income and Wealth, xxx, Columbia U.P., New York and London, 1966).
Navin, T. R., *The Whitin Machine Works since 1831* (Harvard U.P., Cambridge, Mass., 1950).
Nef, J. U., *The Rise of the British Coal Industry* (G. Routledge & Sons, London, 1932).

Payen, J., *Capital et machine à vapeur au XVIII$^e$ siècle* (Mouton, Paris, 1969).
Pigott, S. C., *Hollins: A Study of Industry, 1784–1949* (William Hollins & Co., Nottingham, 1949).
Pollard, S., *The Genesis of Modern Management: A Study of the Industrial Revolution in Great Britain* (Edward Arnold, London, 1965).
Ponting, K. G., *A History of the West of England Cloth Industry* (Macdonald, London, 1957).
—— *The Woollen Industry of South-West England* (Adams & Dart, Bath, 1971).
—— (ed.), *Baines's Account of the Woollen Manufacture of England* (David & Charles, Newton Abbot, 1970).
Pursell, C. W., *Early Stationary Steam Engines in America: A Study in the Migration of a Technology* (Smithsonian Institution Press, Washington, 1969).

Raistrick, A., *Dynasty of Iron Founders: The Darbys and Coalbrookdale* (Longmans, London, 1953; 2nd edn., David & Charles, Newton Abbot, 1970).
Reach, A. B., *Manchester and the Textile Districts in 1849*, ed. C. Aspin (Helmshore, 1972).
Reynolds, J., *Windmills and Watermills* (Hugh Evelyn, London, 1970).
Rimmer, W. G., *Marshalls of Leeds, Flax-Spinners, 1788–1886* (Cambridge U.P., Cambridge, 1960).
Roll, E., *An Early Experiment in Industrial Organisation: being a History of the Firm of Boulton & Watt, 1775–1805* (Longmans, London, 1930).
Rosenberg, N., *Technology and American Economic Growth* (Harper & Row, New York, 1972).

ROSENBERG, N., (ed.), *The American System of Manufactures* (Edinburgh U.P., Edinburgh, 1969).
ROSTOW, W. W., *The Stages of Economic Growth: A Non-Communist Manifesto* (2nd edn., Cambridge U.P., Cambridge, 1971).
— — *How it all Began: Origins of the Modern Economy* (Methuen, London, 1975).
— — (ed.), *The Economics of Take-off into Sustained Growth* (Macmillan, London, 1963).
ROWE, W. J., *Cornwall in the Age of the Industrial Revolution* (Liverpool U.P., Liverpool, 1953).

SCOTT, E. K., *Matthew Murray: Pioneer Engineer* (Edwin Jowett, Leeds, 1928).
SHAPIRO, S., *Capital and the Cotton Industry in the Industrial Revolution* (Cornell U.P., Ithaca, 1967).
SIGSWORTH, E. M., *Black Dyke Mills: A History* (Liverpool U.P., Liverpool, 1958).
SINGER, C. J., HOLMYARD, E. J., HALL, A. R., and WILLIAMS, T. I. (eds.), *A History of Technology* (Clarendon Press, Oxford, 1954–8).
SMELSER, N. J., *Social Change in the Industrial Revolution: An Application of Theory to the Lancashire Cotton Industry, 1770–1840* (Routledge & Kegan Paul, London, 1959).
SWEEZY, P. M., *Monopoly and Competition in the English Coal Trade, 1550–1850* (Harvard U.P., Cambridge, Mass., 1938).

TANN, J., *Gloucestershire Woollen Mills* (David & Charles, Newton Abbot, 1967).
— — *The Development of the Factory* (Cornmarket Press, London, 1970).
TUPLING, G. H., *The Economic History of Rossendale* (Chetham Soc. N.S. lxxxvi, Manchester, 1927).
TURNER, H. A., *Trade Union Growth, Structure and Policy* (Allen & Unwin, London, 1962).

UNWIN, G., HULME, A., and TAYLOR, G., *Samuel Oldknow and the Arkwrights: The Industrial Revolution at Stockport and Marple* (Manchester U.P., Manchester, 1924).
USHER, A. P., *A History of Mechanical Inventions* (McGraw-Hill, New York, 1929; rev. edn., Harvard U.P., Cambridge, Mass., 1954).

WADSWORTH, A. P., and MANN, J. DE L., *The Cotton Trade and Industrial Lancashire, 1600–1780* (Manchester U.P., Manchester, 1931).

Wailes, R., *Windmills in England: A Study of their Origin, Development and Future* (Architectural Press, London, 1948).
—— *The English Windmill* (Routledge & Kegan Paul, London, 1954).
Ware, C., *The Early New England Cotton Manufacture: A Study in Industrial Origins* (Houghton Mifflin, Boston, 1931).
Wood, G. H., *The History of Wages in the Cotton Trade during the Past Hundred Years* (Sherratt & Hughes, London and Manchester, 1910).

(E) ARTICLES IN PERIODICALS

(N.B. Individual articles in the journals mentioned here are not listed separately)

Allen, J. S., "The 1712 and other Newcomen Engines of the Earl of Dudley", *Trans. Newcomen Soc.*, xxxvii, 1964–5.
Andrew, S., "Fifty Years in the Cotton Trade", *Oldham Standard*, 1887.
*Artizan* (1825—).
Asher, E., "Industrial Efficiency and Biased Technical Change in American and British Manufacturing: The case of Textiles in the Nineteenth Century", *J. Econ. Hist.* xxxii, no. 2, June 1972.
Ashmore, O., "Low Moor, Clitheroe: A Nineteenth-Century Factory Community", *Trans. Lancs. & Cheshire Antiq. Soc.* 73–4, 1966.
Ashworth, H. and E., "Number of Steam Engines and Water Wheels in Bolton, February/March 1837", *Rep. Manchester Stat. Soc.*, 1837.
Atkinson, F., "The Horse as a Source of Rotary Power", *Trans. Newcomen Soc.* xxxiii, 1960–1.

Baker, R., "On the Industrial and Sanitary Economy of the Borough of Leeds in 1858", *J. Roy. Stat. Soc.* xxi, 1858.
Bankes, J. H. M., "Records of Mining in Winstanley and Orrell, near Wigan", *Trans. Lancs. & Cheshire Antiq. Soc.* liv, 1939.
Baynes, J., "Experiments on Steam Engine Power", *Mechanic's Magazine*, lxiv, 1846.
Blaug, M., "The Productivity of Capital in the Lancashire Cotton Industry during the Nineteenth Century", *Econ. Hist. Rev.* 2nd Ser. xiii, 1960–1.
Bondt, F. de, "La Dimension du progrès technique dans le textile", *Recherches économiques de Louvain*, Oct. 1970.
Brown, W., "Information Regarding Flax Spinning at Leeds", 1821 (Leeds Reference Library).

Browne's "Engine Reporter", *Royal Cornwall Gazette*, 1847–58.
BURN, R. S., *Commercial Glance*, 1844 edn. (British Museum).

CARNE, J., "On . . . the Improvements which have been Made in Mining", *Trans. Roy. Geog. Soc. Cornwall*, iii, 1828.
CHADWICK, D., "On the Rate of Wages in Manchester and Salford, 1839–59", *J. Stat. Soc.*, 1860.
CHALONER, W. H., "The Cheshire Activities of Matthew Boulton and James Watt, of Soho, near Birmingham, 1776–1817", *Trans. Lancs. & Cheshire Antiq. Soc.*, 1949.
CHAPMAN, D., "The Establishment of the Jute Industry: A Problem in Location Theory?", *Rev. Econ. Studies* vi, 1938–9.
CHAPMAN, S. D., "The Pioneers of Worsted Spinning by Power", *Bus. Hist.*, vii, July 1965.
— — "The Peels in the Early English Cotton Industry", *Bus. Hist.* xi, 1969.
— — "Fixed Capital Formation in the British Cotton Industry, 1770–1815", *Econ. Hist. Rev.* 2nd Ser. xxiii, 1970.
— — "The Cost of Power in the Industrial Revolution", *Midland Hist.*, i, 1970.
*Civil Engineer and Architect's Journal, Scientific and Railway Gazette* (London, 1837–67).
CLAPHAM, J. H., "The Transference of the Worsted Industry from Norfolk to the West Riding", *Econ. J.* xx, No. 78, 1910.
CLAYTON, A. K., "The Newcomen-Type Engine at Elsecar, West Riding", *Trans. Newcomen Soc.* xxxv, 1962–3.

DANIELS, G. W., "Samuel Crompton's Census of the Cotton Industry in 1811", *Econ. Hist.* ii, 1930.
DEANE, P. M., "The Output of the British Woollen Industry in the Eighteenth Century", *J. Econ. Hist.* xvii, 1957.
— — "New Estimates of Gross National Product for the United Kingdom, 1830–1914", *Rev. Income and Wealth*, Ser. 14, June 1968.
DICKINSON, H. W., and LEE, A., "The Rastricks—Civil Engineers", *Trans. Newcomen Soc.* iv, 1923–4.
*The Economist*, (1843—).
*Engineer and Machinist* (London and Manchester, 1850–1).
ENYS, J. S., "Remarks on the Duty of the Steam Engines employed in the Mines of Cornwall at Different Periods", *Trans. Inst. Civil Engineers*, iii, 1842.

FAIRBAIRN, W., "On the Expansive Action of Steam, and a New Construction of Expansion Valves for Condensing Steam Engines", *Proc. Inst. Mech. Engineers*, i, July 1849.

FAIRBAIRN, W., "On the Consumption of Fuel and the Prevention of Smoke", *British Assoc. Advancement of Science*, 1851.
FEATHER, G. A., "A Pennine Worsted Community in the Mid-Nineteenth Century", *Textile Hist.* iii, Dec. 1972.
FORWARD, E. A., "Simon Goodrich and his Work as an Engineer", *Trans. Newcomen Soc.*, iii, 1922–3 and xviii 1937–8.
*Journal of the Franklin Institute* (1828—).


GILBERT, D., "On the Progressive Improvements made in the Efficiency of Steam Engines in Cornwall", *Phil. Trans.* 120, 1830, pt. I.
GLOVER, F. J., "The Rise of the Heavy Woollen Trade of the West Riding", *Bus. Hist.* iv, 1961.
GLYNN, J., "Draining Land by Steam Power", *Trans. Soc. Arts & Manufactures*, ii, Derby, 1836.
GOODCHILD, J., "The Ossett Mill Company", *Textile Hist.* i, 1968.
— — "Pildacre Mill; An Early West Riding Factory", *Textile Hist.* i, 1970.
— — "On the Introduction of Steam Power into the West Riding", *South Yorks. J.*, pt. III, May 1971.
GREENWOOD, T., "On Machinery Employed in the Preparation and Spinning of Flax", *Proc. Inst. Mech. Engineers*, Aug. 1865.
GREG, R. H., *et al.*, "A Return from 412 Cotton Mills in Manchester and Surrounding Districts" (Manchester, 1844; Manchester Reference Library).


HARRIS, J. R., "The Employment of Steam Power in the Eighteenth Century", *History*, lii, no. 175, June 1967.
HARRISON, A., and J. K., "The Horse Wheel in North Yorkshire", *Industrial Archaeology*, x, Aug. 1973.
HEATON, H., "Benjamin Gott and the Industrial Revolution in Yorkshire", *Econ. Hist. Rev.* iii, 1931.
HENWOOD, W. J., "Account of the Steam Engines in Cornwall", *Edinburgh J. of Science*, x, 1829.
— — "On the Expansive Action of Steam in some of the Pumping Engines on the Cornish Mines", *Trans. Inst. Civil Engineers*, ii, 1838.
HILLS, R., and PACEY, A. J., "The Measurement of Power in Early Steam-Driven Textile Mills", *Technology and Culture*, xiii, no. 1, 1972.
HYDE, C. K., "The Adoption of Coke-Smelting by the British Iron Industry, 1709–1790", *Expl. Econ. Hist.* N.S. x, 1973.

— — "Technological Change in the British Wrought Iron Industry 1750–1815", *Econ. Hist. Rev.* 2nd Ser. xxvii, May 1974.

JENKINS, R., "Jonathan Hornblower and the Compound Engine", *Trans. Newcomen Soc.* xi, 1930–1.
— — "A Cornish Engineer: Arthur Woolf, 1766–1837", *Trans. Newcomen Soc.*, xiii, 1932–3.
JEREMY, D. J., "Innovation in American Textile Technology during the Early 19th Century", *Technology and Culture*, xiv, Jan. 1973.
JEWKES, J., "The Localisation of the Cotton Industry", *Econ. Hist.* ii, 1930.

KENNEDY, J., "Observations on the Rise and Progress of the Cotton Trade in Great Britain", *Mem. Lit. and Phil. Soc. of Manchester*, 2nd Ser. iii, 1819.
— — "A Brief Memoir of Samuel Crompton", *Mem. Lit. & Phil. Soc. of Manchester*, 2nd Ser. v., 1831.

LANGTON, W., "Steam Power used in Manufactures in Manchester and Salford", *J. Stat. Soc.* ii, 1839.
LEMON, SIR CHARLES, "The Statistics of the Copper Mines of Cornwall", *J. Stat. Soc.* i, 1838.

MCCULLOCH, J. R., "An Essay on the Rise, Progress, Present State, and Prospects of the Cotton Manufacture", *Edinburgh Rev.* lxvi, 1827.
MCCUTCHEON, W. A., "Water Power in the North of Ireland", *Trans. Newcomen Soc.* xxxix, 1966–7.
MARSHALL, J. G., "Sketch of the History of Flax Spinning in England, especially as developed in the Town of Leeds", *British Assoc. for the Advancement of Science*, 1858.
MARWICK, W. H., "The Cotton Industry and the Industrial Revolution in Scotland", *Scot. Hist. Rev.* xxi, 1924.
*Mechanic's Magazine* (1823–66).
MERTTENS, F. W., "The Hours and Cost of Labour . . .", *Trans. Manchester Stat. Soc.*, 1893–4.
*Mining Journal and Commercial Gazette* (1835–1910).
MITCHELL, G. M., "The English and Scottish Cotton Industries: A Study in Interrelations", *Scot. Hist. Rev.* xxii, 1925.
MOKYR, J., "The Industrial Revolution in the Low Countries in the First Half of the Nineteenth Century", *J. Econ. Hist.* xxxiv, no. 2, June 1974.

MORSHEAD, W. A., jr., "On the Relative Advantages of Steam, Water, and Animal Power", *J. Bath and West of England Soc.* iv, 1856.
MOTT, R. A., "The Newcomen Engine in the Eighteenth Century", *Trans. Newcomen Soc.* xxxv, 1962–3.
MUTTON, N., "Boulton & Watt and the Norfolk Marshland", *Norfolk Archaeology*, xxxiv, 1967.

*Newry Magazine* (1815–16).
NIXON, F., "The Early Steam-Engine in Derbyshire", *Trans. Newcomen Soc.* xxxi, 1957–9.

PANKHURST, K. V., "Investment in the West Riding Wool Textile Industry in the Nineteenth Century", *Yorks. Bull. of Econ. & Soc. Research*, vii, 1955.
PARKES, J., "On Steam Boilers and Steam Engines", *Trans. Inst. Civil Engineers*, iii, 1842.
PERIER, J.-C., "Sur les machines a vapeur", *Bulletin de la Société d'Encouragement pour l'Industrie Nationale*, no. lxxiii, July 1810.
*Philosophical Magazine and Journal* (1814–26).
POLE, W., "On the Pressure and Density of Steam", *Min. Proc. Inst. Civil Engineers*, vi, 1847.

RADLEY, J., "York Waterworks, and other Waterworks in the North before 1800", *Trans. Newcomen Soc.* xxxix, 1966–7.
RAISTRICK, A., "The Steam Engine on Tyneside, 1715–1778", *Trans. Newcomen Soc.* 1936–7.
*Repertory of Arts and Manufactures* (1794–1825).
*Repertory of Patent Inventions* (1825–62).
RIMMER, W. G., "Middleton Colliery near Leeds, 1779–1830", *Yorks. Bull. for Econ. & Soc. Research*, vii, 1955.
RODGERS, H. B., "The Lancashire Cotton Industry in 1840", *Trans. & Papers Inst. British Geographers*, xxviii, 1960.
ROSENBERG, N., "Karl Marx on the Economic Role of Science", *J. Pol. Econ.* 82, no. 4, July–Aug. 1974.
*Reports of the Royal Cornwall Polytechnic Society* (1833—).

SANDBERG, L., "American Rings and English Mules: The Role of Economic Rationality", *Quart. J. Econ.* 83, 1969.
—— "Movements in the Quality of British Cotton Textile Exports, 1815–1913", *J. Econ. Hist.* xxviii, no. 1, Mar. 1968.
SIGSWORTH, E. M., "An Episode in Wool Combing", *J. Bradford Textile Soc.*, 1956–7.
SPENCER, J., "On the Growth of the Cotton Trade", *Trans. Manchester Stat. Soc.*, 1877.

TANN, J., "Some Problems of Water Power", *Trans. Bristol & Gloucs. Arch. Soc.* lxxxiv, 1965.
—— "Richard Arkwright and Technology", *History*, 58, 1973.
—— "Fuel Saving in the Process Industries during the Industrial Revolution: A Study in Technological Diffusion", *Business Hist.* xv, 1973.
TEMIN, P., "Steam and Waterpower in the Early Nineteenth Century", *J. Econ. Hist.* xxvi, June 1966.
—— "Labor Scarcity and the Problem of American Industrial Efficiency in the 1850's", *J. Econ. Hist.* xxvi, Sept. 1966.
THOMAS, J., "Josiah Wedgwood as a Pioneer of Steam Power in the Pottery Industry", *Trans. Newcomen Soc.* xvii, 1936–7.
TITLEY, A., "Trevithick and Rastrick and the Single-Acting Expansive Engine", *Trans. Newcomen Soc.* vii, 1926–7.
—— "Richard Trevithick and the Winding Engine". *Trans. Newcomen Soc.* x, 1929–30.

WICKSTEED, T., "On the Effective Power of High-Pressure Expansive Condensing Steam Engines, commonly in use in Cornish Mines", *Trans. Inst. Civil Engineers*, i, 1836; ii, 1838.
WILLIAMSON, J. G., "Optimal Replacement of Capital Goods: The Early New England and British Textile Firm", *J. Pol. Econ.* 79, Nov–Dec. 1971.
WILSON, P. N., "Water Power and the Industrial Revolution", *Water Power* vi, Aug. 1954.
WOOD, O., "A Cumberland Colliery during the Napoleonic War", *Economica*, N.S. xxi, 1954.
*Woollen, Worsted and Cotton Journal* (1853–4).
WRIGHT, G., "An Econometric Study of Cotton Production and Trade, 1830–1860", *Rev. Econ. Stats.* lii, no. 2, May 1971.

# Index

Sub-entries are arranged with references to power and power source first, and other references to industrial activity in order of productive process thereafter.

Aachen, 280
Abram, W. A., 201n
Adcock, H. and J., 77n, 143n
Agriculture: employment in 1800, 43; prices, 44n; and steam power, 23, 288; and water power, 126, 154n
Aikin, J., 16n, 97n
Air pump, 18, 20, 22, 23, 93, 105
Aire River, 138
Alban, E., 85n
Albion Mills, 122
Allen, G. C., 64n
Allen, J. S., 56n, 144n
Allen, Z., 272n
Alport Mines, 168
Alsace, 166
Andreano, R. L., 285n
Andrew, S., 188n, 237n
Animals: see Horses, Donkeys, Oxen
Arago, D. F., 30, 30n, 166
Arkwright, R.: inventions, 7, 179, 180n, 186, 241–2; use of horses, 118, 160, 176; mill, 119, 128, 130, 155, 177, 182n
Armley, 97n
Armley Mills, 87, 171
Armstrong, R., 70, 87n, 108, 108n
Arrow, K., 259n, 292
Asher, E., 285n
Ashley, J., 219n
Ashley, Lord (Shaftesbury), 217n
Ashmore, O., 88n
Ashton, T. (of Hyde), 203n, 204n, 213
Ashton, Professor T. S., 3, 6n, 148n, 158n, 293
Ashworth, H., 33, 180n, 227n, 228n
Ashworths: mills and water power, 74n, 120n, 128, 130, 132, 135, 215n; rents, 193n; coals, 200n; profits, 213, 215–6; MSS, 72n, 74n, 132n, 135, 193n
Aspin, C., 179n, 184n
Atkinson, F., 117n
Atmospheric Engine: role in Industrial Revolution, 3; on coalfields, 47, 76–7; technical progress in, 17, 19, 21n; number of, 27, 31; distribution of, 105; power of, 26, 29, 68, 143–4, 146; costs of, 75, 77, 94–5; coal consumption of, 67–9, 76, 110–11; scrapping of, 76–7, 78; replacing horses, 120, 155; replacing water-wheels, 287; replacing Watt engines, 156–7, 286; returning engines, 79, 142–3; underground engines, 144n; Bateman & Sherratt engines, 179n. See also Newcomen Engine, Savery Engine, Returning Engine
Attercliffe, 69
Austhorpe, 17
Austin, J. T., (Fowey Consols engine), 79
Auxiliary Engines, 125, 171, 233. See also Returning Engine
Ayr River, 128
Ayrshire, 131

Babbage, C., 189n, 203n, 240n
Backward Linkages: nature and calculation of, 13, 26, 28, 44–5, 98, 100–2, 105–6, 107; duration of, 46, 99–100; and derived innovation, 42; and social savings, 44, 113–15; to iron, 44, 98, 99–104, 105–6, 108–10, 115, 285–6; and total iron output, 103–4, 109, 286; to brass and copper, 99, 104–6, 108; to coal, 45, 99, 110–13, 115, 286; of boilers, 101–2, 108; for textiles, 108–10, 112–13
Bailey, Messrs. (of Staleybridge), 61n
Baines, E., jr: reliability, 7n; water supply, 7, 138, 172; steam power in cotton, 232, 236n; spindleage, 182n, 237n; number of power-looms, 195n; cotton output, 227n, 228n, 230n; wool output, 243n, 251; price of cotton, 180n; price of textile machinery, 272n; wages, 242; employment, 184n, 211n
Baines, E., sr, 32, 33
Baker, R., 34n
Bakewell, 130, 134n, 137n, 162n, 163n, 168
Ballysillan, 167

Bankes, J. H. M., 119n
Banks, J., 126n
Bann River, 128, 172
Barlow, A., 201
Barlow, P., 80, 81n, 119, 120n
Baron, F., 166
Barrow, J., 23n
Bartlett Mills, 268, 270n
Barton, D. B., 58n, 80, 82n, 85n, 89n, 144n, 256n, 261n
Basse-Ransy (Belgium), 277
Batchelder, S., 272n
Bateman & Sherratt, 31, 32, 179n
Bateman, J. F., 128n, 172
Baynes, J., 199, 237n
Beam (of engine): wooden, 16; compound, 18; cast-iron, 21; cost and weight, 94–5; elimination of, 22, 23; and double-acting engine, 20; and side-lever engine, 22; and McNaughted engine, 86; motion of, 18, 142; supporting bobs for stability, 59
Beighton, H., 49n
Belfast, 35, 292
Belgium: lag in industrialization, 276–7; steam engines of, 277, 280, 281; adoption of Cornish engine, 264, 281; coal of, 278, 282; cotton industry, 278–80, 282; woollen industry, 280, 282; and 'comparative history', 266, 281
Benton, T., 162n
Benyon, T. & B., *see* Marshalls
Best-practice: in steam-engine technology, 78, 147, 254; in cotton machinery, 274, 276
Bienefeld, M. A., 71n
Billey, *see* Slubbing Billey
Bingley, 129, 171
Birch, A., 103n
Birkbeck, G., 77n, 143n
Birley & Kirk, 190
Birmingham: steam engines of, 31, 34–5, 61, 62, 89; price of coal at, 64, 96, 97n; price of pig iron at, 44n
Birmingham Canal Coy., 60
Birmingham Philosophical Society, 34
Birmingham Waterworks, 89
Birstal, 33
Black, J., 11
Black Dyke Mills (J. Foster), 66n, 87
Blackburn, 201, 270n

Blast Engine, 61
Blaug, M.: 226; ratio of indicated to nominal horsepower, 27n; cotton spindleage, 182n, 235, 237, 237n; capital, 235–8; quality of cotton, 227, 228n; value of output, 229, 231n; price of cloth, 230, 230n
Blenkinsop, J., 119n, 121
Board of Trade, 227n
Boiler-house, 23, 60
Boilers: number of, 222; technology of, 16n, 18, 90, 220; Newcomen, 75; waggon (Watt), 75, 83, 84, 87, 108, 220, 221, 222–3; cylindrical, 59, 83–4, 87, 89, 220, 221, 222; tubular (Lancashire), 86, 87, 220, 221; internal-flued, 87, 221, 222; water-tube, 89, 253n; of cast and wrought iron, 59, 102, 105, 108; of steel, 24; of copper, 104–5; weight of, 93–4, 100–2, 106; cost of, 50, 51, 54, 72, 83, 84–5, 93–4, 106, 151n; horsepower of, 51, 101, 108; deterioration of, 85; explosions of, 88, 221; management of, 220, 253, 255, 256; conversion from low to high pressures with, 88, 90, 221; for heating, 179; heating surface, 59, 83–4, 85, 221; boiler plates, 84, 88, 90
Bolton: horsepower in, 33; non-condensing engines in, 159n; coal costs in, 197n, 198, 200, 278; medium-fine spinning of, 184, 188n, 215; weaving of, 196–8, 199n, 200, 230n, 290
Bondt, F. de, 280n
Boott Mills, 268, 272
Borinage (Belgium), 277
Boring Mills, 20, 44
Bottlenecks of Industrialization, 6, 7; *see also* Energy Crisis
Boucher, C. T. G., 48n
Boulding, K., 284n
Boulton & Watt Coy.: Soho factory, 35, 42, 136; types of engines built, 23n, 89, 219; number of engines built and sold, 3, 27, 28–9, 31, 32n, 33, 34, 50, 73, 100, 111, 147, 179n, 285, 294; costs of engines, 19n, 21, 22, 50, 51, 53, 59–60, 61, 71–2, 74, 93–4, 106, 151n; prices of engines and pricing policy, 23n, 51–2, 53, 54, 56, 70, 75, 76, 78, 145, 219n, 272;

# Index 323

metal consumption of engines, 99–104, 105, 106, 115n; performance of engines, 25–6, 45, 68, 69, 169; charges for metal, 52n, 56–7, 56n; delays in deliveries, 61–2; representativeness of engines, 52–3, 78, 107; as employers, 23, 293; at Albion Mills, 122; MSS and accounts, 23n, 27n, 29, 33n, 47, 50, 52n, 55n, 56n, 57n, 72n, 97n, 100, 107. *See also* Watt Engine
Boulton, Matthew, 20, 28, 120, 136
Boulton, M. R., 52n, 293
Bourbon (cotton), 180n
Bourne, J., 16n, 18n, 57n, 90n
Bowley, Sir Arthur, 74n
Bowling Ironworks, 138n
Boyson, R., 130n, 135n, 192n, 215n
Braconnier, 277
Bradford, 66, 244, 251, 279
Bradford Soke Mill, 138n
Bran, 121n
Branson, G., 193n, 200n
Brass: in atmospheric engine, 17, 17n, 49, 94, 99, 105–6; in Watt engine, 93, 99, 100–1, 104–5, 105–6; in pirate engines, 106; in textile engines, 108, 109; in Cornish engine, 256; in water-wheels, 164; engine used at battery works, 143; price of, 57, 93, 94
Brazil (cotton), 180n, 181n
Bremner, D., 249n
Breweries, 117, 118n, 119n, 159
Bridgewater Canal, 97n
Briggs, A., 154n
Bristol, 143
Brito, D. L., 266n
Brown, W., 247
Brunel, Sir Marc, 56
Brunner, H., 117n
Brunton, R., 63n, 178n, 186
Brussels, 278
Bruwier, M., 277n
Buchanan, R., 62, 63, 140n
Budge, J., 60, 257n
Bull, E., 22, 41
Burn, R. S., 227n, 228n, 230n
Buschek, C., 23n
Byker, 143n
Bythell, D., 195n, 240

Camborne, 84

Cambridge, 122
Camden Town, 143
Campbell, R. H., 50n, 57n, 125n
Canals, 39, 40, 41, 63–4, 126, 159n
Capacity of Engines, 35, 44, 111, 233, 239n, 296. *See also* Capital Formation
Capital Formation: and steam engines, 8, 84–5, 119, 258, 289; and boilers, 84, 85; and buildings, 207, 208, 213–14; in horses, 119; in waterpower, 171, 177; in cotton, 235–8, 241; productivity of, 81, 203, 236, 238; costs of, 85, 207, 275; price of capital goods, 85, 272–3; saving on, 84, 86, 90, 204, 221, 258–9, 263, 267; fixed and working, 9, 84; in U.S.A., 274, 275; in Belgium, 277. *See also* Capacity of Engines, Cotton, Mill-Owners
Capital Theory, 9, 10, 188–92, 194, 196–8, 293, 298
Carding, 117, 224, 243–4, 289
Cardwell, D. S. L., 11, 11n, 15n, 21n, 127n, 168n
Carlisle, 89
Carloose Mine, 60
Carne, J., 260, 262n
Carnot, S., 294
Carron Coy., 19, 28, 50, 57, 145n, 159
Carshalton, 138n
Cartwright, E., 118, 160, 195, 295
Cast Iron, *see* Iron
Castle Foregate Mill, 33
Castlefield (Manchester), 97n
Cataract, 18, 21
Catling, H. J., 187, 194n, 204, 206n, 242
Catrine Cotton Works, 128, 131, 132, 159
Chacewater Mine, 18, 146
Chaloner, W. H., 7n, 8n, 179n, 227n, 280n
Chapman, D., 250
Chapman, H. S., 242–3, 244
Chapman, S. D.: cotton industry, 3n; mill layout, 71n; costs of waterwheels and water-power, 129–30, 132, 134–5, 137n, 152n, 162n; use of water-power, 138n; jenny, 179n
Chapman, S. J., 188
Charleroi, 278
Charlotte Dundas (steamship), 23
Checkland, S. G., 241

Cheshire, 71, 219, 233
Chester-le-Street, 67
Chevalier, J., 25n
Chicago, 38, 39, 40
Chimney-stacks, 23
City Gas Works (London), 68
Clapham, Sir John, 5, 107, 201n, 245
Clarke, J. A., 122
Clausius, R. J. E., 67
Clayton, A. K., 49n, 67n
Cleland, J., 35, 61n
Climate and Cotton Industry, 276
Cloth, *see* Cotton Cloth, etc.
Coal: role in Industrial Revolution, 6, 45, 99; and railways, 64; output of, 111, 111n, 112–13; transportation of, 63–4, 97n, 278; innovation in modern American industry, 296. *See also* Collieries, Backward Linkages, Forward Linkages
Coal Consumption by Engines: backward linkages, 13, 44–5, 99, 110–15, 286; in Savery engines, 16, 68; in atmospheric engines, 19, 68–9, 70, 73–4, 75, 110–11, 114, 115, 147; in Watt engines, 19, 20, 45, 69–70, 73–4, 75, 76, 81, 86, 110–11, 114, 147, 218, 268, 281; in high-pressure engines, 21, 22, 70, 81, 84, 86, 112, 135–6, 218–19, 220, 256, 281; in portable engines, 23; cost for engines, 119, 135–6, 151n, 161–3, 200n, 278; and duty, 67; and machinery speeds, 45–6, 242; in American engines, 268, 268n, 269; in Belgian engines, 277. *See also* Expansion of Steam
Coal Prices: diversity of, 62–5; geological influences on, 282; and mill location, 65–6; and engine 'thresholds', 76–7, 91, 294; in each county, 148; in Cornwall, 84, 261, 289; in Lancashire, 81; in Yorkshire, 243n; in Durham, 158n; in Birmingham, 64, 96, 97n; in Bolton, 197n, 200n, 278; in Dundee, 208n, 247; in Edinburgh, 96, 97n; in Leeds, 64, 96, 97n, 247, 278; in Liverpool, 96, 97n; in London, 44n, 63, 76, 96, 97n, 121, 158n; in Manchester, 64, 73, 96, 97n, 192, 278; in Newcastle, 158n; in Saddleworth, 64; in Wigan, 64; at collieries, 62, 63, 64, 112; in rural areas, 288; in Massachusetts, 162n, 282; in France, 277–8; in Belgium, 278, 282; compared to oats prices, 121
Coal Quality: diversity of, 62–3; slack, 62–3, 64, 77, 97n, 112, 286; trend in Cornwall, 261; and portable engines, 23
Coalbrookdale, 17, 56, 56n, 143, 172n
Cockerill, J., 277, 278, 279n
Cole, W. A., 147
Collier, F., 163n
Collieries: horse-powered, 5, 117, 119n; atmospheric engines, 18, 68, 77; Watt engines, 143, 143n; whim engines, 83; winding engines, 145n, 254, 281; 'shammalling' of engines, 144, 144n; size of engines, 25, 29, 143–4; burning of slack in engines, 63
Colquhoun, P., 180n, 182n
Combing of Worsteds, 248, 291n
Compounding of Engines: by Hornblower, 24, 41, 253; by Trevithick, 24, 253; by Woolf, 24, 253, 281, 282; by Sims, 253; in London, 87, 88; in Lancashire, 86–7, 223; in Yorkshire, 87, 88; in Newcastle, 87; in Glasgow, 88; in Cornwall, 256; in U.S.A., 269; in Belgium, 281, 282. *See also* Cornish Engine, High-Pressure Engine
Condenser, separate. *See* Separate Condenser
Condensing in Engines: in Savery receiver, 15; in Newcomen cylinder, 16, 18, 19; in compound engines, 24, 253; and expansion of steam, 223; and water requirements, 159, 159n; condensation and steam jackets, 256; condensing *vs* non-condensing engines, 22, 23, 55, 57, 86, 159n, 223, 281. *See also* Separate Condenser, Watt Engine, Water
Conduits, *see* Leats
Cook, T., 171
Copeland, M. T., 270n, 272n
Copper: copper mining and engine technology, 252, 252n; output of, 104, 105, 262n; 'ticketings' (sales), 104; price of, 104, 262, 262n, 263; cost of in engines, 94, 106; weight of in engines, 94, 99, 106, 108, 109;

## Index

use in boilers, 104–5. *See also* Backward Linkages

Corliss Engine: 269

Cornish Engine: reports on, 90n, 253–4, 260, 295; diffusion of, 14, 79, 81, 83–91, 252, 257–64, 289; use outside Cornwall, 87–8, 89, 91, 264, 281; life-history of, 255–6; technical evolution of, 24, 253, 268; costs of, 47, 57–9, 83, 84, 85, 90, 218, 258, 289; coal consumption of, 70, 219, 220; and economy of fuel, 81, 83, 84, 85, 86, 218, 254, 255, 256, 261, 289; duty of, 26, 79, 85, 90, 253–4, 255, 256, 257, 259–60, 262–3, 263n; steam pressures in, 24, 80, 82, 83, 85, 221, 255, 269; horsepower of, 80–1, 254; speed of action of, 79, 79n, 80; motion of, 82, 82n, 83, 89, 90, 255, 289; and double-action, 81, 81n, 83; and disembodied technical progress, 86, 90, 255–6, 257n, 258–61, 261n, 263, 296; and lagging, 87–8, 90, 221, 256, 259; and superheating, 257n, 261; boiler of (cylindrical), 59, 83–4, 85, 87, 88, 90, 220, 221, 256, 259, 289, 296; beams of, 18n, 59, 83, 289; pumps of, 81–2, 255, 256; engine-house of, 58–9, 84. *See also* Diffusion, Expansion of Steam

Cornwall: copper and tin mines of, 104, 252, 262–3, 262n; horsepower in, 36, 148; atmospheric engines in, 28, 60, 68; Watt engines in, 26, 28, 29, 144, 144n, 252; manufacturers of Cornish engines, 58, 58n; rotative motion in, 83; peat used in engines, 62; source of cataract, 18; round-the-clock working, 73, 255; pumps in, 81–2; capital costs in, 85; cost-book system of accounting, 58, 85, 261; engineers' bonuses, 85, 254; price of coal in, 148, 158n, 252, 261n, 264

Cort, H., 7, 7n, 98–9

Costs, *see* Cotton, Horses, Mills and Factories, Steam Engine, Steam Power, Water Power, etc.

Cotton, W., 219n

Cotton: general progress of, 1, 2, 4, 7, 12, 160, 175, 226; output, 209, 210, 226–9; horse-mills, 117–18, 157, 160, 176, 289; use of water-power, 7, 129–30, 139, 150–51, 160, 176–9, 215, 289; costs of water-power in Britain, 134, 151, 157, 177–8, 215n, 284; ditto in U.S.A., 268, 269, 273–4, 275–6, 282; use of steam-power, 7, 8, 12, 66–7, 175, 182–3, 186; costs of steam-power in Britain, 74, 150, 151n, 187, 190, 192–3, 193n, 197n, 198, 200, 201–2, 206–7, 208, 209, 213, 224, 251, 284, 289, 290, 291; ditto in U.S.A., 267–9, 282; power of engines, 71, 218–223, 237n; horsepower in, 32, 33, 108, 179, 179n, 185, 205n, 209, 210, 223, 232–5, 236, 287; metal consumption of engines of, 108, 109; heating of mills, 178; coal consumption of engines, 66–7, 69–70, 218, 219, 220, 268, 278, 282; mill layout, 71, 142, 178, 182n; location of, 66–7, 177n, 184, 184n, 193; number of firms in, 185, 185n; capital in, 235–8, 274; employment in, 185–6, 209, 210, 211n, 223–4, 233, 238–40, 267, 295; profits of, 192n, 213, 215–16, 218–19, 221; costs of production, 208, 209, 211, 212, 213, 214, 216, 218, 224, 230–1, 265, 272, 278, 290; imports of raw cotton, 179n, 180, 182n, 214, 217, 224n, 226–7, 231n, 237; cotton wool, 180, 180n, 182, 214, 216, 217, 231; waste in spinning, 179n, 216, 226; spinning sector, 176n, 177n, 183–4, 210, 230, 233–4, 235, 267, 295; yarn prices, 180, 180n, 181, 212, 216, 251; yarn prices for differing qualities, 181, 186, 190n, 215n, 216, 295; average yarn quality, 182, 182n, 183, 184, 191, 208, 228–9, 237, 250, 290; ditto in U.S.A., 270n, 275–6, 282; ditto in Belgium, 279, 282; range of quality, 183–4; yarn quality and mechanization, 175, 177, 178, 179–80, 181–2, 184, 184n, 185–6, 187–8, 192, 193, 194, 202, 224, 275, 290; output per spindle, 178, 179n, 182, 188, 203, 204, 204n, 205–7, 208, 208n, 209, 211, 212, 214, 218, 224, 225n, 237, 274, 291; ditto in U.S.A., 271, 273, 274, 282; ditto in Belgium, 278, 279, 280, 282; yarn exports, 226–7, 229n, 231n; piecing, 190, 191, 191n, 193, 202, 207, 267; integration of spinning and weaving,

326                                    Index

Cotton: (Cont.)—
  185, 210n, 211n, 233–4; weaving sector, 176n, 183, 210, 230, 233–4, 235, 267, 295; requirements of yarn for cloth, 194, 228, 228n, 271; cloth output, 210, 210n, 226–9, 230n, 230–1; exports of cloth, 215, 227; demand for cloth, 179, 196, 214; cloth imports, 179; rate of production of cloth, 197, 197n, 198, 199, 201, 201n, 202, 203, 204n, 206–8, 211n, 218, 270, 270n, 273; quality of cloth, 179, 186, 196, 196n, 201, 208, 227–8, 250, 290; cloth quality and mechanization of weaving, 196, 197–8, 198–9, 200, 201, 202, 210n, 211n, 290, 295; cost of producing cloth, 211, 230–1, 290; price of cloth, 216, 230; output per loom, 200n; finishing, 32, 230, 230n, 231; cotton smallwares, 32, 185, 223–4; competitiveness of industry, 184, 184n, 189; cotton yarn in worsteds, 251, 291; cotton *vs* woollen industries, 241 245, 291; cotton *vs* flax, 250, 292. *See also* Spindleage, Machinery, Jenny, Mule, Throstle, Power-loom, Textiles, Mills, Millowners, Labour, Working Day, Strikes, Factory Acts, High-Pressure Engine, Capital Formation, Demand
Court, W. H. B., 26n, 34n, 35, 64n, 136n
Coventry, 200n
Craig, R., 12n
Cranbrook, 123
Crank, 20, 20n, 142, 144
Crompton, S., 7, 118, 138, 160, 176, 180, 182n, 203n
Cronstadt, 28, 50
Crouzet, F., 280n
Crump, W. B., 87, 200n
Cubitt, W., 125
Cudworth, W., 87n
Cumberland, 119n
Cumming, T. G., 118n
Cunningham, W., 4, 7, 242n
Curr, J., 50, 69, 76
Cyfarthfa, 169n
Cylinder: diameter and horsepower, 25, 26, 49, 49n, 58, 80, 94; weight and cost of, 56n, 93, 94; cast iron in, 17, 17n, 49, 93; in Newcomen engine,

16, 17, 18, 19, 75, 94, 143; and Watt engine, 19–20, 75, 93, 143; and double-acting engine, 20, 22; and Trevithick engine, 22; and Cornish (Woolf) engine, 82, 85, 253, 255; and compounded engines, 86; cutoff in, 22, 82, 255; clothing of, 18, 87, 88, 90, 256. *See also* Expansion of Steam.

Dandy-loom, 198n, 201, 201n, 202
Danforth, C., 274
Daniels, G. W., 182n
Darby, A., 98, 104
Darby, H. C., 122n
Darlington, J., 57n, 80, 169n
David, P. A., 38n, 42, 74, 75n, 261n
Davies-Shiel, M., 167
Davis, L. E., 43n, 162n, 271
Dead-Spindle, 244, 246, 271, 271n
Dean (parish), 33
Deane, P. M., 3n, 7, 74n, 97n, 121n, 140n, 146n, 147, 149n, 211, 211n, 212n, 262n
Deanston Mills (Perthshire), 131
Deeping Fen, 124n
Delafield, J., 119n, 120
Deliveries of Engines, 48, 61, 93–4
Demand: aggregate, 43, 287–8; for power, 29, 44, 98, 114, 115, 118, 158–9, 274, 284; for water-power, 137–8, 287; for cotton, 4, 179, 185, 196, 214, 215, 224, 224n, 276; for worsteds, 241; for flax, 250. For coal and iron, *see* Backward Linkages.
Demoulin, R., 277n, 280n
Depressions: and rate of innovation, 296; 'locking in' lower prices and costs, 180, 216, 224; and machinery speed-up, 208n, 214, 224; and flax industry, 249; and high-pressure engine, 218–19, 221, 263, 298; and data collection, 232
Derbyshire, 77
Derry, T. K., 113n, 123n
Desaguliers, J. T., 15, 17n, 48n, 155
Devon, 62, 85
Dewsbury, 171
Dhondt, J., 277n, 280n
Dickinson, H. W., 15n; and Savery engine, 48n; and atmospheric engine, 17n, 18n, 120; and Watt engine, 19n, 20n; and high-pressure engine,

43, 55n; and boat engine, 56n; number of horsepower, 27, 28, 29. 29n
Dickinson, W. (of Blackburn), 270n
Diffusion: of Industrial Revolution, 6; of types of engines, 13, 92, 254, 295, 296; of Newcomen engine, numbers, 18, 27; ditto, sizes, 25, 145–6; ditto, periods, 47, 77, 294; ditto, costs, 48–9, 91; ditto, in mining, 77; ditto, in Belgium, 277; of Watt engine, numbers, 3, 27, 28–9, 111; ditto, sizes, 25–6; ditto, periods, 73, 75, 277, 280, 285, 294, 297; ditto, costs, 51–3, 70, 71, 72, 73, 74, 75–9, 91, 93–5, 106, 145, 151n; of Cornish (high-pressure) engine, numbers, 222–3; ditto, sizes, 26; ditto, periods, 79, 83–4, 88, 218, 253, 256, 257–9, 261, 263; ditto, costs, 57–9, 83, 84–5, 86, 90–1, 219–20, 221–2, 258–61, 263, 295; ditto, areas, 81, 83–4, 85–8, 89, 91, 219, 221, 252, 264, 289; ditto, in France, 219, 264, 281; ditto, in Belgium, 264, 277, 281, 282; ditto, in U.S.A., 268–9; ditto and rotative motion, 81–3, 89, 90, 92, 289, 298; ditto, index, 254, 257, 259, 260; of textile machinery, mules, 176–7, 180, 187–8, 193–5, 202, 211, 237, 237n, 243, 297; ditto, frames, 177; ditto, jennies, 242, 243–4, 297; ditto, throstles, 244; ditto, power-looms, 195–6, 197n, 200, 202, 244–5, 245–6, 297; ditto, Jacquard loom, 200n; ditto, flying-shuttle, 244; ditto and reductions of power costs, 14, 193, 200, 202, 211, 243, 245, 246; lag behind invention, 296–7
Dinsdale, W., 222n
Direct-Acting Engine, 23, 23n, 53, 60, 219
Disembodied Technical Progress, *see* Technical Progress, disembodied
Distaff, 241, 295
Dockyards Engines, 59–60, 61, 90
Dodd, G., 88n, 118n
Dolcoath Mine, 60, 169
Domesday Book, 125
Don River, 173
Doncaster, 118
Donkeys, 117–18
Donnachie, I. L., 124

Donnell, E. J., 195n, 203n, 230n
Dorzée, 277
Double-Acting Engine: principle, 20; in Trevithick engine, 22; in high-pressure engine, 80, 81, 81n, 277; and parallel motion, 21; and rotative motion, 83; fuel economy of, 20, 81, 81n, 280; economy of space, 20, 81, 91n
Double Speeder, 275
Dowie, J. A., 38n
Downham, 155
Downs-Rose, G., 168n
Drawing Frame, 212, 248–9
Dressing Machine, 193n, 197n, 198, 201n, 295
Dublin, 35, 61
Duckham, B. F., 148n
Dudley, 25
Du Fay & Coy, 237n
Dundee, 208n, 247, 249–50, 292
Dupin, E. P. C., 30, 30n, 264n
Durham, 158n
Duty of Engines: definition of, 18, 67, 253–4; in atmospheric engines, 18, 67–8; in Watt engines, 67–8, 257; in Cornish engines, 79–80, 85, 220–1, 254, 255, 256–60, 261, 262–3, 294
Dyer, J. C., 203
Dymock, J., 180n

Eagley Brook, 128
East Indies, 180n
Eckersley's (of Wigan), 200n
Economic History: method, 38–43, 265–6, 283–5, 288, 298; and economics, 9, 11, 250n, 283; and economic growth, 11–12, 284, 295–7, 298; and quantification, 8–9, 12, 38, 283, 285; and science, 11, 294, 295–6; and the steam engine, 2–8, 252, 276–7, 292. *See also* Backward Linkages
Economics: capital theory, 9, 188–92, 194, 196–8, 202–4, 293, 298; of innovation, 9–10, 11, 14, 191, 192–3, 194, 195, 198, 200, 201–2, 204, 205–6, 207–9, 211, 250, 251, 265, 291–2, 296–7, 298 (*see also* Technical Progress); general equilibrium, 38, 44, 44n, 158n, 159; production functions, 107, 205, 298; rent theory, 133–4, 136–7, 138–9,

Economics: (*Cont.*)—
170, 178, 286; and engineering, 10–11, 283, 292–5
Economies of Scale, 71, 101, 179, 197n, 224, 254, 255, 261
Edinburgh, 96, 97n
Edmonstone Colliery, 48
Eduction Pipe, 18, 93, 94
Edwards, H., 281
Edwards, M. M., 72n, 152n, 163n, 176n, 179n, 180n, 182n, 205n
Egerton Mill, 130, 132
Electric Motors, 295
Ellison, T., 197n, 199n, 215n, 227, 228n, 230n
Elphinstone Colliery, 97n
Elsecar, 49, 76
Embodied Technical Progress, *see* Technical Progress, embodied
Employment: full employment, 38, 43–4, 287–8; unemployment, 239n (*see also* Capacity of Engines); female, 176, 176n, 223, 243; child, 223, 249; skilled, 177, 249, 266, 266n, 272n, 277, 279; in textile mills, 30, 183–4, 184n, 209–10, 238–40, 238n; in factory spinning, 185, 185n, 209–10, 211n; in non-factory spinning, 176, 176n, 179–80, 223, 242, 242n, 243, 295; on mules, 176–7; in handweaving, 176n, 195–6, 197n, 198, 211n, 240, 240n; in power-weaving, 195, 198, 199, 203, 209–10, 211n, 266; in dandy-loom weaving, 202; per horsepower, 117, 185–6, 211n, 235, 239; and speed-up of machinery, 207–8, 211n, 217–18, 224, 225, 267, 272, 274, 291; total power supplied, 295; in woollen mills, 247n; in windmills, 123; in agriculture, 43; in Armed Forces, 43, 287; in watermills, 130–1; labour conditions in mills, 178. *See also* Labour Productivity, Wages, Working Hours
Encyclopaedia Britannica, 165n
Energy Crisis, 6, 8, 44, 46, 149, 156, 284, 288
Engels, F., 154n
Engerman, S. L., 189n, 272n
Engine, *see* Steam Engine, Waterpressure Engine, etc
Engine-house: costs of, 50, 60, 72, 95; for Cornish engine, 58–9, 59n, 84; eliminated in portable engine, 23
Engineering: and innovation, 10, 21, 292–4; and science, 294; and steam engine, 67; and high-pressure engine, 80, 85, 89, 254, 255–6, 260, 261, 289; and alternatives to steam, 116; and water-power, 141; and mining, 77, 85; trends in, 69, 205, 293; in London, 87; in France, 165; in Belgium, 278
Enginemen, *see* Wages
English, W., 274n
Enos, J. L., 297n
Entrepreneurs, *see* Mill-owners
Enys, J. S., 82n, 85n
Equalizing Chain, 21
Erection of Engines: costs of, 48, 49, 59–60, 72, 144, 151n; re-erection, 27–8
Evans, O., 268, 269n
Exchange Rates, 162n, 267, 272, 278, 280, 280n
Expansion of Steam: in atmospheric engine, 21n; in Watt engine, 21, 21n, 76, 110, 223, 252; in Hornblower engine, 21n, 253; in Cornish engine, 82, 218–19, 253, 255–6; coal consumption with, 218–19; dissemination of, 219, 223, 256, 257, 261; degree of, 223, 255, 269; in waterpressure engine, 21n, 168. *See also* Steam Pressures
Explosions, *see* Boilers
Exports of Engines, 27, 28, 111
Extensor, 275

Factories, *see* Mills
Factory Acts, 172, 204n, 216–17, 217n, 225
Fairbairn, P., 249
Fairbairn, Sir William: mill-buildings, 61n; power of engines, 26, 86, 86n; introduction of high-pressure engines, 218–19, 259n; coal consumption of engines, 70, 218–19; boilers, 88n, 220n; Lancashire boiler, 84n, 86, 221; water-wheels, 126n, 127, 128, 128–9, 130, 131, 132n, 140n, 164, 165, 167, 169n; turbines, 166, 167, 168n; water schemes, 172.
Falck, Dr. N. D., 20
Farey, J., jr: *Treatise*, 2, 7n; power of engines, 26n; number of engines, 31,

## Index

32, 33, 34, 36n; Savery engine, 48, 68; atmospheric engine, 19n, 21n, 48, 49n, 50, 70n, 75, 77, 94–5, 143n; Smeaton, 18n, 70n, 75, 156; Watt engine, 2, 19n, 21n, 69, 70n, 75, 110n, 156, 159n, 219n, 293n; double-acting engine, 20n; Cornish engine, 80n, 81–2, 81n, 82n, 83, 83n, 88n, 259n, 260n, 262n; returning engine, 142n, 143n, 145n; 'shammals' engines, 144n; Thompson's engine, 179n; steam pressures, 16n, 17n, 21n, 110n; water-tube boiler, 88–9; engine-houses, 60; erection of engines, 144n; horse mills, 118n; water-pressure engine, 168n; patenting, 292, 293n, 297
Farey, J., sr, 77, 154n
Fenlands, 59, 122–4, 125n, 155, 157
Fenton, Murray & Wood, 34, 54, 55n
Fielden, J., 217n
Finch, W. C., 122, 123n, 124n
Fire Engine, *see* Atmospheric Engine
Firth of Forth, 97n
Fishlow, A., 38n, 45, 46n
Fitton, R. S., 163n, 180n
Flax: horsepower in, 32, 33, 108, 109, 139; technology, 248, 248n, 249, 250; heckling, 248, 249, 249n; gill frames, 248–9; spindleage, 247, 247n, 248, 249–50; wet spinning, 248, 248n, 249; mechanization and yarn quality, 208, 248, 250, 251, 291–2; productivity, 247–8, 250; weaving, 251n; location, 247, 250, 292; and jute, 250; and cotton, 250, 292. *See also* Demand, Depressions, Mill-owners
Flintshire, 138
Floud, R. C., 9, 12n
Flour Mills, 21, 122, 166
Flux, A. W., 64n
Fly Frame, 274–5, 276
Flying Shuttle, 176n, 244
Flywheel, 20, 20n, 82n, 142
Fogel, R. W., 38, 38n, 39, 40n, 41, 189n, 284
Fong, H. D., 182n, 247n
Forbes, H., 246
Forbes, R. J., 34
Forward, E. A., 22n, 55n
Forward Linkages: from steam-power to industries adopting, 14, 98, 152, 159, 175, 285, 289; from steam engine to machinery, 10, 13, 45–6, 175, 202, 205–9, 224, 265, 289, 291–2, 298; 'spin-off' to other innovation, 42, 44, 45, 46, 159, 160, 176, 289; timing of, 46, 292; from coal industry, 113. *See also* Steam-Power, Machinery, Textiles
Foster, J., 66. *See also* Black Dyke Mills
Fothergill, J., 136
Foundling Hospital (London), 97n
Fourneyron, B., 166, 167
Fowey Consols Mine, 79, 168
Framework: for steam engines, costs of, 59, 72, 95, 144, 161, 289; ditto, materials, 59, 84; ditto, eliminated in portable engines, 22, 23; for water-wheels, 130, 161
France: and Watt engine, 280; and high-pressure engine, 219, 264, 281; costs of steam-power in, 161–2, 280; costs of water-power in, 161–2; coal prices in, 264, 277–8; source of Jacquard loom, 200; engineers of, 165; Napoleonic Wars, 43–4, 44n, 155, 280
French, G. J., 143n
Frome River, 138
Fuel, *see* Coal, Lubricants, Peat
Fulling Mills, 133n, 172
Furness, 138

Galloway, E., 22n, 88n
Galloway, R. L., 28n
Gamble, J., 56n
Garforth & Sedgwicks, 129
Gaudry, J., 24n
Gauldie, E., 250n
Gayer, A. D., 227n
Ghent, 278, 280, 282
Ghorbal, S., 87
Gibb, G. S., 272n
Gilbert, D. (Giddy), 68n
Gillispie, C. C., 294n
Glasgow: engines and horsepower in, 31, 35; engine builders in, 61; typical steam mill of, 206; compounding engines in, 24, 88; horse mills of, 117; water schemes of, 172; coal quality of, 63; double-decking of mules, 188n; price of cotton, 181n; costs of hand weaving, 197n
Gloucestershire, 138

# 330  Index

Glynn, J., 125n, 126, 140n, 166n, 168, 169n, 172n
Goodchild, J., 33, 134n
Goodrich, S.: and Trevithick engine, 22n; estimates of engine costs, 54, 55n, 56n, 59n, 60, 61, 62, 72n, 86n; estimates of boiler costs, 85–6
Goodwyn's Brewery, 119n
Gordon, Barron, 170
Gordon's Mills (William Pirie), 172–3
Gott, B., 73, 87, 171
Gould, J. D., 42n
Governor, 21, 164
Graham, W., 194
Grain, 39, 119–21, 155, 157, 280, 286, 296
Gray, A., 117n
Great Exhibition, 55n, 86n, 87n, 130, 162n, 166, 200n, 221n
Greenock, 167, 172, 172n
Greenwich, 88
Greg, R. H., 137, 197n, 239n, 275–6
Greg, S., 130, 131, 135–6, 163n, 167, 171. *See also* Quarry Bank Mill
Greg, W. R., 239n
Gribble, J., 261
Grier, W., 117n
Griff Colliery, 155
Griliches, Z., 257n, 258, 263, 263n, 296
Grinding Windmills, 122
Grose, S., 256
Grouvelle, P., 162n
Guest, R., 203n
Gunderson, G., 38n

Habakkuk, Sir John, 3n, 195n, 210n, 267, 267n, 272n, 273–4
Halifax, 66
Hall, A. R., 34n
Hall & Son, 83n, 87
Hamilton, H., 35n, 252n
Hamilton, S. B., 25
Hammond, J. L. and B., 5
Hampshire, 43
Hand-loom, *see* Power-loom, Employment, Labour Productivity, Wages
Harcourt, G. C., 9n
Hargr(e)aves, J., 118, 160, 176, 179, 242, 297
Harley, C. K., 46
Harris, J. R., 27, 28
Harris, T. R., 87n, 253n, 257n
Harrison, A. and J. K., 119n

Harte, N. B., 60n, 69n, 119n, 176n
Hartwell, R. M., 199n, 254n
Harvey, W. S., 168n
Harveys, 58, 89, 261n
Hassan, J. A., 148n
Hawke, G. R., 12, 41, 64, 64n, 113, 114, 258n, 284
Hawkesbury Colliery, 25, 48
Hawkhurst, 124
Hayle, 58
Headingley, 97n
Heaton, H., 244
Heaton Colliery, 25
Hemp, 19, 74n
Henderson, L. J., 294
Henderson, W. O., 227n, 281n
Henshall, W., 74n, 152n
Henwood, W. J., 68n, 79n, 254n
Herne Bay Coy, 23n
Hetherington, J., 221
Hetton, 97n
Hewes, T. C., 130, 164, 164n
Hick, B., 222
High-Pressure Engine: and Savery, 15; and Watt, 21, 89, 252–3; and Trevithick, 21–2, 24, 41, 43, 55, 253, 256; and Woolf, 24, 83, 87, 90, 253, 256, 281, 282; compound engine, 24, 253, 269; non-condensing engine, 159n, 268–9. 277, 281; Uniflow engine, 24; and steamboats, 90; diffusion of, 79, 85, 86–8, 89, 90–1, 143n, 252, 277, 294; performance of, 27, 83, 263; costs of, 57, 70, 83, 86, 162n, 219; and rotative motion, 83, 90, 92, 264, 298. *See also* Cornish Engine, Diffusion
Highs, T., 176
Hills, J. G. B., 67, 68
Hills, R. L.: horse mills, 117, 118n; windmills, 123, 124; water-wheels, 129n, 132n, 164n, 165n; water-power, 138n, 172n, 177n; returning engine, 142n; Wrigley engine, 16n; double-acting engines, 20n; prices of engines, 52n; coal prices, 124n, 132n; costs of engines, 72n, 177n; delivery time for engines, 62n; textile engines, 118n; spindles per horsepower, 177n
Hitchin, J., 197n
Hobsbawn, E. J., 154n, 175
Hoëne-Wroński, J. M., 264n

## Index

Hoffmann, W. G., 227n
Holbeck, 33, 34
Holdsworths, 69
Hollander, S., 12n
Holmyard, E. J., 34n
Holroyds Mill, 69
Hoole, H., 97n, 186n
Hopkins, R. T., 138n
Hornblower, J., 21n, 24, 41, 253, 293
Horner, Inspector L.: and horsepower, 233, 234n, 235n, 239; and high-pressure engine, 219, 221; and fine-spinning, 185; yarn required for cloth, 228n; and employment, 238n, 239n
Horrocks, H. (patentee of power-loom), 195
Horrockses, 87, 217n
Horsehay, 25, 143
Horsepower Definition, 26–7, 29n, 118, 233, 237n
Horsepower Installed, *see* Steam-Power
Horses: use in mills, etc., 5, 6, 117, 118, 118n, 155, 157, 286; costs of, 14, 116, 118–22, 155, 156, 157, 162n, 286; strength of, 117, 118, 119n; life-expectancy of, 118; gins for, 119, 119n
Hosiery, 241
Hot Well, 18, 20, 93, 94–5
Hours of Work, *see* Working Hours
Huddersfield, 87, 244
Hueckel, G., 44n
Hughes, J. R. T., 12, 12n, 162n, 235n
Hull, 61, 69
Hulme, A., 59n
Hunslet, 34, 69
Hyde, C. K., 7n, 12, 57, 98, 99, 99n, 103n, 286
Hyde, J., 204n

India, 179, 181n
Industrial Revolution: invention in, 1, 2, 5–6, 7, 42, 156, 294, 297; coal in, 6; iron in, 7; under-employment in, 288; bottlenecks to, 6, 7; dating of, 3, 5; studies of, 2, 7n, 8. *See also* Energy Crisis, Machinery, National Income, Diffusion
Industrialization and steam-power, 1–8, 36, 42, 44, 70, 116, 152, 156, 287–8. *See also* Steam Power, Social Savings, Watt Engine

Injection Cock, 16
Innovation, *see* Machinery, Economics, Diffusion, Industrial Revolution, Iron, Atmospheric Engine, Best-practice, Boilers, Copper, Cornish Engine, Depressions, Mill-owners, Engineering, Forward Linkages, Jenny
Interest Rate, 72n, 73, 146n, 192
Intermediate Technology, 200–2, 242
Ireland: horsepower in, 35, 139, 148; peat used in, 62; price of coal, 148; rain fall of, 153; water-wheels of, 126; water resources of, 153–4; linen industry of, 139, 247
Iron: output, 103, 103n, 104, 109, 285; innovation, 1, 7, 98–9, 104, 105, 296; prices, 52n, 53, 56–7, 85, 93; and water-wheels, 98, 130, 132, 141, 143, 164, 166; and steam-power, 12, 44, 67, 98–110, 112, 114, 115, 285–6; cylinders of, 17, 17n, 49, 93; beam of, 18, 21; steam case of, 75, 86n, 93; boilers of, 59, 93. *See also* Backward Linkages
Irwell River, 138, 159n, 172
Isle of Man, 126

Jacquard Loom, 200–1, 200n, 202
James, C. T. ('Justitia'): 206n; mill, 269n; steam and water-power costs, 162n, 267, 268; improvements in steam engines, 268, 268n; spindles per horsepower, 270; output per spindle, 271; working year, 271
James, J., 246, 251
Jarrett, H., 284n
Jaunez, A., 162n
Jeffery, R., 261
Jemappes, 277
Jenkins, D. T., 66n, 119n, 126n, 232n
Jenkins, J. G., 133n, 201n, 242n, 244n
Jenkins, R., 19n, 20n, 253n
Jenny: origin of, 242; improvements in, 176, 242; diffusion of, 242, 297; for domestic use, 118, 176, 242; role in cotton, 2, 179; role in woollens, 242, 243–4, 291; compared to mule, 242–3, 246, 295; cost reductions from, 243–4, 291, 295; output of, 243; sizes of, 176, 242n, 243. *See also* Wages
Jeremy, D. J., 266, 267

Jevons, W. S., 64n
Johansen, L., 259
Jones, E. L., 43n, 149n
Joule, J. P., 294
Justitia, *see* James, C. T.
Jute, 250

Kane, Sir Robert, 148n, 153, 153n, 172n
Kay, J., 176n
Kay, R., 199
Kay-Shuttleworth, Sir James, 185n
Keir, P., 143
Keirstead, B., 296
Kelly, W., 176
Kempson, J., 197n
Kendal, 167, 172n
Kennedy, C., 11n
Kennedy, J.: sun and planet in textiles, 20n; coal consumption per hour, 70; mule spindleage, 180, 182n; spinning productivity, 203n; power-looms, 195. *See also* McConnel & Kennedy
Kent, 122, 123, 124
Ketley, 67
Kilnhurst Forge, 159
Kinneil, 297
Kuznets, S., 1, 12

Labour Productivity: in spinning, 209–10, 243, 243n; in hand-weaving, 196, 210n, 211n; on dandy-looms, 201, 201n; in power-weaving, 197–8, 197n, 199, 203, 209–10, 210n, 211n, 240, 245, 270n; labour-saving mechanization, 5, 16, 43–4, 176, 194, 195, 199, 207, 217, 224–5, 249, 249n, 266–7, 274, 276, 295. *See also* Employment, Wages, Working Hours
La Houlière, M. de, 25n
Lambeth, 69
Lanarkshire, 184n
Lancashire: horsepower in, 32, 148, 232, 233, 236; atmospheric engines, 143; Watt engines, 31, 148, 179n, 268; compounding, 24, 282; high-pressure engines, 84, 87, 219, 221; double-acting engines, 81; piston speeds, 219; engineers of, 221; coal consumption of engines, 70, 268; coalfields of, 64; coal prices of, 84, 148; water resources of, 7, 153n; boilers of, 83–4, 86, 87, 221; mill building in, 7, 125; spinning machinery of, 206, 242, 265, 273, 274; costs of spinning in, 265; power-weaving system of, 199; wages in, 199n; employment in, 238, 239–40; working day in, 71
Landes, D. S., 5, 6n, 155n, 260
Lanes (of Stockport), 142
Langton, J., 97n
Latent Heat, 11, 11n
Lawson, J., 33
Lazenby, W., 130n, 131n, 164n
Lead, 94, 105–6
Lean, T. and J., 90n, 253, 253n, 254
Learning by Doing, 10, 248, 255, 295
Leather, 19, 94
Leats and Conduits, 127, 128, 131–2, 161–3, 173
Lebergott, S., 39n
Lebrun, P., 279n
Lee, A., 55n
Lee, C. H., 59n, 72n, 118n
Lee, E., 125
Lee, G., 180n, 182n
Leeds: horsepower, 31, 33–4, 34n; cost of engines, 61; compounding, 88; price of coal, 64, 96, 97n, 247, 278; flax industry, 247–50, 291–2; spindles per horsepower, 247, 249; output per spindle, 250, 291–2; spinning wages, 242
Leith, 97
Lejeune & Billard, 277
Lemon, Sir Charles, 262n
Leupold, J., 22
Lewinski, J., 277n
Liège, 277, 278
Lillie, J., 130
Lincoln, 122
Lindley, W., 34
Linen, *see* Flax
Linkages, *see* Backward Linkages, Forward Linkages
Live-Spindle, *see* Dead-Spindle
Liverpool: horsepower, 32; engine costs, 219n; delivery route for engines, 61, 219n; coal prices, 96, 97n; bar iron prices, 44n
Lloyd-Jones, R., 176n
Location of Mills, *see* Mills

# Index

Locomotive Engines, 12, 23, 119n, 121, 253, 293. *See also* Railways
London: horsepower, 31, 148; atmospheric engine, 77, 143; premium on Watt engines, 52, 53, 76; price of Watt engines, 53; Cornish engine, 83n, 87, 89, 91; returning engine, 143; engine delivery charges, 61; steam flour-mill, 122; coal prices, 44n, 63, 76, 87, 91, 96, 97n, 121n, 148, 158n; coal transportation charges, 63; costs of horses, 119n, 120
Long Benton Mine, 18, 67, 69, 75, 145n
Loom, *see* Power-Loom
Lord, J., 8, 27, 28, 29, 42, 179, 179n
Lotka, A. J., 257n
Lowell, 162n, 268, 269, 270n, 273, 282
Lubricants: oil, 19, 62, 74n, 120, 151n, 161, 163n, 190n, 193n, 197n, 198, 242, 255; tallow, 74n, 120, 161, 163n, 193n
Lynch & Inglis, 55n

McClelland, P., 39n
McCloskey, D. N., 46n, 272n
McConnel & Kennedy, 59, 62n, 72n, 118n, 184, 184n, 205n
McCulloch, J. R., 70n
McGouldrick, P., 266n, 269n, 272n, 273
McNaughted Engine, 70, 86–8, 136, 221, 223, 289
Machinery: role of mechanization in Industrial Revolution, 1, 5, 7, 217, 224; forward linkages from, 4; mechanical problems, 5–6, 208, 218; ditto with power, 159, 295; ditto in spinning, 142; ditto with preparing frames, 203, 248–9, 274–5; ditto with jenny, 176, 179, 242; ditto with hand and self-acting mule, 176–7, 180, 186, 187–8, 188n, 191, 193, 194, 204, 208–9, 211, 218, 242, 242n, 279, 290; ditto with water-frame and throstle, 177, 179, 186, 241, 242, 246, 248, 274, 279, 291; ditto with power-loom, 195, 196, 199, 204n, 218, 245, 246, 291; ditto with Jacquard loom, 201; ditto with dandy-loom, 201, 201n; ditto in worsteds, 241, 244, 245, 246, 291; ditto in woollens, 241–2, 245, 291; ditto in flax, 248–9; contingency of advance on steam-engine, 2, 6, 7, 8, 12, 44, 45, 179, 179n, 181, 183, 184, 186, 187, 194, 224, 243, 274, 289, 295; entrepreneurial expectations about, 189, 189n, 257; 'semi-reserve capacity', 178, 296; and cost reductions, 4, 180, 199, 207, 209, 224, 243–4, 246, 249n, 265; price of, 10, 272–3, 272n, 279; speed-up of, 13, 45–6, 178, 202, 203–4, 204n, 206–8, 213–14, 217, 217n, 218, 224, 246, 266–7, 270n, 273–4, 276, 279, 291, 292; and quality of production, 175, 179–80, 186, 193, 196, 199, 200, 201, 202, 208, 251, 267, 275, 276, 290; 'challenge and response', 176n, 267, 295; replacement of, 191, 196, 202, 203, 242–3, 244–6, 248, 251, 272, 274, 276, 290, 291; substitution for labour, *see* Labour Productivity
Major, J. K., 117n
Manchester: horsepower in, 31, 32–3, 223; types of engines, 223; atmospheric engine, 69, 72, 73–4; Watt engine, 72, 74; Cornish engine, 84; boilers, 84, 108, 222; costs of engines, 71, 72, 74, 78, 84, 139, 150–1, 151n, 161–2, 193, 193n, 200n, 282; Thompson engine, 179n; water-wheels, 129–30; water-power costs, 137n, 150–1; water schemes, 172; location of mills close to rivers, 159n; price of yarn, 181n; costs of spinning, 278–9, 290; speed of spinning, 279; yarn quality, 182, 184, 185, 190; costs of weaving, 193n; cloth quality, 196
Manchester Statistical Society, 32
Manchester Steam Users' Association, 222, 222n
Mann, J. A., 188n, 215n, 227n, 230n, 231n
Mann, J. de L., 64n, 242n
Mansfield, E., 296, 297n
Mantoux, P., 6n
Manufacturers, *see* Mill-owners
Marine Engines, 12, 22, 23, 53, 56, 85, 90, 91n, 142. *See also* Steamships
Marshall, A., 30
Marshall, H. C., 249
Marshall, J., 227n
Marshall, J. D., 167
Marshall, J. G., 217n, 247n

Marshall, John, 66, 129, 249n

Marshalls (of Leeds): factory location, 66; engines at, 33, 73n, 87–8, 90; speeding engines and machinery, 88, 217n; clothing of cylinders, 87–8, 90; hours of work of, 250; productivity of, 250, 250n; profits, 250; heckling machines, 249, 249n; wet spinning, 248n, 249, 249n, 250, 250n; yarn quality, 250, 250n, 292; employing Farey, 34; MSS, 55n, 64n, 68n, 70n, 73n, 87, 88n, 129n, 140n, 143n, 249n, 250n

Marsland, J., 222

Marx, K., 8, 8n, 14, 216–18, 223, 224–5, 233

Massachusetts, 162n, 268, 269, 271, 282

Mathias, P., 42n, 67n, 118, 118n, 119n, 256n

Matschoss, C., 20n, 22n, 29n, 68n

Matthews, R. C. O., 178, 204n, 227, 228n, 229n, 296

Maudslay, H., 22

Medlock River, 159n

Meikle, A., 125

Meikleham, R. S., 22n

Mersey River, 159n

Metal Industries, 44, 106, 108. *See also* Iron, Brass, Copper, Lead, Tin

Methodology, *see* Economic History

Metz, 162n

Mildenhall, 124, 155

Millington, J., 126n

Mill-owners: choice of location for mill, 66; raising fixed capital, 85; decision-making over innovation, 188–9, 189n, 194, 217–18, 226, 257, 296, 298; choice between Watt and atmospheric engine, 76–7, 79, 144–6; choice and neglect of high-pressure engine, 85, 90, 91, 282, 298; and steam pressures, 220–1, 223; and reducing fuel consumption, 218–19, 282; and coal costs, 278, 281–2; costs of engines for, 59, 77–8; control over engineers, 85, 254; working engines in opposition, 142–3; and auxiliary steam engine, 171; and returning engine, 142–3, 144–5; choice between steam and other power sources, 116, 152, 158; competition for water rights, 132–3, 134, 136, 137–8; leases, 132–3, 134, 134n, 172–3; control over water-power, 127, 152, 171–3, 286; costing of water-power, 135–6, 137, 170, 286; profits and losses on water-power, 134, 136–7, 170, 172–3, 177, 178, 286; profits and losses in cotton, 192, 192n, 196, 213, 215–16, 219; responses to price changes, 212, 215, 217, 290; decision to speed up machinery, 212, 213, 215, 217, 298; introduction of hand-mule, 243, 291; introduction of self-acting mule, 188, 188n, 189, 190, 191, 194, 290; use of throstle, 192; introduction of power-loom, 195, 196–7, 197n, 241, 290; renting out of looms, 200, 202; worsted *vs* woollen manufacturers, 241, 244–5, 291; Leeds *vs* Dundee flax-spinners, 250, 291–2; American *vs* British manufacturers, 266–7; profits of Cornish mining, 261–3; and Factory Acts, 172, 204n, 216–17, 217n, 225; and strikes, 194, 217. *See also* Cotton, Machinery, Mills, Capital Formation, Mule, Mule, Self-Acting, Throstle, Power-loom, Woollen Industry, Flax

Mills: construction of, 7, 234n, 235n; average size of, 71, 118–19, 155, 182n; costs of, 45, 61n, 133, 134n, 154, 171, 198, 206n, 207–8, 209, 213–14, 235, 272–3, 279, 282, 291; semi reserve capacity, 178, 296; factory system, 4, 5, 134, 159, 289; horses in, 5, 117–18, 155, 176; windmills, 122, 155; water-wheels in, 129–30, 134, 154, 159n, 171, 176; steam-engine in, 21, 177, 178; water-power gearing, 140n, 164; transmission and gearing, 3, 20, 82, 142–3, 144, 218, 220, 264, 289; heating of, 178; factory villages, 66, 133, 154; location of, 66, 125, 133, 154, 158, 159, 159n, 160, 161–3, 278, 282; boring, 20, 44; breweries, 117, 118n, 119n, 159; carding, 117, 224, 243–4, 289; flour, 21, 166; scribbling, 243–4; starch, 69; sugar, 22; threshing, 119n, timber, 122

Mining and Power, 28, 29, 29n, 30, 35, 60, 252, 255–6. *See also* Pumping Engines

Mitchell, B. R., 74n, 97n, 121n, 140n, 146n, 149n, 211n, 262n

Mokyr, J., 277n
Mons, 277, 278
Montgomery, J.,: horsepower at Lowell, 270; costs of power, 273; mill of, 206, 206n, 207; spindles per horsepower, 270, 270n, 271n, 275; speed of spindles, 206, 207, 273; costs of self-acting mules, 190n, 191, 191n; costs of spinning, 206–7; yarn required for cloth, 271; looms per horsepower, 210n, 270; piecing wages, 191, 191n.
Morin, A. J., 165, 166n, 264n
Morshead, W. A., jr: horses, 120; water-wheels, 131, 140n, 164n; ditto in agriculture, 126n, 127; reaction wheels, 167n; water-pressure engines, 168n; portable $vs$ fixed engines, 22–3, 23n, 55n
Motion of Steam-Engines: irregularity of, 16–17, 18, 21, 142–3, 143n, 255, 289; rotative, 3, 16, 20, 21, 29, 52, 71, 72, 74, 75–6, 80, 83, 90, 92, 156
Mott, R. A., 48n, 67, 77, 103, 143n
Muggeridge, R., 197n
Mule: role in cotton, 2; role in woollens, 242; man-powered, 118, 160, 176; animal-powered, 117; water-powered, 176, 177n, 178; steam-power and, 177, 177n, 178, 179n, 183, 187, 224, 242; pairing of mules, 176–7; cost of, 190n, 214, 279; quality of in Belgium, 278–9; number of spindles, 180, 205n, 237, 237n; spindles per mule, 176, 218; spindles per horsepower, 177, 179n, 186, 190n, 204–5, 205n, 270, 270n; output per spindle, 179n, 203, 204, 206, 212–13, 243, 273, 279; and fine yarns, 177, 180, 188n, 294–5; wages on, 242, 243; profitability cf. jenny, 242–3, 246, 291; conversion into self-actor, 191. *See also* Cotton, Labour Productivity, Wages
Mule, Self-Acting: role in cotton, 2; evolution of, 187, 188, 191, 209; size of, 189n; product of, 192, 194; cost of, 190n, 273, 279; number of spindles, 187, 237; spindles per horsepower, 186, 190n; output per spindle, 188, 190, 209, 218, 237, 279; and coarse yarns, 186, 188, 190–1, 192, 193, 202, 290; profitability cf. hand-mule, 189–93, 194, 224, 290;

additional breakages, 190, 191, 193, 202; introduction in strikes, 194; and throstle, 192; and power-loom, 194–5
Mulhall, M., 29, 29n, 30, 35, 36, 287
Murdock, W., 23, 81, 81n, 293
Murray, A. and G., 184n
Murray, M., 22, 33, 34, 54, 55n, 61, 249
Mushet, D., 109
Musson, A. E., 8, 29n, 31, 32n, 48, 78n, 143, 179, 179n, 294
Mutton, N., 59n

Nasmyth, J., 219
National Income, 147, 149, 152, 156, 211, 286–7, 288
Nef, J. U., 99, 111, 111n, 112
New Constitution Mill, 69
New Eagley Mill, 74n, 135n
New Lanark, 134n, 159
New Orleans, 38n
New Tarbet Mills, 132
New York, 38, 39, 40
Newburyport, 162n, 268, 269n
Newcastle, 18, 48, 87, 97n, 145n, 158n
Newcomen, T., 17, 252
Newcomen Engine: origin of, 17; patent, 27, 49, 294; manufacturing of, 17, 47, 277, 294; improvement of, 17–18, 18n, 69, 70; duty of, 18, 67–8; fuel consumption of, 19, 68–9, 70, 75, 76, 114, 147–8, 158; size and power of, 25, 29, 49n, 75, 117, 143–6; number of engines, 27, 28; in Belgium, 277; on coalfields, 77, 112, 143–4, 158, 277; costs of, 47–50, 72, 73–4, 75, 76, 94–5, 105–6, 145, 151n; repairs of, 76, 79; weight of components, 94–5, 105–6; cylinder of, 16, 17, 17n, 18, 19; beam of, 16, 18; equalizing chain, 21; irregularity of motion of, 16–17, 79; steam pressures in, 16, 17, 17n, 18–19; and latent heat, 111n; profitability cf. Watt engine, 75–9, 91, 146–8, 156, 158n; and iron industry, 98, 115n. *See also* Atmospheric Engine, Pumping Engines, Returning Engines
Newton (Massachusetts), 269n
Nixon, F., 67, 69n, 77n
Nordhaus, W. D., 11n, 284n, 292
Northumberland, 28, 168

Oats, 121, 121n
Oil, see Lubricants
Old Machar (Aberdeenshire), 170
Oldham, 117, 184, 196
Oldknow, S., 59
Ormrod & Hardcastle, 196, 196n
Orrell's Mill, 204n, 210n
Oscillating Engine, 23, 23n
Ossett, 134n
Owen, R., 134n
Oxen, 117n, 118

Pacey, A. J., 177n
Paisley, 117
Palmer, G. H., 90n
Pankhurst, K. V., 291n
Papin, D., 20
Papplewick, 132, 143, 164
Parallel Motion, 21, 41, 93
Parent, A., 127, 127n
Parker, W. N., 297n
Parkes, J., 59n, 82n, 83n, 84, 88n
Parsons, E., 34, 34n
Partington, C. F., 30, 57n, 68n, 123, 124, 155
Patents, 292–4. *See also* Newcomen Engine, Watt Engine, Savery Engine
Pattison, J., 87
Paul, L., 118, 160, 176
Payen, J., 281n
Peat, 62–3, 280
Pecqueur, C., 30, 30n
Peile & Williams, 210n
Penns (of Greenwich), 88
Perier, J.-C., 280
Perkins, J., 24
Perthshire, 131
Petroleum, 296
Philadelphia, 269n
Phillips, J. A., 57n, 80n, 169n
Pickard, J., 20
Pigott, S. C., 180n
Pipes, Steam, 18, 20, 21n, 60–1, 61n, 72, 88, 90, 93, 94, 221, 256. *See also* Eduction Pipe
Pirate Engines, 19, 52, 78, 105, 106, 147, 179n, 286, 293. *See also* Separate Condenser
Pirie, William, & Coy., 172–3
Piston: in Newcomen engine, 16–19, 21n, 94; in Watt engine, 19–21, 21n, 93; in double-acting engine, 20, 21, 81; in Trevithick engine, 22; length of stroke, 18, 22, 69, 79n, 80, 88; load on, 18, 20, 82, 82n, 94, 255; speed of, 16–17, 79, 79n, 80, 81, 82, 218, 219–20, 253; cost of, 93–4; clothing of, 18; sealing of, 17, 19
Poldice Mine, 68
Pole, W., 85n
Pollard, S., 9n, 197n, 204n
Pollution, 154n, 285, 288
Poncelet, J. V., 165
Ponting, K. G., 60n, 69n, 119n, 176n, 241n, 242n, 243n, 245n, 251n
Portable Engines, 22, 22n, 23, 23n, 55, 145n
Porter, G. R., 97n
Portsmouth Dockyard, 59, 60, 61, 62
Power, see Steam Power, Water Power, etc.
Power-loom: technical evolution of, 118, 195, 196, 199, 203, 270n, 290, 295; number and diffusion of, 195, 205n, 224, 241, 244; delays to diffusion of, 197n, 200, 201n, 245, 295; power source of, 118, 295; horsepower in, 205n, 210; looms per horsepower, 206, 210n, 245–6, 251, 270–1, 291; cost of, 197n, 198, 201n, 202, 203, 279; cost of power for, 193n, 198, 200, 201–2, 245–6, 251, 295; cost of plant per loom, 197n, 198, 235; rental for room and turning, 193n; dimensions of, 193n, 197n, 201n, 245; speed of (picks), 193n, 204n, 211n, 218, 245, 246, 266, 270, 270n; output per loom, 197–8, 197n, 200n, 201n, 202n, 203, 245, 246, 266, 266n, 273; quality of cloth, 196, 196n, 197n, 198–9, 200, 202, 210n, 211n, 245, 251, 267, 290, 295; turnover of capital with, 197n; lubrication of, 197n, 198; consumption of flour in, 197n, 198; looms per weaver, 195, 199, 202, 203, 211n, 266, 266n; employment on, 210, 211n; output per weaver, 210, 270n; productivity of hand-looms, 211n; output of hand-looms, 210, 210n; disappearance of hand-loom weavers, 195, 196, 198, 202, 211n, 240; profitability of power-loom over hand-loom, 195, 196–200, 202, 210n, 245, 246, 251, 290; profitability over Jacquard loom, 200; profitability

over dandy-loom, 201; in worsteds and woollens, 241, 244–7, 291; and American spinning, 267; and self-acting mule, 194–5. *See also* Employment, Wages, Diffusion
Prices, *see* Agriculture, Cotton, Iron, Steam Engine, etc.
Production functions, 107, 205, 298
Productivity, *see* Labour Productivity, Machinery, Steam Power
Profits, *see* Mill-owners
Pumping Engines: Newcomen, duty of, 68; ditto, costs of, 72, 74; Watt, compared to Newcomens, 77, 146, 157; ditto, size of, 29, 71; ditto, pricing of, 52, 75, 76, 145; ditto, coal consumption of, 76, 110, 147; ditto, expansion in, 76, 110; pirated, 147; Cornish, 79, 83, 254, 289; hours of work of, 73, 110, 147; use of engines for pumping, 15–16, 79, 82n, 89, 142–3. *See also* Mining and Power
Pumps: 253, 256; lifting pumps, 81; plunger-pumps, 81–2; width of, 18; depth of, 143n, 255; angle of, 256; leaks in, 255; weight of water in, 82n; pump rods, 82, 82n, 143n, 255; Savery engine as, 15–16

Quarry Bank Mill: costs of mill and machinery, 273n; water-wheel at, 130, 135, 162n, 164; costs of water-power at, 130, 132, 135–6, 152n, 159, 162n, 163n, 171n; floods and droughts at, 171; engine costs at, 74n; coal consumption of, 70, 135–6; price of coal at, 136; compounding of engine, 70, 136; steam pipes of, 61n; wages at, 163n; cottages at, 163n; MSS, 61n, 130n, 132n, 136n, 162n, 167n, 171n, 222n
Queensbury, 66, 66n

Radcliffe, W., 201, 201n, 295
Radley, J., 63n
Railways, 4, 12, 23, 38–41, 46, 64, 119n, 284, 293. *See also* Locomotive Engines
Rainfall, 153
Raistrick, A., 17n, 49n, 55n, 56n, 60n, 143n, 172n

Rankine, W. J. M., 26, 27n, 67, 118, 140n, 165, 294
Rate of Return, 78, 189–92, 196–200, 242, 290, 293
Raybould, T. J., 63n
Reach, A. B., 184n
Receiver, 15, 16
Reciprocating Engine, *see* Pumping Engines
Reddaway, W. B., 191n
Redford, A., 5
Redlich, F., 265
Rees, A., 26n, 81n, 140n, 177n
Regulator, 16
Reid, H., 18n
Reiter, S., 12, 12n
Renfrewshire, 184n
Rents, *see* Economics, Water-power
Renwick, J., 81
Research and Development, 12
Reservoirs, 127, 131–2, 135, 143, 173
Returning Engines: with Savery engine, 16, 143; with Newcomen engine, replacing Watts, 114n, 142–3, 144–5, 145n, 146, 149, 156; with Watt engine, 143; with Cornish engine, 82n; with water-pressure engine, 169; early use of, 143, 143n; in textile mills, 142, 143, 143n; loss of power in, 144–5, 145n, 146
Réunion, 180n
Reynolds, J., 125n, 126n
Rhode Island, 266n
Ricardo, D., 133n
Riche de Prony, G. C. F. M., 140n
Rickards, Inspector R., 232, 233
Rimmer, W. G., 97n, 121n, 248
Ring Spinning, 274, 282
Roberts, R., 187, 191, 194, 279
Robinson, E., 8, 31, 32n, 48, 78n, 143, 179, 179n, 294
Robison, J., 165
Rodgers, H. B., 185n
Rogerson & Coy., 69
Roll, Sir Eric, 5, 8, 19n, 28n, 52, 52n, 60n, 61, 70n, 72n, 76n, 78n
Rolt, L. T. C., 67
Rosenberg, N., 8n, 122n, 194n, 263n, 266n, 272n
Rostow, W. W., 3–4, 3n, 6–7, 13, 43, 46, 99, 156, 175, 227n, 292
Rotative Engines, *see* Motion of Steam-Engines, Watt Engine

Rotherham, 158n
Rousseaux, P., 139–40
Roving Frames, 203, 249
Rowe, W. J., 252, 262
Royal Commission on Coal Supply, 111, 111n, 112

Saddleworth, 64
St Blasien, 166, 167
St Davids (Scotland), 97n
St Louis, 38n
St Pancras, 77n
Salford, 31, 32, 33
Saltaire, 26
Salter, W. E. G., 191n, 259n
Samuelson, P. A., 11n
Sandberg, L. G., 228, 228n, 229n, 274n
Saunders, Inspector R. J., 232n
Savery, T., 62, 144n, 252
Savery Engine: building and use of, 27, 47–8, 77n, 143; coupling of, 144, 144n; size of, 25, 48, 49, 68, 91; costs of, 47–8, 49, 75, 91, 106, 145; fuel consumption of, 68, 158; steam pressures of, 15–16; functioning as pump, 15–16; as returning engine, 16, 48, 143; patent, 27, 49
Savory, E., jr (of Downham), 155
Sawing Mills, 122
Schemnitz (Chemnitz), 169n
Schon, D., 297n
Schulze-Gaevernitz, G. von, 197n
Schumpeter, J. A., 175
Schwartz, A. J., 227n
Scientific Discovery, 8, 10–11, 79, 83, 294
Scotland, 124, 139, 167, 247, 270
Scribbling Mills, 133n, 172, 243–4
Sea Island (cotton), 180n
Sedgwick & Coy., 129
Select Committee on Exportation of Machinery, 186, 281
Select Committee on Patents, 292
Self-Acting Mule, *see* Mule, Self-Acting
Separate Condenser: and science, 11, 11n; patent, 19, 293; weight and cost of, 93; materials of, 75, 93, 105; location of, 20; and atmospheric engine, 19, 19n, 146n; and Hornblower compound engine, 24, 41–2, 293–4; and pirates, 147, 293; and social saving, 42, 294
Seraing, 277

Shadwell Waterworks, 110n
Shaftesbury, Lord (Ashley), 217n
Shaw's Waterworks (Sir Michael Shaw Stewart, Greenock), 172
Sheepscar, 69
Sheffield, 31
Shepton Mallet, 242
Sherratt, *see* Bateman & Sherratt
Shrewsbury, 87, 90, 248
Shudehill, 128
Side-Lever Engine, 22
Sigsworth, E. M., 66n, 201n, 245n
Silk, 32, 88, 108–9, 139, 200
Simmons, P. L. 61n
Simpson & Coy., 69, 180n
Sims, J., 83, 253
Singer, C., 34n, 120n, 140n
Sirhowy, 56n, 93–4
Skilton, C. P., 123n
Slubbing Billey, 242
Smallwares, *see* Cotton
Smeaton, J.: use of models, 17–18, 125, 126, 127; and horses, 119, 120, 155; and windmills, 125; and waterwheels, 126, 127, 140, 164, 165, 166; and costs of water-power, 132; and returning engine, 142, 143; and water-pressure engine, 168n; improvements to atmospheric engine, 18, 18n, 19n, 20, 21, 69, 70, 142, 156, 252; cylinder-boring, 19; weight and cost of engine components of, 94–5, 105; duty of engines of, 18, 67; power of engines of, 75, 146; coal consumption of engines of, 67, 69; and coal quality, 63, 63n; and Cronstadt engine, 28; improvements to boiler, 18; and Watt engine, 19n, 70, 70n; and portable engine, 22, 22n; and Cornish mining, 252; use of iron parts, 18, 125, 164
Smelser, N., 176n
Smethwick, 60
Smiles, S., 2
Smithfield, 149
Social Savings: definition of, 38–9, 141, 152; dating of, 36, 43, 287; on railways, 13, 38–41, 64, 64n, 284; differences in this book, 13, 41, 283–4; on Watt engine, 14, 41–2, 43, 114, 116, 141–9, 150, 156–7, 158, 252, 284, 286, 288, 293; on pirate engines, 147; on all steam engines,

14, 114–15, 116–17, 121–2, 141–2, 149–56, 157–60, 170, 173, 284–5, 287–8; contemporary estimates of ditto, 30n, 118n; and actual power supplied, 26, 28, 146; and unemployment, 38, 43–4, 287–8; and technical progress, 39, 41–2, 42n, 116, 141, 158, 158n, 159–60, 170, 293; and threshold model of diffusion, 116; and backward linkages, 44, 114; and 'energy crisis' notion, 44, 156, 288; and income distribution, 152–3, 157; and social rates of return, 293
Society of Arts, 246
Soho, 22, 26, 31, 35, 61, 97n, 110n, 136, 159
Solow, R. M., 259n
Sorocold, G., 159
Speed of Machinery, *see* Cotton, Depressions, Employment, Machinery, Power-loom
Speed of Power Supply, *see* Cornish Engine, Piston, Water-wheels
Spencer, J., 228n
Spindleage, cotton: number, 182, 205n, 209, 225n, 237n; number in fine-spinning, 185; number of mule spindles, 180, 183, 237; number of self-acting spindles, 187, 237; number of throstle spindles, 237; water-frame spindleage per mill, 182n; spindleage per water-wheel, 129, 166–7; spindles per mule, 176, 190n; spindles per horsepower, 10, 45, 110n, 205, 209, 212; ditto, for mules, 177, 179n, 186, 186n, 204, 204n, 206, 270n, 279; ditto, for self-acting mules, 186; ditto, for water-frames, 177, 186; ditto, for throstles, 166–7, 186n, 206, 269–70, 271, 271n, 279; for preparatory frames, 275; for fine and coarse spinning, 177n, 186, 204n; and output per spindle, 205–6, 208, 211, 271–2; comparison with U.S.A. and Belgium, 282. *See also* Machinery
Spindleage, flax, 247–8, 249–50
Spinning, *see* Cotton, Employment, Labour Productivity, Machinery, Flax, Woollen Industry, Wages
Spinning Wheel, 241–2, 246, 295
Staffordshire, 63
Staleybridge, 61n

Stamping Engines, 83, 254
Stanway, S., 184n, 190n, 197n, 199n, 211n, 238
Starch Mills, 69
Steam Case, 19, 90, 93, 105, 147, 256, 257, 259. *See also* Iron
Steam Engines. *See* Atmospheric Engine, Auxiliary Engines, Blast Engine, Compounding of Engines, Condensing in Engines, Corliss Engine, Cornish Engine, Direct-Acting Engine, Dockyards Engines, Double-Acting Engine, High-Pressure Engine, Locomotive Engines, McNaughted Engine, Marine Engines, Newcomen Engine, Oscillating Engine, Pirate Engines, Portable Engines, Pumping Engines, Returning Engines, Savery Engine, Side-Lever Engine, Stamping Engines, Table Engine, Uniflow Engine, Watt Engine, Whims, Winding Engines
Steam Engines, Components of. *See* Air Pump, Beam, Boiler-House, Boilers, Brass, Cataract, Chimney-Stacks, Copper, Crank, Cylinder, Eduction Pipe, Engine-House, Expansion of Steam, Flywheel, Framework, Governor, Hemp, Hot Well, Injection Cock, Iron, Lead, Leather, Motion of Steam-Engines, Pipes, steam, Piston, Pumps, Receiver, Regulator, Separate Condenser, Steam Case, Stuffing, Valves
Steam for Heating, 23, 60–1, 70, 161, 178–9, 268, 268n, 286
Steam Passages, *see* Pipes, steam
Steam Power: horsepower installed, 13, 27–30, 31, 35–7, 111, 151, 287; ditto in Watt engines, 27, 28–9, 148; ditto in textiles, 179, 205n, 210, 222n, 234, 236, 247, 247n; distribution of horse-power, 100, 102, 105, 107–8, 146; power of engines, 25–7, 28–9, 29n, 118, 143–4, 145–6, 233, 237n; life-history of engines, 79, 255–6; costs of, and industrialization, 4, 6, 10, 11, 11n, 13–14, 152–3, 157; ditto, *vs* manpower costs, 198, 200, 202; *vs* horse-mill costs, 119–20, 121, 155, 156, 157; *vs* windmill costs, 155–6, 157; *vs* water-power costs, 134, 135–6, 138–9, 161–3, 177–9, 269,

Steam Power: (*Cont.*)—
284, 286; ditto, depending on coal prices, 22, 64, 141–2, 158–9; ditto, per unit of output, 45–6, 207–9, 224, 266; ditto and machinery driven, 177, 187, 190, 193, 193n, 198, 200, 202, 241, 243, 247, 251, 266, 267–8, 273–4, 276, 289–91, 295, 298; ditto ditto in woollen industry, 242, 243, 245–6; ditto ditto in flax industry, 249; ditto, between U.S.A. and Britain, 161–2, 267–8, 269, 273–4; ditto, in France, 161–2; ditto and technology chosen, 46, 274, 276; ditto and economies of scale, 71, 224; ditto, periods of change, 14, 70–1, 73, 139–40, 150, 193, 200, 202, 213, 224, 286, 287, 289, 290; ditto, in Manchester, 71–4, 77–8, 150, 193; ditto in Bolton, 200; mobility of, 4, 27–8, 35–6, 125, 159–60, 159n. *See also* Steam Engine, Agriculture, Backward Linkages, Belfast, Belgium, Best-practice, Birmingham, Bolton, Bottlenecks to Industrialization, Boulton & Watt Coy., Capacity of Engines, Capital Formation, Coal Consumption by Engines, Collieries, Cornwall, Cotton, Dean, Deliveries of Engines, Demand, Depressions, Diffusion, Dublin, Duty of Engines, Economies of Scale, Energy Crisis, Engineering, Erection of Engines, Exports of Engines, Flax, Forward Linkages, France, Glasgow, Industrialization and steam-power, Ireland, Iron, Lancashire, Leeds, Liverpool, London, Lubricants, Machinery, Manchester, Massachusetts, Mill-owners, Mills, Mining and Power, Mule, Mule, self-acting, Patents, Peat, Pollution, Power-loom, Railways, Sheffield, Social Savings, Spindleage, Textiles, Windmills, Water Power, Woollen Industry, Wages

Steam Pressures: in Savery engine, 15–16; in atmospheric engine, 16–17, 18–19; in Watt engine, 20, 21, 252; in Trevithick engines, 21–2, 24, 253; in Uniflow engine, 24; in compound engine, 24, 86, 223, 253, 269; in Cornish engine, 24, 80, 82, 253, 269; in marine engines, 90; after condensation, 18–19, 20; distribution of engines by, 222–3; costs of increasing, 222, 258–9, 261, 296; superheating, 257, 257n, 261; attitudes to higher pressures, 15–16, 21, 24, 77, 79, 90, 219, 220–1, 223, 252, 264n, 293; ditto, in U.S.A., 268–9; ditto, in Belgium 281. *See also* Cornish Engine, High-Pressure Engine

Steamships, 12, 21, 22, 23, 46, 90, 218
Stern, W. M., 2, 280n
Stettler, H. L., III, 271
Stewart, N. K., 124
Stewart's Wallsend (coal), 97n
Stirrat, *see* Whitelaw & Stirrat
Stockport, 32, 138, 142, 184, 193n, 196, 196n, 204n, 239n, 240n
Stoddart (Cornish engineer), 26, 26n
Stowers, A., 140n, 164n
Strikes, 194, 217
Stroke of Engines, *see* Piston
Stuffing, 74n, 161, 163n
Styal, 130, 162n, 163n, 171. *See also* Quarry Bank Mill
Sugar Mills, 22
Sun and Planet, 20, 20n, 179n
Sun Fire Insurance, 129
Sunderland, 97n
Svedenstierna, E. T., 164n
Sykes, J., 148n
Symington, W., 23
Symons, J. C., 279n

Table Engine, 22
Tallow, *see* Lubricants
Tann, J.: horse-mills, 117, 117n, 119n, 121n; water-mills, 128n, 129, 129n, 132n, 164n; returning engines, 143; costs of Savery engines, 48, 143; threshold for Watt engine, 75n; erection costs, 60; engine-house costs, 60; coal consumption of engines, 69, 111n; steam heating, 178–9
Tariffs, 272, 276, 279
Taylor, A. J., 184n, 194n
Taylor, G., 59n
Taylor, J., 82, 89
Taylor, R. C., 109
Taylor, W. C., 159n
Technical Progress: disembodied, 11–12, 18, 86, 221–2, 258–61, 261n, 263,

296; embodied, 221, 259, 260, 261n. See also Forward Linkages, Social Savings, Machinery, Steam Engines, Water Power
Temin, P., 162n, 217n, 267, 268, 268n, 269, 269n, 270n
Textiles: location of, 66–7; horsepower in, 29n, 30, 32, 34n, 35–6, 37, 108–10, 113n; water horsepower in, 138, 139–40, 151, 159, 171; horsemills, 118n, 286; water and steam costs for, 161–3, 171–2, 179, 286; forward linkages from high-pressure engine to, 14, 45, 85; machinery and costs of power, 251; social savings in, 30n, 118n; and sun and planet motion, 20; delicacy of work in, 79, 83, 142–3; hours of work in, 71, 73. See also Cotton, Flax, Silk, Woollen Industry, Employment, Working Hours
Thackray & Whiteheads, 129
Thomas, J., 63n
Thomas, M. W., 217n
Thomasson, T., 200n
Thompson, F., 144n, 179n
Thomson, J., 167
Thorp, J., 274
Threshing Mills, 119n
Threshold Model: 13, 91; for Watt and Newcomen engines, 74–9; for high-pressure and low-pressure engines, 83–5, 90–1, 92; for auxiliary steam-engine, 171; for steam heating, 178–9; for self-actor and hand-mule, 192; and social savings, 116
Throstle: number of, 237; driven by water-power, 178, 286; driven by turbine, 166, 167; and hand-mule, 192, 270, 270n; and self-actor, 192; and coarse yarns ('water twist'), 192, 274, 279n; speed of, 204n, 246; spindles per horsepower, 166, 167, 186, 206, 269–70, 270n, 271, 271n; cost of plant per spindle, 209, 272–3; and worsted spinning, 241, 244, 277, 291; in U.S.A., 269; in Belgium, 278–9; cost of, 279; Danforth throstle ('cap frame'), 274, 276. See also Water-frame
Thrutchers, 86
Tin, 252, 262, 262n, 263
Titley, A., 22n, 24n

Toynbee, A., 2, 3, 12
Transmission, see Mills
Transportation of Materials, 66, 93–4, 119n, 133, 154, 158, 161–3, 176n, 276. See also Coal
Tredgold, T., 140n
Trevithick, F., 55n, 60n, 169n
Trevithick, R., jr: puffer engine, 21–2, 22n; plunger-pole engine, 24; Cornish engine, 253; and compound engine, 24, 253; price of engines of, 22, 55, 55n; efficiency of engines of, 22, 22n, 55n; and high pressures, 21, 24, 41, 43, 253; cylindrical boiler, 89, 221; water-pressure engine, 169
Trevithick, R., sr: 60
Truman, B., 118n
Tupling, G. H., 117n, 138n, 177n
Turbine, 166–8
Turner, H. A., 194n
Turner, T., 55n
Tyneside, 49, 60n, 77, 120, 143n

Uniflow Engine, 24
Union Mills (Cranbrook), 123
United Mines, 26
United States of America: high-pressure non-condensing engine, 162, 268–9, 282; compound engine, 269; costs of steam-power, 161–2, 179, 267–8, 268n, 269, 269n; cost of fuel, 162, 281–2; costs of water-power, 137, 161–2, 179, 267–8, 269, 269n, 273, 275, 276, 282; power costs and machinery, 267, 272, 273, 274, 275, 276, 282; spindles per horsepower, 270, 270n, 271, 271n, 272, 282; output per spindle, 271, 282; relative speed of machinery, 266–7, 270, 270n, 271, 273, 274, 276, relative price of machinery, 272, 272n, 275, 282; differences in textile technology, 274–5, 276, 282; introduction of power-loom, 267, 270n; looms per horsepower, 270, 271; plant costs, 272, 272n, 273, 275, 282; labour and power-looms, 211n, 266, 270, 270n; skilled labour, 266, 266n, 272n; labour scarcity, 266, 272, 272n, 274, 275, 276, 282; working year, 271; response to British competition, 267; tariff, 272, 276; textile products, 273, 274, 275–6, 282; nature of

United States of America: (*Cont.*)—
demand, 267, 276, 282; supply of raw cotton, 217, 224n, 275; profitability of slavery, 189n; diffusion of technology, 261n, 296; and economic conditions in Britain, 136; schools of economic history, 283

Unwin, G., 59n, 134n, 138, 177n, 180n

Ure, A.: cost of horse-feed, 120n; spindleage, 182n, 187n, 237, 237n; spindles per horsepower, 186n; handmule *vs* self-actor, 191, 194n; speeds of spinning, 204n, 274n; looms per horsepower, 210n; quality of power-loom cloth, 196n; cotton prices, 190, 192n; steam pipes, 61n

Usher, A. P., 49n, 156n, 166n

Valves, 17, 20, 22, 83, 88, 221, 258

Verviers, 278, 280

Von Tunzelmann, G. N., 254n, 257n, 258n, 261n, 285n

Vortex Wheel, 167–8

Wadsworth, A. P., 180n

Wages: price of labour, 38; average wages of labourers, 120, 208n, 238, 238n, 272n, 288; in U.S.A., 161–2, 266n, 266–7, 272n, 274, 275, 282; in France, 161–2; in Belgium, 278–9, 282; labour costs for engines, 50, 74, 74n, 95, 120, 151n, 161–2, 193n, 287; ditto, for erection, 60; labour costs for horse-mills, 119, 120; labour costs for windmills, 124; labour costs for water-mills, 130, 131, 132, 161, 163, 168; labour costs in spinning, 216, 230n; ditto, for jennies, 176, 242, 243; ditto, for mules, 190, 190n, 191n, 242; labour costs in handloom-weaving, 196, 197n, 198, 230n, 290; labour costs in power-weaving, 197n, 216, 230n, 266; labour costs in heckling, 249n; labour costs as a restraint on mechanization, 45, 202, 207–8, 209, 272, 274, 282; raising wages in strikes, 194, 194n. *See also* Labour Productivity

Wailes, R., 122

Wakefield, 69

Walker Colliery, 25, 145n

Walkers (of Rotherham), 158n

Wallis, G., 266, 270

Wallsend, 97n

Waltham, 270

Wandle River, 138n

Wasborough, M., 20, 20n

Wasseige, 277

Water, for Steam-Engines, 15, 18, 19, 159, 159n

Water-frame: role in cotton industry, 2; in standard mill, 118; power source for, 118, 176, 177–8, 177n; spindles per horsepower, 126, 177, 186; and yarn quality, 179; in worsteds, 241; in woollens, 242; in flax, 248. *See also* Throstle

Water Power: extension in Industrial Revolution, 4, 134, 136, 138, 158, 172, 224; limits to ditto, 6, 7, 133, 136–7, 138, 138n, 154, 158–9, 170–3; horsepower and use of, 125, 137–8, 139–40, 151, 172, 172n, 177–8, 284; ditto in Ireland, 153–4; ditto in Lancashire, 153n; ditto in Cornwall, 168; ditto in Bolton, 33; ditto in U.S.A., 269n, 273, 275; ditto in Belgium, 280, 282; harnessing of, 126–8, 131–2, 135, 140, 154, 172–3; water rights, 119n, 132–3, 134, 134n, 136, 152, 162n; restricting factory mobility, 125, 159–60; and factory system, 159, 224, 289; relative costs cf. animals, 155, 157, 286; relative costs cf. steam-power, 4, 14, 130, 133–4, 135–6, 137, 138–40, 141, 149–51, 157, 161–3, 170, 171–2, 172n, 173, 177–9, 177n, 284–5, 286, 287; ditto, in U.S.A., 161–2, 267–8, 269, 269n, 282; ditto, in France, 161–2; costs of, 128, 130–41, 150, 150n, 151, 152, 152n, 153, 154, 156, 157, 161–3, 165, 172, 177, 178, 179, 286, 287; capital-intensity of, 43, 269; driving machinery, 160, 176–7, 177n, 215, 244, 273, 275–6, 286, 289; improvements in technology of, 140–1, 154, 164–9, 287; reliability of, 6, 127, 135, 154, 170–3. *See also* Auxiliary Engines, Returning Engines, Water-wheels, Cotton, Iron, Woollen Industry

Water-pressure Engine, 21n, 141n, 168–9

Waterways, *see* Canals

Wood, G. H., 74n, 211n, 230n, 238, 238n, 239, 239n, 240
Wood, O., 119n, 148n
Woodbury, L., 273
Woolf, A.: significance, 253, 292; and high-pressure steam, 24, 82, 87, 253, 281; compound engine, 24, 59, 87n, 253, 281, 282; coupling of engines, 83; rotative engines, 83, 90; engines in London, 87, 129n; strokes per minute of small engines, 80; consumption of coal in engines, 220; water-tube boiler, 89. *See also* Cornish Engine
Woollen Industry: entrepreneurs of, 241, 291; technology of, 241–5; domestic *vs* factory system, 242, 246, 291, 295; horsepower of, 32, 33, 246–7, 247n; steam-power and, 108–9, 243, 295; water-power and, 139, 171–3; employment, 247n; cost reductions in, 243–4, 246, 291, 295; raw wool, 66, 241, 242; productivity of spinners, 243–4; spinning *vs* weaving, 244; productivity of weavers, 245; hand *vs* power weaving, 245–6, 291; looms per horsepower, 245–6, 246n, 251; output, 247n; yarn quality, 245, 291; cloth quality, 245, 251; woollens *vs* worsteds, 241, 244–7, 247n, 291; worsteds, entrepreneurs of, 241; ditto, technology of, 241–2, 244, 246, 248, 274, 291, 295; ditto and steam-power, 108–9, 244, 295; ditto and water-power, 139, 244; ditto, capital of, 241; ditto, demand for, 241; ditto, price of, 291, 291n; ditto, combing of, 248, 291n; ditto, spindles per horsepower, 251; ditto, spinning productivity, 246, 251, 291; ditto, looms per horsepower, 246, 251; ditto, weaving productivity, 245, 246, 291; ditto, cloth quality, 251, 291; ditto in Belgium, 279; ditto, competition from cotton, 251, 291; location of, 66, 278, 280, 282; hosiery, 241. *See also* Jenny, Mule, Throstle, Water-Frame, Machinery, Mill-owners, Mills
Working Hours: hours per year of mill engines, 71, 73, 110, 111, 112, 147; hours per year of pumping engines, 73, 110, 147; hours per day in Cornwall, 73, 255; hours per day for relays of horses, 118; hours per day for windmills, 123, 124; nightwork in water-mills, 127, 172–3; losses of hours per year for water-mills, 170–1; hours per day in textile mills, 71, 73, 112, 170, 223; hours per week in flax mills, 208n, 250; working days per year, 76; ditto in France, 162n; ditto, in U.S.A., 271
Worsted, *see* Woollen Industry
Wright, G., 217n, 285n
Wrigley, J., 16, 16n, 68, 129, 143n
Wyatt, J., 118, 160, 176

Yarn, *see* Cotton, Flax, Woollen Industry
Yatestoop Mine, 144n
York Buildings Waterworks, 15–16
York Waterworks, 63n
Yorkshire: horsepower, 31, 33; prices of horse gins, 119n; water resources, 125, 133n; returning engines, 143; engine speeds, 219; coal of, 63, 64; jenny, 242, 243n; Danforth throstle, 274; flax spindleage, 247n; flying shuttle, 244; looms per horsepower, 245–6, 246n; Jacquard loom, 200n
Young, T., 120
Young, T. M., 273

# Index

Water-wheels: dimensions, 117, 128–9, 166, 167; power, 128–9, 129n, 130, 135, 165; speed, 129, 140, 140n, 164, 165, 167–8; efficiency, 127–8, 140, 144–5, 145n, 146, 164–5, 166–7; relative costs cf. steam-engines, 114–15, 116–17, 130, 141, 161–3; costs, 129–30, 131, 140–1, 161–3, 164–5, 167, 168–9; life-expectancy, 131; erection, 130–1, 168; horizontal, 125–6, 166; undershot, 126–8, 129n, 164, 165; breast, 126–7, 164, 170; overshot, 126–7, 131, 140, 140n, 143, 164, 170–1; floats, 126, 165; buckets, 126, 129, 164, 165, 166; shroudings, 164; arms, 164; axles, 164; governor, 164; pentrough, 127; wheelhouse, 131; use of iron in, 130, 132, 141, 143, 164, 166. *See also* Turbine, Vortex Wheel, Water-pressure Engine

Watt, G., 77, 293

Watt, J.: abilities, 17, 21, 78, 89, 205, 293; influences, 11, 11n, 252, 252n; and latent heat, 11, 11n; and sun and planet, 11, 20, 20n; and parallel motion, 21; patents, 19–21, 292–3; measure of Horsepower, 26, 67, 118. *See also* Boulton & Watt Coy., Watt Engine

Watt Engine: role in industrialization, 1–3, 6, 7, 41–2, 79, 91, 98, 116, 149–50, 151n, 152, 154n, 155, 156–60, 284–5, 286–7, 288, 292–3; ditto, if replaced by Newcomen Engine, 114, 116, 141–9, 156, 252, 286; ditto, in Belgium 277, 280; ditto, in France, 280; diffusion, 74–9, 91, 297; at collieries, 29, 77, 143; and iron industry, 98–9, 114–15, 285–6; costs, 20, 72, 74, 75, 76, 77, 81, 91, 93–4, 120, 151n, 161; prices, 53–7, 76, 77, 145; performance and power, 26, 29, 68, 69, 70, 71, 75, 77, 78, 117, 143–4, 147, 169, 205, 256, 257; coal consumption, 45, 69–70, 110–12, 114–15, 116, 269n, 286; steam pressures, 20, 24, 27, 252–3; expansion, 21, 21n, 76, 110, 110n, 252; compounded, 24; converted to high pressures, 258–9, 261; rotative motion (driving machinery), 2–3, 6, 20, 82, 82n, 142, 143; parallel motion, 21; cast-iron beam, 21; sun and planet, 20, 20n; double-action, 20, 21, 81; boring of cylinder, 19, 19n, 20, 79; piston and packing, 19, 79, 81; steam case, 19, 20; separate condenser, 11, 19, 20, 24, 41–2, 293; governor, 21, 169. *See also* Boulton & Watt Coy., Steam Power

Weaving, *see* Cotton, Flax, Silk, Woollen Industry

Wednesbury, 25n, 48

Weiszäcker, C. C. von, 11n

Westgarth, W., 168

Wheal Abraham, 80

Wheal Fortune, 25n

Wheal Maid, 25, 81n

Wheal Prosper, 253

Wheal Vor, 83

Wheat, 39

Whims, 83

Whitbreads (breweries), 119n

Whitehaven, 25, 25n, 144n

Whitelaw & Stirrat, 167

Whitmore, W., 62n

Wicksteed, T., 69–70, 89, 143, 145n

Wigan, 64

Wild, M. T., 133n, 244n

Wilkinson, J., 20, 44

Williams, T. I., 34n, 113n, 123n

Williamson, J. G., 266n, 272, 272n, 273

Williamson Bros., 167

Wilson, P. N. (Lord Wilson of High Wray), 125n, 129n, 132, 172n

Winding Engines, 145n, 254, 281

Windmills: types, 123; improvements in, 124–5; dimensions, 117, 123; power, 117, 123, 123n, 155; number and horsepower, 122–3; uses, 122; location, 122; inputs to, 114–15; costs, 116, 123–4, 155–6, 157; costs cf. animals, 117, 123, 124, 155; costs cf. steam-engines, 117, 124, 125, 155–6, 157–8, 287; capital-intensity, 43–4; life-expectancy, 124; governor, 164; reliability, 124, 125, 155

Wise, J., 257n

Withers, G., 278, 280n, 281

Witney, 192

Wood: shortage, 7, 7n; water-wheels, 132, 141, 164; framing, 59, 94; sawing mills, 122

Wood, D., *see* Fenton, Murray, & Wood